PETROLEUM SCIENCE AND TECHNOLOGY

RECENT ADVANCES IN PRACTICAL APPLIED WELL TEST ANALYSIS

PETROLEUM SCIENCE AND TECHNOLOGY

Additional books in this series can be found on Nova's website
under the Series tab.

Additional e-books in this series can be found on Nova's website
under the e-book tab.

PETROLEUM SCIENCE AND TECHNOLOGY

RECENT ADVANCES IN PRACTICAL APPLIED WELL TEST ANALYSIS

FREDDY HUMBERTO ESCOBAR

New York

Copyright © 2015 by Nova Science Publishers, Inc.

All rights reserved. No part of this book may be reproduced, stored in a retrieval system or transmitted in any form or by any means: electronic, electrostatic, magnetic, tape, mechanical photocopying, recording or otherwise without the written permission of the Publisher.

We have partnered with Copyright Clearance Center to make it easy for you to obtain permissions to reuse content from this publication. Simply navigate to this publication's page on Nova's website and locate the "Get Permission" button below the title description. This button is linked directly to the title's permission page on copyright.com. Alternatively, you can visit copyright.com and search by title, ISBN, or ISSN.

For further questions about using the service on copyright.com, please contact:
Copyright Clearance Center
Phone: +1-(978) 750-8400   Fax: +1-(978) 750-4470   E-mail: info@copyright.com.

NOTICE TO THE READER

The Publisher has taken reasonable care in the preparation of this book, but makes no expressed or implied warranty of any kind and assumes no responsibility for any errors or omissions. No liability is assumed for incidental or consequential damages in connection with or arising out of information contained in this book. The Publisher shall not be liable for any special, consequential, or exemplary damages resulting, in whole or in part, from the readers' use of, or reliance upon, this material. Any parts of this book based on government reports are so indicated and copyright is claimed for those parts to the extent applicable to compilations of such works.

Independent verification should be sought for any data, advice or recommendations contained in this book. In addition, no responsibility is assumed by the publisher for any injury and/or damage to persons or property arising from any methods, products, instructions, ideas or otherwise contained in this publication.

This publication is designed to provide accurate and authoritative information with regard to the subject matter covered herein. It is sold with the clear understanding that the Publisher is not engaged in rendering legal or any other professional services. If legal or any other expert assistance is required, the services of a competent person should be sought. FROM A DECLARATION OF PARTICIPANTS JOINTLY ADOPTED BY A COMMITTEE OF THE AMERICAN BAR ASSOCIATION AND A COMMITTEE OF PUBLISHERS.

Additional color graphics may be available in the e-book version of this book.

Library of Congress Cataloging-in-Publication Data
Recent advances in practical applied well test analysis / editor, Freddy Humberto Escobar (Universidad Surcolombiana, Huila-Colombia, South America).
    pages cm
  Includes index.
  ISBN 978-1-63483-473-5 (hardcover)
  1. Oil wells--Testing. I. Humberto Escobar, Freddy, editor.
  TN871.R3643 2014
  622'.33820287--dc23
                                2015031074
Published by Nova Science Publishers, Inc. † New York

This book is dedicated to…

God for being my Creator and for all the blessings I have received throughout my entire life, and also to the Most Holy Virgin Mary, Mother of God, who helps me follow Jesus, the only Son of God.

Dr. Djebbar Tiab for being my mentor, friend and professor. He created the TDS technique, which is the basis of this work.

My beautiful children, Jennifer Andrea, Freddy Alonso and Maria Gabriela, for being my pride and a great blessing from God to my life.

My parents, Sotero and Delfina; my brothers, Sotero Alonso (RIP) and Leonardo Fabio; and my sister, Dayra Stella; to all my nephews, nieces, cousins and relatives; to my godchildren, brothers-in-law, sisters-in-law and mother-in-law and friends.

My son-in-law Carlos Andres Perez, his father Jose Omar Perez—who is also my relative— and Fanny Novoa (Jose Omar's wife) for being part of my family and loving, caring for and helping my oldest daughter.

Last but not least, this book is dedicated to the most important human being in my universe and its surroundings: my beloved wife Matilde Montealegre.

CONTENTS

Acknowledgments		ix
Foreword		xi
Introduction		xiii
Part One: Homogeneous and Heterogeneous Reservoirs		1
Chapter 1	Fundamentals of Transient Pressure Analysis in Homogeneous Reservoirs	3
Chapter 2	Double-Porosity, Single-Permeability Reservoirs	29
Chapter 3	Transition Period off from Radial Flow Regime	53
Chapter 4	Double-Porosity and Double-Permeability Reservoirs	95
Chapter 5	Triple-Porosity. Single-Permeability Reservoirs	105
Chapter 6	Triple-Porosity, Double-Permeability Reservoirs	125
Part Two: Non-Newtonian Fluids		147
Chapter 7	Infinite Reservoirs	149
Chapter 8	Non-Newtonian/Newtonian Interface	187
Chapter 9	Finite Systems	203
Chapter 10	Naturally Fractured Reservoirs and Bingham Plastic Fluids	215
Part Three: Diverse Modern Topics		247
Chapter 11	Transient-Rate Analysis	249
Chapter 12	Conductivity Faults	295
Chapter 13	Variable Temperature	319
Chapter 14	Bottom Aquifers	351
Chapter 15	Shale Reservoirs	375
Index		419

ACKNOWLEDGMENTS

I am so thankful to the people and institutions that provided support and help for the completion of this book: Universidad Surcolombiana for granting me a sabbatical period for the purpose of writing this work; Dr. Chandra Rai and the Mewborne School of Petroleum Engineering (MP&GE) at the University of Oklahoma (OU) for the support and acceptance of me to spend my sabbatical there as a visiting professor; and Ms. Sonya D. Grant, assistant to the director of the MP&GE department, for his kind cooperation in my sabbatical work at OU.

I thank all my students, coauthors and friends who have conducted research with me.

I thank Dr. Sergio Berumen for writing this book's foreword.

I thank my wife and children for being my support.

FOREWORD

The publication of this book, *Recent Advances in Practical Applied Well Test Analysis* by Freddy Humberto Escobar, will be of great interest to reservoir engineers, reservoir characterization geologists, teachers and advanced students in these fields, and all others who are interested in well test design and analysis.

Having the opportunity to share with this passionate researcher the wealth of information he has acquired and developed during a long career devoted to a concentrated scholarly study and practical investigation of well test analysis is, indeed, a privilege for me.

This book aims at providing advanced applications for well testing under different well and reservoir geological boundaries and production conditions. Since the discovery of the pressure derivative by Tiab and Kumar in the middle 70s, no other significant advances have revolutionized the practical aspects of reservoir engineering and well testing as the pressure derivative did. Further creative developments by Dr. Tiab, the father of the pressure derivative, led to a more precise and simplified method of diagnostics in well testing, called Tiab's direct synthesis technique, or TDS. This technique provides a powerful tool in the interpretation of pressure data from well tests and confers a tremendous certainty in detecting the flow model and reservoir properties. In addition, the beauty of TDS is that it deepens the understanding of both pressure and rate-transient response and behavior of the analytical solutions. The interpretation of well tests using conventional techniques compares favorably with the TDS in the extensive examples used throughout the entire book. It validates and demonstrates the simplicity and creativity of TDS and provides a superior and simpler way to understand oil and gas reservoirs as well.

One of the interesting shifts the modern practice of well testing analysis has taken in the last decade is that it has become truly interdisciplinary. Reservoir engineers, geologists and geophysicists integrate together the results of pressure well test interpretation into building geological models. Throughout this book, there is a detailed discussion of analytical solutions for several conditions, followed by well test examples illustrating the data interpretation results and the benefits of the TDS technique.

The Escobar volume is a clear, practical, readable account, but, most importantly, it is adequately seasoned with multiple examples of well tests. Every single example in the book illustrates and shows how the TDS technique simplifies the interpretation of several complex combinations of geological setting and well flow condition: Well pressure data sets associated to heavy oil; sealing faults, conductive faults, and/or parallel faults; channels, fractured fractal porous media, etc., enrich the understanding of every chapter. A wealth of figures clarifies the text and is a mine of reference information. The author treats the topics in a logical and reasonable way, and a particularly strong point is the attention the book gives to the practical procedure of TDS and its valuable contribution to the modern interpretation of well pressure data.

SERGIO BERUMEN[*]
UNIVERSITY OF MÉXICO

---

[*] Sergio Berumen is a petroleum engineer with over 34 years of experience in the oil and gas industry in projects in reservoir engineering, oil & gas field development, reservoir engineering and geomechanics, well test analysis, reservoir simulation and microseismic technology. He obtained his B.Sc. degree (1980) from National Polytechnic Institute, M.Sc. degree from the University of Mexico (1987) and Ph.D. degree (1995) from the University of Oklahoma. All his degrees are in petroleum engineering. His dissertation work was supervised by Dr. Djebbar Tiab. He is a registered professional at SPE and the Mexican Association of Petroleum Engineers, and he has published more than 40 technical papers. Sergio is also professor of the Graduate School of Engineering at the University of México (UNAM). He was Manager of Technology at Pemex E&P at the time of his retirement from that institution in 2008. Then, in 2009, he served as senior technology advisor to the president of the Mexican Petroleum Institute. From 2010 to 2014 was Director of Diavaz Geosciences, where he was responsible of the development of the heavy oil reserves of naturally fractured heavy oil fields at Tampico-Misantla basin. In October 2014, he was appointed as Exploration and Production technology advisor at Diavaz E&P in México City.

INTRODUCTION

Contrary to many fields in science, petroleum engineers neither design nor see the system with which they are working. Actually, it is through the well that petroleum engineers make contact with the reservoirs; using such indirect measurements as sound traveling, spontaneous potential, hydrogen content, formation and fluid resistivities, densities, pressure wave traveling, etc., petroleum engineers can obtain the desired reservoir parameters.

Well testing is a valuable and economical tool in the oil and gas industry. Thanks to the advances in mathematical modeling, measurement devices and computer capabilities, well testing remains a continuously growing subject. The information obtained from well testing is analyzed with the purpose of obtaining important reservoir information useful for hydrocarbon field management. The intricate mathematical models developed by numerous researchers during past and recent years attempt to show the benefit of mathematics in well test interpretation.

The conventional straight line method was the first tool introduced to interpret well pressure transient behavior. This is based on plotting either pressure or pressure drop (the reciprocal rate can be also used) against a function of time, which depends upon the given flow regime. Then, plots of pressure versus logarithm of time, square root of time or fourth root of time, among others, are meant to be applied for radial, linear and bilinear flow regimes, respectively. This method, however, has two major drawbacks: (1) the difficulty in defining flow regimes and (2) the inability to verify the obtained parameters. Type-curve matching was thus introduced to improve well test interpretation; however, it is both tedious and basically a trial-and-error procedure.

Nowadays, most petroleum engineers use computer software for modeling pressure test behavior. The softwares use either analytical or numerical models that automatically match the well test data by means of nonlinear regression analysis, which is subject of none uniqueness of the solution. The appropriate model selection and the initial input values are the key for a successful interpretation. Some other engineers misuse the software modeling, performing an inverse problem by using models randomly with the aim of matching the well test data to provide the output solution. This procedure is wrong since the engineer must choose the reservoir model.

It is the author's opinion that there have been three milestones that have revolutionized the science of well test interpretation: (1) the introduction of the pressure derivative function, which was first formally used by Tiab in 1976 and spread by Bourdet in the middle 80s; (2)

deconvolution, which allows going deeper in the reservoir; and (3) the *TDS* technique invented by Tiab in 1993.

This book revolves around the *TDS* technique. This revolutionary method is strongly based on the logarithm pressure derivative versus time log-log plot. It is applied to specific regions, features and flow regimes that can be easily identified in the pressure derivative curve so several analytical expressions are obtained for a practical, easy and exact way of conducting a well test interpretation. This tool is powerful and also allows verifying most of the estimated parameters. All the known commercial softwares include it without referring to it by its original name. Over several years of providing training to numerous engineers and companies in Latin America, the author has noticed that whoever knows and uses the *TDS* technique will end up taking it as their favorite interpretation method. Then, they take the outcomes from *TDS* to computer software to set the bounds for faster and less risky modeling.

The book contains the latest application of the *TDS* technique to several important reservoir/fluid scenarios. Several step-by-step examples are given for a better understanding of the interpretation methodology. The book is divided into 15 chapters distributed in three parts. The chapters are not expected to be read in chronological order. However, for beginners with the *TDS* technique or junior engineers, it is strongly recommended to read chapter 1, which deals with infinite, finite and elongated reservoir cases. When possible, conventional analysis is also included in the chapters. Chapter 2 contains a characterization of double-porosity reservoirs and the determination of the average reservoir pressure by the *TDS* technique. Chapter 3 continues working on naturally fractured reservoirs but for cases when the transition period does not fall in the radial flow regime but before fractured wells or after elongated systems. Chapter 4 is focused on the application of the TDS technique to double-permeability reservoirs. And chapters 5 and 6 deals with triple-porosity, single-permeability and triple-porosity, double-permeability reservoirs, respectively. Then, the first part of the book covers homogeneous and heterogeneous reservoirs.

Part 2 deals with non-Newtonian fluids in which heavy oil may fall and some enhanced oil recovery and stimulation projects. Chapter 7 studies infinite systems, and interpretation by conventional analysis is also included. This chapter includes spherical flow, which has a unique pressure derivative behavior, and hydraulically fractured vertical wells. Chapter 8 presents the pressure transient and interpretation of the injection of foams, gels or any non-Newtonian fluid in a well whose reservoir contains conventional Newtonian oil. Wrong interpretation may result from not handling such cases appropriately. The chapter includes the injection of either a dilatant or pseudoplastic power-law non-Newtonian fluids. Chapter 9 considers both bounded and constant-pressure reservoirs with the purpose of determining the well-drainage area. Only the *TDS* technique exists for such situations. Part 3 ends with chapter 10, which presents both the TDS technique and conventional analysis for naturally fractured reservoirs. Bingham fluids in vertical wells are also included in this chapter.

Part 3 is concerned with modern topics recently studied by the author. It starts with chapter 11, which deals with transient rate analysis in homogenous and heterogeneous finite and infinite oil and gas reservoirs. Hydraulically fractured wells producing under constant-pressure conditions and also treated there along with elongated gas reservoirs are also considered. Chapter 12 presents both the conventional analysis and the *TDS* technique for finite-conductivity faults with or without mobility contrast. Chapter 13 is dedicated to composite reservoirs resulting from steam injection, hot water injection or in situ combustion-enhanced oil recovery projects. Only the *TDS* technique is applied in such cases. Chapter 14 includes a characterization of the leakage factor in coalbed methane reservoirs and the spherical stabilization flow regime. Both *TDS* and conventional techniques are used in chapter 14. Finally, chapter 15 deals with the most popular subject in today's oil and gas industry: shale reservoirs. Only two models are considered in this chapter, and the conventional technique exists for only one of them.

PART ONE: HOMOGENEOUS AND HETEROGENEOUS RESERVOIRS

Chapter 1

FUNDAMENTALS OF TRANSIENT PRESSURE ANALYSIS IN HOMOGENEOUS RESERVOIRS

INFINITE BEHAVIOR

Well testing fundamentals are originally based on a book on heat transfer written by Carslaw and Jaeger (1959), who provided the differential equation for heat flow in cylindrical and spherical polar coordinates. Since then, many researchers, starting with Matthews and Russell (1967), have used the famous line-source solution.

The one-dimensional radial flow partial differential equation in dimensionless for is then:

$$\frac{1}{r_D}\frac{\partial}{\partial r_D}\left(r_D\frac{\partial P_D}{\partial r_D}\right) = \frac{\partial P_D}{\partial t_D} \tag{1.1}$$

Where the dimensionless quantities in oil-field units are defined by:

$$t_D = \frac{0.0002637kt}{\phi\mu c_t r_w^2} \tag{1.2}$$

$$P_D = \frac{kh(P_i - P)}{141.2q\mu B} \tag{1.3}$$

$$r_D = \frac{r}{r_w} \tag{1.4}$$

Neglecting both skin factor and wellbore storage effects, the line-source solution of Equation (1) for the constant-rate production well inside an infinite reservoir. is given by:

$$P_D(r_D, t_D) = -\frac{1}{2} \int_x^\infty \frac{e^{-x}}{x} dx \tag{1.5}$$

being x:

$$x = -\frac{r_D^2}{4t_D} = -\frac{948\phi\mu c_t r^2}{kt} \tag{1.6}$$

The integral expression provided in Equation (5) is better known as the exponential integral. Then, Equation (5) becomes:

$$P_D(r_D, t_D) = -\frac{1}{2} Ei\left(-\frac{r_D^2}{4t_D}\right) \tag{1.7}$$

Normally, at very short production times ($x \leq 0.0025$) the Ei(-x) can be taken as ln $(1.781x)$ at the wellbore, and considering the skin effects, Equation (1.7) can be approximated as:

$$P_D + s = -\frac{1}{2}\ln(1.781x) \tag{1.8}$$

After substituting into Equation (8) the definition of dimensionless pressure given by Equation (3) and Equation (6), it yields:

$$P_{wf} = P_i - \frac{162.6q\mu B}{kh}\left[\log\left(\frac{kt}{\phi\mu c_t r_w^2}\right) - 3.2275 + 0.8686s\right] \tag{1.9}$$

The above expression suggests a semilog plot of flowing pressure versus time will provide a linear behavior during radial flow regime whose slope, m, and intercept, $P1hr$, allow obtaining:

$$T = \frac{kh}{\mu} = \left|\frac{162.6qB}{m}\right| \tag{1.10}$$

$$s = 1.1513\left[\frac{P_{1hr} - P_i}{m} - \log\left(\frac{k}{\phi\mu c_t r_w^2}\right) + 3.2275\right] \tag{1.11}$$

CLOSED-BOUNDARY SYSTEMS

Jones (1956) introduced the reservoir limit test. If the external infinite-boundary condition is changed to a closed-boundary condition, $\partial_{PD}/\partial_{rD}\,|\,r_e = 0$, then the late time pseudosteady-state solution is obtained (Ramey and Cobb 1971):

$$P_D = 2\pi t_{DA} + \frac{1}{2}\ln\left(\frac{A}{r_w^2}\right) + \frac{1}{2}\ln\left(\frac{2.5458}{C_A}\right) \tag{1.12}$$

From this, once the dimensionless quantities are replaced above, the Cartesian slope, m^*, and intercept, P_{INT}, during the late pseudosteady-state period are defined as:

$$m^* = -\frac{0.23395qB}{\phi c_t Ah} \tag{1.13}$$

$$P_{INT} = P_i - \frac{70.6q\mu B}{kh}\left[\ln\frac{A}{r_w^2} + \ln\left(\frac{2.2458}{C_A}\right) + 2s\right] \tag{1.14}$$

Solving the Dietz shape factor, C_A, from Equation (1.14) yields:

$$C_A = 5.456\frac{m}{m^*}e^{2.303\frac{P1hr-p_{INT}}{m}} \tag{1.15}$$

The above expression implies the reservoir drainage area can be estimated from the slope and the shape factor from the intercept of a Cartesian plot of pressure versus time during the late pseudosteady-state period.

The former expressions apply only to drawdown tests; for pressure buildup tests, time superposition applies, as do with Equations (1.3) and (1.8), leading to:

$$P_{ws} = P_i - \frac{162.6q\mu B}{kh}\log\left(\frac{t_p + \Delta t}{\Delta t}\right) \tag{1.16}$$

where the slope is similar to Equation (1.10). The skin factor is calculated from:

$$s = 1.1513\left[\frac{P_{1hr} - P_{wf}}{m} - \log\left(\frac{k}{\phi\mu c_t r_w^2}\right) + 3.2275\right] \tag{1.17}$$

$$s = 1.1513\left[\frac{P_{1hr} - P_{wf}}{m} + \log\left(1 + \frac{1}{t_p}\right) - \log\left(\frac{k}{\phi\mu c_t r_w^2}\right) + 3.2275\right]; \text{if } t_p < 1hr \tag{1.18}$$

TDS TECHNIQUE FOR INFINITE AND CLOSED-BOUNDARY SYSTEMS

So far, these equations constitute part of the conventional straight line method, which is based upon building a pressure or pressure drop plot against a function of time, the slope and intercept of which will help to determine important reservoir parameters.

In 1975, Tiab invented the pressure derivative function Tiab (1975), Tiab and Kumar (1980a, 1980b). He took the derivative to the E_i function using Leibitz's rule and obtained:

$$\frac{\partial P_D}{\partial t_D} = \frac{1}{2t_D} e^{-x}$$

(1.16)

The above expression, known as the Cartesian or arithmetic derivative, uncovered the radial flow regime as a negative one slope in the log-log arithmetic derivative versus time. Later on, this flow regime was recognized as a zero slope (flat line) on the natural log pressure derivative log-log plot, which is today better known as the pressure derivative.

$$t_D * \frac{\partial P_D}{\partial t_D} = \frac{\partial P_D}{\partial \ln t_D} = \frac{1}{2} e^{-x}$$

(1.17)

The pressure derivative has played a remarkable role in well test analysis since flow regimes are better identified according to the appropriate behavior (slope) each of them follows. This function was the basis for the invention of Tiab's direct synthesis technique (*TDS* technique). The pioneer paper was first published in 1993 in an SPE Production and Operations Symposium in Oklahoma City (Tiab 1993a), and then it was accepted as a peer-reviewed paper in a very prestigious journal (Tiab 1995).

Based upon Equation (1.16), Tiab (1993a, 1995) found the formation permeability could easily be estimated from the pressure derivative during radial flow regime, $(t*\Delta P')_r$:

$$k = \frac{70.6 q \mu B}{h(t * \Delta P')_r}$$

(1.18)

Tiab (1993a, 1995) used suffixes to point out the precedence of the parameter. For instance, letter r in Equation (1.18) denotes "radial." Notice that a zero slope straight line intercepting the y axis in 0.5 should be drawn.

He also demonstrated the wellbore storage coefficient can be estimated from a point, N, on the early unit-slope line of either the pressure or pressure derivative:

$$C = \left(\frac{qB}{24}\right) \frac{t_N}{\Delta P_N} = \left(\frac{qB}{24}\right) \frac{t_N}{(t * \Delta P')_N}$$

(1.19)

During the radial flow regime, the governing dimensionless pressure equation is given by:

Fundamentals of Transient Pressure Analysis in Homogeneous Reservoirs 7

$$P_{Dr} = 0.5\left\{\ln(t_D / c_D)_r + 0.80907 + \ln[c_D e^{2s}]\right\} \tag{1.20}$$

By dividing the dimensionless pressure expression, Equation (1.20), by the pressure derivative, Equation (1.17), which actually takes the value of 0.5, and solving for the skin factor, Tiab (1993a, 1995) also found:

$$s = 0.5\left(\frac{\Delta P_r}{(t * \Delta P')_r} - \ln\left[\frac{k\, t_r}{\phi\mu c_t r_w^2}\right] + 7.43\right) \tag{1.21}$$

Where t_r is an arbitrary time during the radial flow regime at which ΔP_r was read.

By drawing a unit-slope line along the wellbore storage-dominated period and intercepting that with the zero-slope radial straight line, Tiab obtained two more expressions to either estimate permeability or verify wellbore storage:

$$k = \frac{70.6q\mu B}{h(t * \Delta P')_i} \tag{1.22}$$

$$k = \frac{1695\mu C}{h t_i} \tag{1.23}$$

where the suffix i stands for intersection.

Especially for short tests in which the radial flow regime may not be seen, Tiab (1993a, 1995) also developed correlations based upon the coordinates of the maximum pressure derivative, t_x and $(t*\Delta P')_x$, taking place at the "hump" of the pressure derivative curve during the transition from wellbore storage to radial. These expressions are shown below:

$$k = \left(\frac{70.6q\,\mu B}{h}\right)\frac{1}{(0.014879qB / C)t_x - (t * \Delta P')_x} \tag{1.24}$$

$$k = 4745.36\frac{\mu C}{h t_x}\left[\frac{(t * \Delta P')_x}{(t * \Delta P')_r} + 1\right] \tag{1.25}$$

$$s = 0.171\left(\frac{t_x}{t_i}\right)^{1.24} - 0.5\ln\left(\frac{0.8935C}{\phi h c_t r_w^2}\right) \tag{1.26}$$

$$s = 0.921\left[\frac{(t * \Delta P')_x}{(t * \Delta P')_i}\right]^{1.1} - 0.5\ln\left(\frac{0.8935C}{\phi h c_t r_w^2}\right) \tag{1.27}$$

$$C = \frac{0.014789qBt_x}{(t*\Delta P')_x + (t*\Delta P')_r} \quad (1.28)$$

Later, Tiab (1993b, 1994) used the interception point, t_{rpi}, of the derivative taken from Equation (1.12) and Equation (1.17) to obtain an accurate and practical formula to calculate the well drainage area:

$$A = \frac{kt_{rpi}}{301.77\phi\mu c_t} \quad (1.29)$$

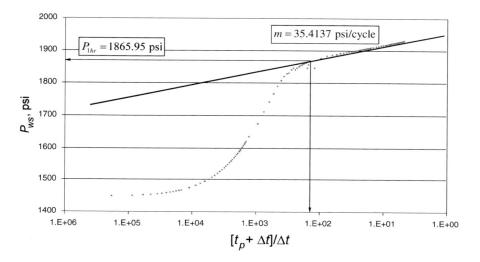

Figure 1.1. Horner plot for example 1.1.

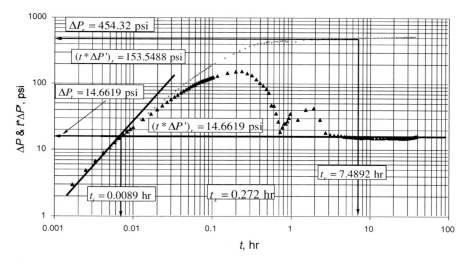

Figure 1.2. Pressure and pressure derivative log-log plot for example 1.1.

Fundamentals of Transient Pressure Analysis in Homogeneous Reservoirs   9

TDS technique equations for gas wells can be found in the work by Nunez, Tiab and Escobar (2003) or, including the pseudotime function, in a paper presented by Escobar, López and Cantillo (2007c).

It is worth mentioning that four SPE monographs have been written in the field of well test analysis. The third one (volume 23) was written in 2009, but it does not mention anything related to the *TDS* technique. The last one, by Spivey and Lee (2013), uses Equation (1.18) exactly and a variation of Equation (1.21) on page 91. Although, the authors called it "manual log-log analysis," the procedure given on page 91 is basically the same as originally reported by Tiab (1993a, 1995). Also, most commercial well test softwares use the *TDS* technique without referring to it as such.

Example 1.1

An oil well produced for 134.38 hr, and it was then shut in to run a pressure buildup test. Relevant information concerning this test is given below, and pressure versus time data are given in Table 1.1. It is required to find the total mobility and skin factor by conventional analysis and by the *TDS* technique.

$\phi = 20\%$	$h = 40$ ft	$c_t = 9\text{x}10^{-6}$ psi^{-1}
$B_o = 1.133$ bbl/STB	$B_w = 1.014$ bbl/STB	$B_g = 0.000979$ ft^3/STB
$\mu_w = 0.419$ cp	$\mu_o = 19.4$ cp	$\mu_g = 0.02075$ cp
$q_o = 895.23$ BPD	$q_w = 1.76$ BPD	$q_g = 451882$ scf/D
$r_w = 0.375$ ft	$T = 153$ °F	$R_s = 279$ scf/STB
$P_i = 1440.891$ psi		

Solution by Conventional Analysis

A Horner plot was built and provided in Figure 1.1. The pressure and pressure derivative log-log plot is also provided in Figure 1.2. Notice that at about 0.7 hr an anomalous behavior is observed. This ends at about 3 hr and masks the start of the radial flow regime. Phase segregation may be responsible for such behavior.

Since this is a multiphase test, the total flow rate must then be estimated for the application of Perrine's approximation (Perrine 1956) by means of the following equation:

$$q_t = q_o B_o + (q_g - q_o R_s)B_g + q_w B_w \tag{1.30}$$

Then, the application of Equation (1.30) is:

$$q_t = 895.23(1.133) + [451882 - 895.23(279)]0.000979 + 1.76(1.014) = 1213 \text{ STB/D}$$

Table 1.1. Pressure data for example 1.2

t, hr	ΔP, psi	$t*\Delta P$, psi	t, hr	ΔP, psi	$t*\Delta P$, psi	t, hr	ΔP, psi	$t*\Delta P$, psi
0.0008	0.889	1.161	0.0933	159.506	119.373	5.1092	450.451	15.702
0.0017	2.125	2.945	0.0967	163.946	121.099	5.5058	451.611	15.542
0.0025	3.557	4.827	0.1	168.353	122.711	5.9025	452.636	15.472
0.0033	5.33	6.719	0.1033	172.428	124.391	6.2992	453.735	15.367
0.0042	6.809	8.869	0.115	186.717	130.876	6.6958	454.584	15.313
0.005	8.375	10.598	0.1558	226.389	144.487	7.0925	455.598	15.222
0.0058	10.17	12.781	0.1967	262.977	153.403	7.4892	456.311	15.162
0.0067	12.12	14.630	0.2375	293.451	154.169	7.8858	457.063	15.151
0.0075	13.608	16.516	0.2783	318.913	143.625	8.2825	457.868	15.238
0.0083	15.496	18.387	0.3192	342.192	133.446	8.6792	458.6	15.232
0.0092	17.622	20.098	0.36	358.995	115.135	9.0758	459.177	15.219
0.01	19.16	21.761	0.4008	372.075	102.917	9.4725	459.906	15.224
0.0133	26.051	28.477	0.4417	380.685	93.547	9.8692	460.467	15.210
0.0167	32.833	34.580	0.4825	382.965	82.683	10.818	461.885	15.120
0.02	39.572	40.588	0.5233	393.307	72.567	12.171	463.596	15.090
0.0233	46.007	46.364	0.5642	398.306	46.870	13.524	465.296	15.122
0.0267	52.432	51.969	0.605	392.728	46.132	14.878	466.744	15.127
0.03	58.719	57.299	0.6458	398.344	35.463	16.231	468.042	15.169
0.0333	64.811	62.325	0.6867	402.496	26.507	17.584	469.202	15.059
0.0367	70.707	67.569	0.7275	405.915	18.550	18.938	470.357	15.105
0.04	76.904	72.093	0.7683	407.552	21.521	20.291	471.378	15.101
0.0433	82.705	76.707	0.8092	410.864	24.780	21.644	472.373	14.986
0.0467	88.462	80.947	0.85	413.887	24.786	22.998	473.315	14.988
0.05	94.138	85.069	0.8908	413.49	27.014	24.351	474.114	14.878
0.0533	99.554	89.229	0.9317	397.959	30.497	25.704	474.978	14.847
0.0567	105.219	92.992	0.9725	408.368	35.025	27.058	475.776	14.846
0.06	110.723	96.714	1.2133	398.582	28.544	28.411	476.429	14.911
0.0633	116.088	99.804	1.5392	428.967	40.550	29.764	477.1	14.977
0.0667	120.925	102.828	1.9358	434.253	42.733	31.118	477.787	15.045
0.07	126.16	105.944	2.3325	437.591	28.621	32.471	478.411	15.188
0.0733	131.3	107.784	2.7292	440.294	17.984	33.824	479.04	15.341
0.0767	136.096	109.625	3.1258	442.582	16.860	35.178	479.57	15.506
0.08	140.599	111.512	3.5225	444.544	16.449	36.531	480.14	15.691
0.0833	145.788	113.454	3.9192	446.311	16.188	37.884	480.727	15.888
0.0867	150.556	115.512	4.3158	447.824	16.105	39.752	481.37	16.277
0.09	154.959	117.470	4.7125	449.281	15.804			

With a slope of 35.4137 psi/cycle seen in Figure 1.1, the total fluid mobility can be estimated from Equation (1.10):

$$\frac{k}{\mu}\bigg|_t = \frac{162.6q_t}{mh} = \frac{162.6(1213)}{(35.4137)(40)} = 139.23 \text{ md/cp}$$

Fundamentals of Transient Pressure Analysis in Homogeneous Reservoirs 11

The skin factor is estimated using Equation (1.17):

$$s = 1.1513\left[\frac{1865.95 - 1449.89}{35.4137} - \log\left(\frac{139.23}{(0.2)(5.95\times10^{-6})(0.375^2)}\right) + 3.2275\right] = 7.26$$

Solution by TDS technique The following information was obtained from Figure 1.2:

$t_r = 7.4892$ hr $(t^*\Delta P')_r = 14.6619$ psi $\Delta P_r = 454.32$ psi

$t_i = 0.0089$ hr $\Delta P_i = 14.6619$ psi $t_x = 0.272$ hr

$(t^*\Delta P')_x = 153.5488$ psi

The wellbore storage coefficient can be found using Equation (1.19):

$$C = \left(\frac{q_t}{24}\right)\frac{t_i}{\Delta P_i} = \left(\frac{1213}{24}\right)\frac{0.0089}{14.6619} = 0.03079 \text{ bbl/psi}$$

The total mobility is estimated with Equations (1.18), (1.23) and (1.24):

$$\frac{k}{\mu}\Big|_t = \frac{70.6(1213)}{40(14.6619)} = 146.0203 \text{ md/cp}$$

$$\frac{k}{\mu}\Big|_t = \frac{1695C}{ht_i} = \frac{1695(0.03079)}{(40)(0.0089)} = 146.072 \text{ md/cp}$$

$$\frac{k}{\mu}\Big|_t = \left(\frac{70.6(1213)}{40}\right)\frac{1}{(0.014879(1213)/0.03079)0.272 - 153.5488} = 154.0816 \text{ md/cp}$$

The skin factor is estimated with Equations (1.21) and (1.26):

$t_r = 7.4892$ hr $(t^*\Delta P')_r = 14.6619$ psi $\Delta P_r = 454.32$ psi

$t_i = 0.0089$ hr $\Delta P_i = 14.6619$ psi $t_x = 0.272$ hr

$(t^*\Delta P')_x = 153.5488$ psi

$$s = 0.5\left(\frac{454.32}{14.6619} - \ln\left[\frac{(146.02)(7.4892)}{(0.2)(5.95\times10^{-6})(0.375^2)}\right] + 7.43\right) = 5.6055$$

$$s = 0.171\left(\frac{0.272}{0.0089}\right)^{1.24} - 0.5\ln\left(\frac{0.8935(0.03079)}{(0.2)(40)(5.95\times10^{-6})(0.375^2)}\right) = 5.3365$$

Table 1.2. Comparison of results from example 1.1

Parameter	Value	Method
$(k/\mu)t$, md/cp	146.0203	*TDS*
$(k/\mu)t$, md/cp	146.072	*TDS*
$(k/\mu)t$, md/cp	154.0816	*TDS*
$(k/\mu)t$, md/cp	139.23	Conventional
s	5.6055	*TDS*
s	5.3365	*TDS*
s	7.26	Conventional

As observed in table 1.2, the results obtained from the *TDS* technique are comparable to those of conventional analysis, with one remark: *TDS* allows the determination of the same parameter —in this case mobility—from more than one source, indicating the information can be verified.

CONSTANT-PRESSURE BOUNDARIES

Since Equations (1.13) and (1.29) only applied to closed-boundary reservoirs of any shape, Escobar, Hernandez and Tiab (2010) implemented the *TDS* methodology for constant-pressure boundaries for different reservoir geometry. Escobart et al. (2010) noticed that for long and narrow reservoirs, which will be briefly studied next, the applied equation depends upon the position of the well along the reservoir if at least one of the lateral boundaries is at constant pressure. Their developed equations are reported in Table 1.3. Dashed lines indicate a constant-pressure boundary.

As mentioned earlier, the Cartesian plot slope, $m*$, cannot be used in Equation (1.13) because it has a zero value in strong water influx reservoirs; therefore, Escobar et al. (2015) developed conventional analysis for such a situation. Their calculation is based upon reading the starting time of the steady-state period, which takes place once a constant-zero slope on the semilog or Cartesian plot of pressure versus time is developed. In a case similar to the one presented by Escobar et al. (2010), the developed equations are summarized in Table 1.4. Since pressure is not involved in the expressions, the equations from Table 1.4 are used for both oil and gas reservoirs. An example of these cases is given in the next section.

LONG AND NARROW SYSTEMS—*TDS* TECHNIQUE

Transient linear flow may be presented once the radial flow regime vanishes in elongated systems. Systems with approximately parallel no-flow boundaries arise in three geological situations: (a) terrace faulting or wells between two faults, (b) fluvial deposition (deltaic) and (c) carbonate reefs. The second case is common in the Magdalena River Basin in Colombia, as well as in some places in Venezuela and Canada.

Fundamentals of Transient Pressure Analysis in Homogeneous Reservoirs 13

Table 1.3. Summary of equations for area determination using the *TDS* technique; adapted from Escobar et al. (2010)

Constant, Ξ	Equation	Equation Number	Reservoir Geometries
	$A = \dfrac{kt_{rpssi}}{283.66\phi\mu c_t}$	(1.31)	
4066			
482.84	$A = \sqrt[3]{\dfrac{kt_{ssri}Y_E^4}{\Xi\phi\mu c_t}}$	(1.32)	
7584.2			
2173.52	$A = \left(\dfrac{kt_{sslri,ss2ri}}{\Xi\phi\mu c_t}\right)^{2/3}\dfrac{Y_E^{5/3}}{b_x}$	(1.33)	
6828.34			

Table 1.4. Summary of equations for area determination using conventional analysis; adapted from Escobar et al. (2015)

Constant, Ξ	Equation	Reservoir Geometries
869.9		
23366.7		
37034	$A = \dfrac{kt_{sss}}{\Xi\phi\mu c_t}$	
41414	(1.34)	
7389.1		
46263.1		

Not many authors have devoted their research to elongated systems. In a book chapter, Escobar (2008) presented a comprehensive study of long reservoirs. It started from an SPE paper presented in 2004 (Escobar et al. 2004) in which the *TDS* technique was introduced for such systems. An improved publication was conducted by Escobar, Hernandez and Hernandez (2007a). Then, Escobar et al. (2005) introduced the parabolic flow, its governing equation and the conditions for this to show up. A study of the geometrical skin factors occurring by convergence from radial flow to linear flow and from linear flow to hemilinear flow was presented by Escobar and Montealegre (2006). Also, Escobar, Tiab and Tovar (2007b) took advantage of the presence of the linear flow regime to determine areal anisotropy whenever the reservoir width is known. The most important equations for characterizing long and narrow reservoirs are presented here.

For mathematical modeling, the geometry of these reservoirs is assumed to be rectangular. The dimensionless quantities normally used in the well test analysis literature are given as follows:

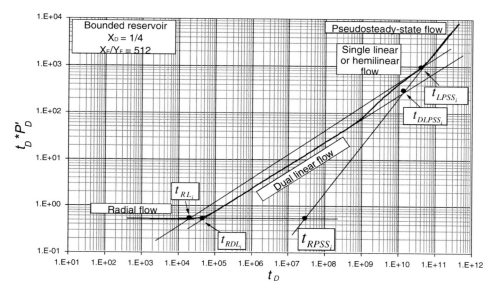

Figure 1.3. Pseudosteady-state line intersection with either linear, dual-linear or radial flow lines; adapted from Escobar et al. (2007a).

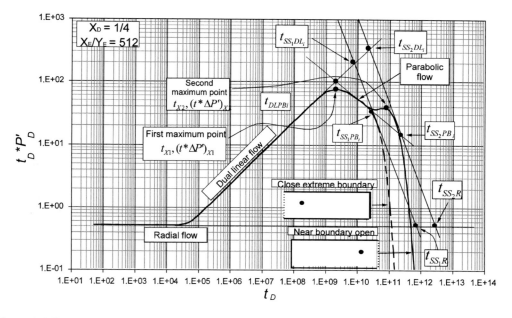

Figure 1.4. Pressure derivative behavior for a well (a) both extreme boundaries are open, (b) open near boundary and close far boundary; adapted from Escobar et al. (2007a).

$$t_{DA} = \frac{0.0002637kt}{\phi\mu c_t A} \tag{1.35}$$

$$t_{D_L} = \frac{t_D}{W_D^2} \tag{1.36}$$

$$W_D = \frac{Y_E}{r_w} \tag{1.37}$$

$$X_D = \frac{2b_x}{X_E} \tag{1.38}$$

$$Y_D = \frac{2b_y}{Y_E} \tag{1.39}$$

$$t_D * P_D' = \frac{kh(t*\Delta P')}{141.2q\mu B} \tag{1.40}$$

As observed in Figure 1.3, a transient linear or hemilinear flow regime is observed when the well is located at one of the reservoir extremes. This means a single-linear flow takes place along one side of the reservoir. This is governed by the following constant-rate equation (Escobar et al. 2004, 2007a):

$$P_D = 2\pi\sqrt{t_{D_L}} + s_L = \frac{2\pi\sqrt{t_D}}{W_D} + s_L' + s_{DL}' \tag{1.41}$$

As long as the well is located far away from one of the extreme boundaries, a true linear flow regime is developed; for purposes of practical identification, Escobar et al. (2004, 2007a) called this as dual-linear flow, which is developed before the single-linear flow. In this case, the governing equation is:

$$P_D = 2\sqrt{\pi t_{D_L}} + s_{DL} = \frac{2\sqrt{\pi t_D}}{W_D} + s_{DL}' + s \tag{1.42}$$

As observed in Figure 1.4, parabolic flow, characterized by a slope of -½ on the pressure derivative curve, develops as a result of the simultaneous effect of an open boundary near the well and the expected linear flow regime along the far lateral side of the reservoir (Escobar et al. 2005). The pressure equation for this flow regime is:

$$P_D = -(W_D)(X_D)^2\left(\frac{X_E}{Y_E}\right)^2 t_D^{-0.5} + s_{PB} = -(W_D)(X_D)^2\left(\frac{X_E}{Y_E}\right)^2 t_D^{-0.5} + s_{PB}' + s_{DL}' \quad (1.43)$$

Suffixes *L*, *DL* and *PB* in Equations (1.42) through (1.44) stand for linear or single-linear or hemilinear, dual-linear and parabolic flow regimes. Their pressure derivatives are:

$$(t_D * P_D')_L = \frac{\pi\sqrt{t_D}}{W_D} \quad (1.44)$$

$$(t_D * P_D')_{DL} = \frac{\sqrt{\pi t_D}}{W_D} \quad (1.45)$$

$$t_D * P'_D = \frac{W_D}{2}(X_D)^2\left(\frac{X_E}{Y_E}\right)^2 t_D^{-0.5} \quad (1.45)$$

Since reservoir permeability is taken from radial flow (assuming areal isotropy), reservoir width can then be solved from the above equations once the dimensionless quantities are replaced:

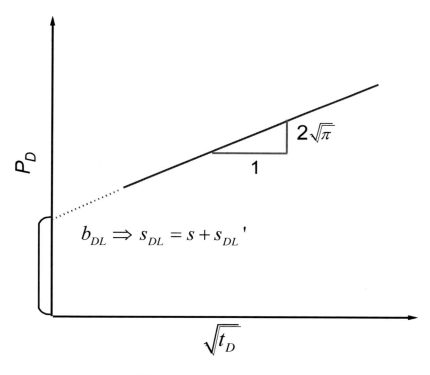

Figure 1.5. Cartesian plot of *PD* vs. $t_D^{0.5}$ during transient dual-linear flow regime

$$Y_E = \frac{7.2034qB}{h(t*\Delta P')_L}\sqrt{\frac{\Delta t_L \mu}{k\phi c_t}} \tag{1.47}$$

$$Y_E = \frac{4.064qB}{h(t*\Delta P')_{DL}}\sqrt{\frac{\Delta t_{DL} \mu}{k\phi c_t}} \tag{1.48}$$

Notice that for simplicity the pressure derivative value could be read at the time, $\Delta t = 1$ hr. The skin factor caused by the convergence of radial flow into linear flow is determined by dividing the dimensionless pressure equation by the dimensionless pressure derivative equation. The same procedure is performed for the skin factor due to the convergence of lineal to dual-linear, lineal to radial or dual-linear to parabolic flow. After replacing the dimensionless quantities in these and solving for the skin factor, the following are obtained:

$$s_L = \left(\frac{\Delta P_L}{(t*\Delta P')_L} - 2\right)\frac{1}{34.743Y_E}\sqrt{\frac{kt_L}{\phi\mu c_t}} \tag{1.49}$$

$$s_{DL} = \left(\frac{\Delta P_{DL}}{(t*\Delta P')_{DL}} - 2\right)\frac{1}{19.601Y_E}\sqrt{\frac{kt_{DL}}{\phi\mu c_t}} \tag{1.50}$$

$$s_{PB} = \left(\frac{\Delta P_{PB}}{(t*\Delta P')_{PB}} + 2\right)\left(\frac{123.16b_x^2}{Y_E}\right)\sqrt{\frac{\phi\mu c_t}{kt_{PB}}} \tag{1.51}$$

where Δ_L and $(t*\Delta P)_L$ are the pressure and pressure derivative values, respectively, of the linear flow regime during any convenient time, t_L. This is similar for the dual-linear and parabolic cases. The total skin factor results as the summation of the geometric and mechanical skin factors. Replacing the dimensionless parameters in Equation (1.47) will result in a relationship for obtaining well location:

$$b_x^2 = \frac{17390}{k^{1.5}Y_E}\left[\frac{q\mu B}{h(t*\Delta P')_{PB}}\right]\left[\frac{\phi\mu c_t}{t_{PB}}\right]^{0.5} \tag{1.52}$$

where $(t*\Delta P)_{PB}$ is the pressure derivative during parabolic flow at any convenient time, t_{PB}.

Escobar et al. (2004, 2007a) reported more useful equations using the intersection formed between the different straight lines drawn along the diverse flow regimes and the maximum points as well.

LONG AND NARROW SYSTEMS—CONVENTIONAL ANALYSIS

Escobar and Montealegre (2007) presented the equations for the interpretation of pressure tests by conventional analysis, expressing Equation (1.43) in dimensional form:

$$\Delta P = \frac{8.1282}{Y_E} \frac{qB}{kh} \left(\frac{\mu}{\phi c_t k} \right)^{0.5} \sqrt{t} + \frac{141.2 q \mu B}{kh} s_{DL} \tag{1.53}$$

which indicates a Cartesian plot of pressure versus the square root of time (or the tandem square root for buildup tests)—see Figure 1.5—will provide a straight line whose slope and intercept are used to estimate reservoir width and geometrical skin factor, respectively:

$$Y_E = 8.1282 \frac{qB}{m_{DLF} h} \left[\frac{\mu}{k \phi c_t} \right]^{0.5} \tag{1.54}$$

$$s_{DL} = \frac{khb_{DLF}}{141.2 q \mu B} \tag{1.55}$$

The governing equation for hemilinear flow will also provide the means to estimate reservoir width and geometrical skin factor:

$$\Delta P_{wf} = \frac{14.407}{Y_E} \frac{q \mu B}{kh} \sqrt{\frac{kt}{\phi \mu c_t}} + \frac{141.2 q \mu B}{kh} s_L \tag{1.56}$$

$$Y_E = \frac{14.407}{m_{LF}} \frac{qB}{h} \left(\frac{\mu}{\phi c_t k} \right)^{0.5} \tag{1.57}$$

$$s_L = \frac{khb_{LF}}{141.2 q \mu B} \tag{1.58}$$

Finally, the governing equations for parabolic flow introduced by Escobar et al. (2004, 2005, 2007a) for drawdown and buildup tests in dimensional form are given below:

$$\Delta P_{wf} = -\frac{34780.8 q B b_x^2 \sqrt{\phi c_t}}{h Y_E} \left(\frac{\mu}{k} \right)^{1.5} \frac{1}{\sqrt{t}} + \frac{141.2 q \mu B}{kh} s_{PB} \tag{1.59}$$

$$\Delta P_{ws} = -\frac{34780.8 q B b_x^2 \sqrt{\phi c_t}}{h Y_E} \left(\frac{\mu}{k} \right)^{1.5} \left(\frac{1}{\sqrt{t_p + \Delta t}} - \frac{1}{\sqrt{\Delta t}} \right) \tag{1.60}$$

So, a Cartesian plot of pressure drop versus either $1/t^{0.5}$ or $1/(t_p + \Delta t)^{0.5} - 1/\Delta t^{0.5}$ will develop a straight line (during parabolic flow) whose slope, m_{PB}, and intercept, b_{PB}, allow us to obtain well location along the x direction (well eccentricity inside the reservoir), b_x, and geometric skin factor, s_{PB}, respectively.

Fundamentals of Transient Pressure Analysis in Homogeneous Reservoirs   19

$$b_x = \sqrt{-\frac{m_{PB} h Y_E}{34780.8 q B \sqrt{\phi c_t}} \left(\frac{k}{\mu}\right)^{1.5}} \qquad (1.61)$$

$$s_{PB} = \frac{kh}{141.2 q \mu B} b_{PB} \qquad (1.62)$$

The intercept in Equations (1.55), (1.58) and (1.62) refers to a pressure drop, ΔP; then, if a plot of pressure versus the appropriate function of time is built, that intercept value must be converted to pressure drop.

Example 1.2

Escobar et al. (2007) presented a synthetic example of a pressure test in which the well is off-centered along a long reservoir (see Table 1.5). The near boundary is closed and the other one is at constant pressure. Reservoir and well parameters are given below, pressure versus the square root of time is given in Figure 1.5 and pressure and pressure derivative data versus time are given in Figure 1.7. Characterizing this reservoir using both conventional and *TDS* techniques is required:

q = 500 BPD	h = 100 ft	c_t = 1x10-5 psi-1
r_w =0.3 ft	ϕ = 10%	B =1.25 bbl/STB
μ =5 cp	Y_E = 800 ft	X_E = 5000 ft
k = 100 md	b_x = 1000 ft	P_i = 3000 psi

Solution by conventional analysis

The slope and intercept (m_{DL} = -14.27 psi/hr0.5, bDL = 266.4 psi) obtained from Figure 1.6 were applied to Equations (1.54) and (1.55) to yield an estimation of Y_E = 796.34 ft and s_{DL} = 6.04, respectively. Also, from Figure 1.5, the values of m_{LF} = 25.47 psi/hr0.5 and b_{DLF} = 96.3 psi were applied to Equation (1.56) to estimate a Y_E value of 790.63 ft, and Equation (1.57) was used to estimate an s_L of 2.18. A summary of the results are given in Table 1.5. A very good agreement is observed between the calculations by both the conventional analysis and the *TDS* technique against the input Y_E value.

Table 1.5. Pressure data for example 1.2

t, hr	ΔP, psi	$t*\Delta P$, psi	t, hr	ΔP, psi	$t*\Delta P$, psi	t, hr	ΔP, psi	$t*\Delta P$, psi
0.01	23.560	22.415	1.600	267.413	25.519	201.485	488.035	142.964
0.02	43.823	38.298	2.015	273.120	24.557	253.654	523.264	169.572
0.03	61.540	49.973	2.537	278.699	24.027	319.332	565.032	200.287
0.04	77.122	58.156	3.193	284.148	23.578	402.015	614.328	235.293
0.05	90.881	63.675	4.020	289.505	23.278	501.069	669.497	272.729
0.06	103.075	67.309	5.061	294.801	23.208	601.069	721.636	305.976

Table 1.5. (Continued)

t, hr	ΔP, psi	t*ΔP, psi	t, hr	ΔP, psi	t*ΔP, psi	t, hr	ΔP, psi	t*ΔP, psi
0.07	113.924	70.007	6.372	300.091	23.464	701.069	770.694	331.550
0.08	123.609	70.995	8.021	305.463	24.155	801.069	816.992	358.111
0.09	132.289	71.880	10.098	311.040	25.377	901.069	860.753	377.570
0.113	149.317	71.244	12.713	316.962	27.186	1001.069	902.147	395.791
0.143	165.923	68.536	16.005	323.376	29.597	1351.069	1030.245	431.239
0.180	181.562	64.126	20.148	330.423	32.618	1701.069	1135.661	442.796
0.226	195.828	58.310	25.365	338.233	36.282	2051.069	1222.274	440.776
0.285	208.512	51.957	31.933	346.943	40.705	2601.069	1328.265	409.910
0.358	219.606	45.746	40.202	356.726	46.104	3301.069	1424.252	357.186
0.451	229.260	40.189	50.611	367.820	52.807	4001.069	1490.248	307.394
0.568	237.718	35.568	63.715	380.560	61.217	4851.069	1543.915	247.728
0.715	245.248	32.025	80.213	395.392	71.770	5901.069	1585.546	190.453
0.900	252.096	29.196	100.982	412.870	84.881	7051.069	1612.865	153.140
1.133	258.461	27.268	127.128	433.646	100.914	8251.069	1629.887	90.389
1.271	261.510	26.661	160.045	458.440	120.180	9851.069	1642.202	121.857

Figure 1.7 is used for the estimation of area. The time value, t_{sss}, at which the pressure stabilizes is about 8300 hr; then, Equation (1.34) can be applied:

$$A = \frac{kt_{sss}}{37034\phi\mu c_t} = \frac{100(7400)}{41414(0.1)(5)(1\times10^{-5})} = 4008306 \text{ ft}^2$$

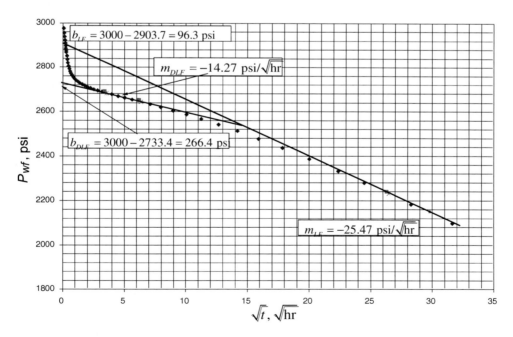

Figure 1.6. Cartesian plot of P_{wf} vs. $t^{0.5}$ for example 1.2

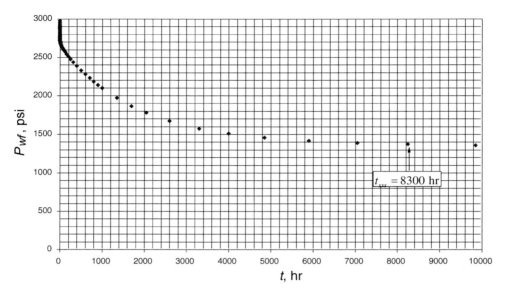

Figure 1.7. Cartesian plot of P_{wf} vs. t for example 1.2

Solution by TDS Technique

The following parameters were read from the pressure and pressure derivative versus time log-log plot given in Figure 1.8:

t_{DL} = 28.46 hr $(t*\Delta P')_{DL}$=37.83 psi ΔP_{DL}=342.47 psi
t_L=801.07 hr $(t*\Delta P')_L$= 360.39 psi ΔP_L= 817 psi
t_{SS1Ri} = 60000 hr

Equations (1.47) and (1.48) provide the estimation of reservoir width values of 810 and 790.6 ft, respectively. The geometric skin factors were calculated with Equations (1.49) and (1.50) to be s_L = 1.23 and s_{DL} = 10.72, respectively. To find reservoir area, Equation (1.33), which is a function of the intersection time between the radial flow and the negative-unit-slope line drawn on the steady-state period, was used:

$$A = \sqrt[3]{\frac{kt_{ssri}Y_E^4}{7584.2\phi\mu c_t}} = \sqrt[3]{\frac{(100)(60000)(801^4)}{7584.2(0.1)(5)(1\times10^{-5})}} = 4016772 \text{ ft}^2$$

A summary of the results is given in Table 1.6, where a good match between the methods is observed. There exist some differences in the geometric skin factor, which the author considers to be acceptable. It is worth remarking that two more values for reservoir width and three values for drainage area can be computed by the *TDS* technique from the intercepts of the linear flow regimes with both the radial flow regime straight line and the negative late time unit-slope line, as well as from the maximum point formed before the steady-state period starts. Those expressions are not reported here, but the reader can refer to Escobar et al. (2004, 2007a) and Escobar (2008).

It is also worth mentioning some research conducted on elongated systems in which either the reservoir width or permeability (due to possible changes in facies) changes. Conventional analysis and TDS methodologies were presented, respectively, by Escobar, Montealegre and Carrillo-Moreno (2009) and Escobar, Montealegre and Carrillo-Moreno (2010).

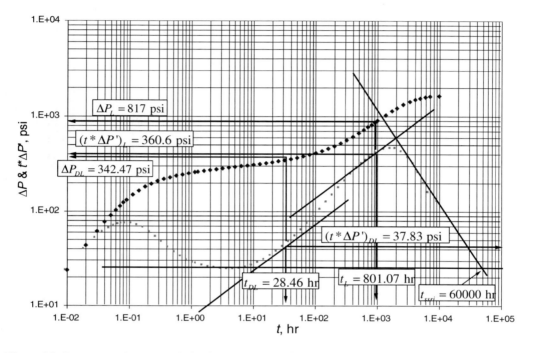

Figure 1.8. Pressure and pressure derivative log-log plot for example 1.2.

Table 1.6. Comparison and summary of results for example 1.2

Parameter	Value	Flow Regime	Method	Absolute Error
Y_E, ft	796.34	Dual-linear	*Conventional*	0.46
Y_E, ft	790.63	Linear	*Conventional*	1.19
Y_E, ft	810	Dual-linear	*TDS*	1.23
Y_E, ft	790.6	Linear	*TDS*	1.19
s_{DL}	6.04	Dual-linear	*Conventional*	-
s_{DL}	10.72	Dual-linear	*TDS*	-
s_L	2.18	Linear	*Conventional*	-
s_L	1.23	Linear	*TDS*	-
A, acres	3996328		*Conventional*	0.09
A, acres	4008306		*TDS*	0.21

INFINITE BEHAVIOR—GAS RESERVOIRS—CONVENTIONAL ANALYSIS

$$m(P_i) - m(P_{wf}) = \frac{1637.74qT}{kh}\left[\log\left(\frac{kt}{\phi\bar{\mu}_g c_t r_w^2}\right) - 3.23 + 0.8686(Dq + s)\right] \quad (1.63)$$

$$k = \frac{1637.74qT}{mh} \quad (1.64)$$

$$s = \left[\frac{m(P_i) - m(P_{1hr})}{m} - \log\left(\frac{k}{\phi(\mu c_t)_i r_w^2}\right) - 3.227 + 0.8686\right] \quad (1.65)$$

CLOSED GAS RESERVOIRS—CONVENTIONAL ANALYSIS

$$A = \frac{2356q\,T}{\phi h m^*(\mu c_t)_i} \quad (1.66)$$

INFINITE BEHAVIOR—GAS RESERVOIRS—TDS TECHNIQUE

$$C = \left(0.419\frac{q_{sc}T}{\mu_i}\right)\left(\frac{t}{t*\Delta m(P)'}\right) \quad (1.67)$$

Table 1.7. Summary of equations for area determination in gas reservoirs using the *TDS* technique

Constant, Ξ	Equation	Equation	Reservoir Geometries			
0.23395	(1.71)	$A = \dfrac{\Xi qBt_{pss}}{h\phi c_t(t*\Delta P')_{pss}}$	•	⊙	•	•
20026.05	(1.72)	$A = \dfrac{k^2h(t*\Delta P')_{ss}t_{ss}}{\Xi qB\phi\mu^2 c_t}$	•	⊙		
34088.25	(1.73)	$A = \dfrac{k^2hr_w^2Y_E(t*\Delta P')_{ss}t_{ss}}{\Xi qB\phi\mu^2 c_t X_E^3}$		•		
287058.94	(1.74)			•		
535456.96	(1.75)			•		
153451.04	(1.76)	$A = \dfrac{k^2hr_w^2Y_E(t*\Delta P')_{ss}t_{ss}}{\Xi qB\phi\mu^2 c_t(X_E b_x)^{1.5}}$	•			
482080.64	(1.77)			•		

$$k = \frac{711.26 q_{sc} T}{h(t * \Delta m(P)')_r} \qquad (1.68)$$

$$s' = 0.5 \left[\frac{(\Delta m(P))_r}{(t * \Delta m(P)')_r} - \ln\left(\frac{k t_r}{\phi(\mu c_t)_i r_w^2} \right) + 7.43 \right] \qquad (1.69)$$

CLOSED AND CONSTANT-PRESSURE BOUNDARY GAS RESERVOIRS—*TDS* TECHNIQUE

$$A = \frac{k t_{rpi}}{301.77 \phi(\mu c_t)_i} \qquad (1.70)$$

LONG AND NARROW GAS SYSTEMS— CONVENTIONAL ANALYSIS

$$Y_E = \frac{81.897 q T}{h m_{DLF}} \sqrt{\frac{1}{k \phi(\mu c_t)_i}} \qquad (1.78)$$

$$s_{DL} = \frac{k h b_{DLF}}{1422.67 q T} \qquad (1.79)$$

$$Y_E = \frac{145.16 q T}{h m_{LF}} \sqrt{\frac{1}{k \phi(\mu c_t)_i}} \qquad (1.80)$$

$$s_L = \frac{k h b_{LF}}{1422.67 q T} \qquad (1.81)$$

$$b_x = \frac{1}{591.98} \sqrt{\frac{Y_E h m_{PB}}{q T}} \sqrt[4]{\frac{k^{2.5}}{\phi(\mu c_t)_i}} \qquad (1.82)$$

$$s_{PB} = \frac{k h b_{PB}}{1422.67 q T} \qquad (1.83)$$

Fundamentals of Transient Pressure Analysis in Homogeneous Reservoirs   25

LONG AND NARROW GAS SYSTEMS—*TDS* TECHNIQUE

$$Y_E = \frac{72.571qT}{h(t*\Delta m(P)')_{L1}}\sqrt{\frac{1}{k\phi(\mu c_t)_i}} \qquad (1.84)$$

$$Y_E = \frac{40.94qT}{h(t*\Delta m(P)')_{DL1}}\sqrt{\frac{1}{k\phi(\mu c_t)_i}} \qquad (1.85)$$

$$Y_E = \frac{175200qTb_x^2}{h\sqrt{k^3}(t*\Delta m(P)')_{PB}}\left(\frac{\phi(\mu c_t)_i}{t_{PB}}\right)^{0.5} \qquad (1.86)$$

Skin factors can be found with oil equations if the viscosity-compressibility product is referred to initial conditions.

Nomenclature

A	Area, ft2
B_g	Gas volume factor, ft^3/STB
B_o	Oil volume factor, bbl/STB
B_w	Oil volume factor, bbl/STB
b_x	Distance from closer lateral boundary to well along the x direction, ft
b_y	Distance from closer lateral boundary to well along the x direction, ft
C_A	Reservoir shape factor
c_t	Total or system compressibility, 1/psi
h	Formation thickness, ft
k	Permeability, md
m	Slope of P-vs-log t plot
$m*$	Slope of P-vs-t plot
P	Pressure, psi
$\overline{P}$	Average reservoir pressure, psi
$P_D{}'$	Dimensionless pressure derivative
P_D	Dimensionless pressure
P_i	Initial reservoir pressure, psia
P_e	External reservoir pressure, psia
P_{wf}	Well flowing pressure, psi
P_{ws}	Well shut-in or static pressure, psi
q	Flow rate, bbl/D; for gas reservoirs. the units are Mscf/D
r_D	Dimensionless radius
r_e	Drainage radius, ft
r_w	Well radius, ft

s	Skin factor
s_t	Total skin factor
T	Reservoir temperature, °R; transmisibity, md-ft/cp
t	Time, hr
t_p	Production (Horner) time before shutting in a well, hr
$t*\Delta m(P)'$	Pseudopressure derivative function, psi2/cp
t_D	Dimensionless time based on well radius
t_{DA}	Dimensionless time based on reservoir area
X_E, x_e	Reservoir length, ft
x_f	Fracture half-length, ft
Y_E	Reservoir width, ft

Greek

Δ	Change, drop
Δt	Shut-in time, hr
ϕ	Porosity, fraction
μ	Viscosity, cp

Suffixes

D	Dimensionless
DL	Dual linear
DA	Dimensionless ith respect to area
g	Gas
i	Intersection or initial conditions
L	Linear or hemilinear
$LPPSi$	Intercept of linear and pseudosteady-state lines
o	Oil
p	Pseudosteady
PB	Parabolic
SS	Steady
$DLPSSi$	Intersection of pseudosteady-state line with dual- linear line
$LPSSi$	Intersection of pseudosteady-state line with lineal line
rpi	Intersection of pseudosteady-state line with radial line
r	Radial flow
$SS2$	-1 slope line formed when the parabolic flow ends and steady-state flow regime starts; well is near the open boundary and the far boundary is closed
$SS1Ri$	Intersection between the radial line and the -1 slope line (SS1)
w	Well, water
x	Maximum point (peak) during wellbore storage

REFERENCES

Carslaw, H. S. & Jaeger, J. C. (1959). *"Conduction of Heat in Solids."* Oxford University Press. Second Edition. 510p.

Djebrouni, A. (2003). *"Average Reservoir Pressure Determination Using Tiab's Direct Synthesis Technique"* M.S. Thesis, *The University of Oklahoma*, Norman, OK, U.S.A.

Escobar, F. H., Saavedra, N. F., Hernández, C. M., Hernández, Y. A., Pilataxi, J. F. & Pinto, D. A. (2004). *"Pressure and Pressure Derivative Analysis for Linear Homogeneous Reservoirs without Using Type-Curve Matching."* Paper SPE 88874, Proceedings, 28th Annual SPE International Technical Conference and Exhibition to be held in Abuja, Nigeria, Aug. 2-4, 2004.

Escobar, F. H., Muñoz, O. F., Sepulveda, J. A. & Montealegre, M. (2005). *"New Finding on Pressure Response In Long, Narrow Reservoirs."* CT&F – Ciencia, Tecnología y Futuro. Vol. *2*, No. 6.

Escobar, F. H. & Montealegre, M. M. (2006). *"Effect of Well Stimulation on the Skin Factor in Elongated Reservoirs."* CT&F – Ciencia, Tecnología y Futuro. Vol. *3*, No. 2. p. 109-119. Dic. 2006.

Escobar, F. H., Hernandez, Y. A. & Hernandez, C. M. (2007a). *"Pressure transient analysis for long homogeneous reservoirs using TDS technique."* Journal of Petroleum Science and Engineering, Volume *58*, Issue 1-2, pages 68-82.

Escobar, F. H., Tiab, D. & Tovar, L. V. (2007b). *"Determination of Areal Anisotropy from a single vertical Pressure Test and Geological Data in Elongated Reservoirs."* Journal of Engineering and Applied Sciences, *2*(11). p. 1627-1639. 2007.

Escobar, F. H. (2008). *"Petroleum Science Research Progress."* Nova Publishers. Edited by Korin L. Montclaire. Chapter title *"Recent Advances in Well Test Analysis for Long an Narrow Reservoirs."* *412*p. 2008.

Escobar, F. H., Lopez, A. M. & Cantillo, J. H. (2007c). *"Effect of the Pseudotime Function on Gas Reservoir Drainage Area Determination."* CT&F – Ciencia, Tecnología y Futuro. Vol. *3*, No. 3. p. 113-124. ISSN 0122-5383. Dec. 2007.

Escobar, F. H. & Montealegre, M. (2007). *"A Complementary Conventional Analysis For Channelized Reservoirs."* CT&F – Ciencia, Tecnología y Futuro. Vol. *3*, No. 3. p. 137-146. ISSN 0122-5383. Dec.

Escobar, F. H., Montealegre-M. M. & Carrillo-Moreno, D. (2009). *"Straight Line Methods for Estimating Permeability or Width for Two-Zone Composite Channelized Reservoirs."* CT&F – Ciencia, Tecnología y Futuro. Vol. *3*, No. 5. p. 107-124. Dec.

Escobar, F. H., Montealegre-M, M. & Carrillo-Moreno, Daniel. (2010) *"Pressure and Pressure Derivative Transient Analysis Without Type-Curve Matching For Elongated Reservoirs With Changes in Permeability or Width."* CT&F – Ciencia, Tecnología y Futuro. Vol. *4*, No. 1. p. 75-88. June.

Escobar, F. H., Hernandez, Y. A. & Tiab, D. (2010). *"Determination of reservoir drainage area for constant-pressure systems using well test data."* CT&F – Ciencia, Tecnología y Futuro. Vol. *4*, No. 1. p. 51-72. June.

Escobar, F. H., Fahes, M., Gonzalez, R. & Pinchao, D. M. (2015) and Zhao, Y. L.. *"Determination of Drainage Area for constant pressure systems by conventional analysis.*" *Journal of Engineering and Applied Sciences*, Vol. *10*. Nro. 12. p. 5193-5199. July 2015.

Jones, P. (1956). "*Reservoir Limit Test.*" *Oil and gas journal.* June 18. p. 184-196.

Nunez-Garcia, Walter, Tiab, D. & Escobar, F. H. "*Transient Pressure Analysis for a Vertical Gas Well Intersected by a Finite-Conductivity Fracture.*" Paper SPE 80915, Proceedings, SPE Production and Operations Symposium held in Oklahoma City, Oklahoma, U. S.A., 23–25 March 2003.

Matthews, C. S. & Russell, D. G. (1967). "*Pressure Buildup and Flow Tests in Wells.*" *Society of Petroleum Engineers of AIME.* 163p.

Perrine, R. L. (1956). "*Analyses of Pressure Buildup Curves.*" *Drilling and Production Practices*, API (1956) 482.

Ramey, H. R., Jr. & Cobb, W. M. (1971). "*A general Buildup Theory for a Well in a Closed Drainage Area.*" JPT Dec. *1971.* 1493-1505.

Spivey, J. P. & Lee, W. J. (2013). "*Applied Well test Interpretation.*" SPE Textbook Series Vol. *13.* 374p. Ricardson, Texas, U.S.A.

Tiab, D. (1993a, January 1). "*Analysis of Pressure and Pressure Derivatives Without Type-Curve Matching: I-Skin and Wellbore Storage.*" Society of Petroleum Engineers. This paper was prepared for presentation at the Production Operations Symposium held in Oklahoma City, OK, U.S.A., March 21-23. doi:10.2118/25426-MS.

Tiab, D. (1993b, January 1). "*Analysis of Pressure and Pressure Derivative without Type-Curve Matching - III. Vertically Fractured Wells in Closed Systems.*" Society of Petroleum Engineers. This paper was originally presented as SPE 26138 at the 1993 SPE Western Regional Meeting, held May 26-28, Anchorage, Alaska. Doi. :10.2118/26138-MS.

Tiab, D. (1994). "*Analysis of Pressure Derivative without Type-Curve Matching: Vertically Fractured Wells in Closed Systems.*" *Journal of Petroleum Science and Engineering.* 11. 323-333.

Tiab, D. (1995). "*Analysis of Pressure and Pressure Derivative without Type-Curve Matching: 1- Skin and Wellbore Storage.*" *Journal of Petroleum Science and Engineering*, Vol *12*, 171-181.Also Paper SPE 25423, Production Operations Symposium held in Oklahoma City, OK. 203-216.

Chapter 2

DOUBLE-POROSITY, SINGLE-PERMEABILITY RESERVOIRS

BACKGROUND

Barenblatt and Zheltov (1960) introduced the basic equations for flow in dual-porosity, naturally fractured reservoirs. They applied continuum mechanics to mathematically describe the media (fractures and matrix) and flow parameters. The transfer of fluid between the two media is maintained in a source function, where the flow is assumed to be in a pseudosteady state in the matrix system. Later, Warren and Root (1963) used this approach to develop a comprehensive and applicable solution to characterize the discussed systems.

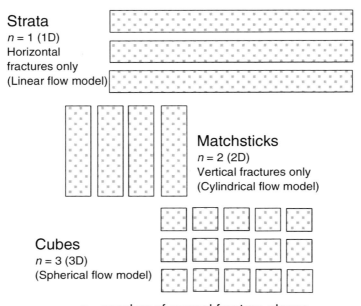

Figure 2.1. Summary of orthogonal geometries; adapted from Stewart (2011).

For describing the fracture network, Warren and Root (1963) considered three idealized block geometries: horizontal slabs, vertical matchsticks and cubes. A clear explanation of these models was graphically presented, see Figure 2.1, by Stewart (2011) in his excellent book on well testing, which presented a graphical explanation of each model.

Although, the above are ideal cases, Stewart (2011) also pointed out that the matchstick model could be the most relevant model for a naturally fractured reservoir. This only allows vertical fractures and $n = 2$, as sketched in Figure 2.1. The model of cubes has a regular three-dimensional fracture network with $n = 3$ and represents a perfect fracture system, although reservoirs actually have irregular geometries with random block size and shape distribution. Due to the high vertical stress in deep reservoirs, horizontal fractures are less likely to occur than vertical ones. Therefore, the matchsticks model works better.

Two key parameters were derived by Warren and Root (1963) to characterize naturally fractured formations: the dimensionless storage coefficient, also called the storativity ratio, ω, and the interporosity flow parameter, or flow capacity ratio, λ. ω provides an estimate of the magnitude and distribution of matrix and fracture storage, and λ is a measure of the mass transfer rate from the matrix to the fractures and therefore describes the matrix flow capacity available to the fractures.

$$\omega = \frac{(\phi c_t)_f}{(\phi c_t)_f + (\phi c_t)_m} \tag{2.1}$$

$$\lambda = \frac{4n(n+2)k_m r_w^2}{k_f h_m^2} \tag{2.2}$$

in which n takes the value of 1, 2 or 3 depending upon the flow model, as described in Figure 2.1.

In double-porosity systems, flow is only expected to occur in the more permeable zone (fracture network). Thus, the model assumes there exists a two-layer system in which one layer possesses a flow capacity 100 times higher than the other one. However, there are cases in which such a situation does not fit, and thus, radial flow in the less permeable layer has to be considered. Such a system is referred to as a double-porosity system, and its model was proposed by Wijesinghe and Culham (1984). Their model assumes a two communicating layer or double-permeability situation (double-porosity, single-permeability reservoirs) in which two zones with different flow capacities, kh, are interacting, separated by a semipermeable barrier with negligible storage capacity. A new term, called capacity flow ratio, is then defined:

$$\kappa = \frac{k_{fb} h_f}{k_{fb} h_f + k_m h_m} \tag{2.3}$$

For the two-layer system assumed, ω and κ are measures of the contrast between the layers. Reservoirs having naturally occurring fractures or naturally fractured reservoirs can be classified as double porosity (or double porosity, single permeability); double permeability (or double porosity, double permeability); triple porosity (or triple porosity, single permeability) and triple porosity, double permeability. These systems will be graphically explained later, and each case will be studied in a separate chapter.

In the case where the matrix permeability is negligible, then $\kappa \approx 1$ and the system is admitted as a double-porosity system, and this is graphically presented in Figure 2.2.a. For such a case, the well is fed only by the fracture network. If the matrix has a significant flow contribution, then κ is smaller than one and the well is first fed by the fractures; once fluid depletion prevels, both the well and fractures start being fed by the matrix. Such a system is referred to as double permeability or double porosity, double permeability. Its behavior is sketched in Figure 2.2.b.

Camacho-Velazquez et al. (2002) presented a triple-porosity model, which includes the interaction between matrix, fractures and vugs, as shown in Figure 2.3. Likewise, primary flow may occur through both the fractures and vugs networks. For such cases, six parameters are involved in this model compared to the two originally introduced by Warren and Root (1963). These parameters are: permeability ratio between fractures and vugs, three interaction parameters and two storativity ratios.

In the case where there is only primary flow through the fracture network, then $\kappa \approx 1$, and it is acknowledged to be a triple-porosity, single-permeability system. On the other hand, when vugs also contribute to primary flow, $\kappa < 1$, meaning primary flow occurs through both the fractures and vugs networks; thus, a triple-porosity, double-permeability system is dealt with.

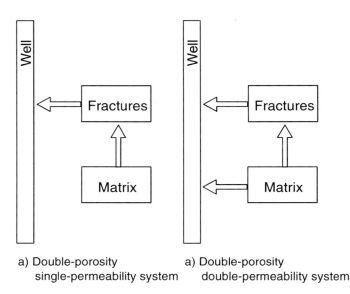

a) Double-porosity single-permeability system   a) Double-porosity double-permeability system

Figure 2.2. Schematical representation of double-porosity and double-permeability systems.

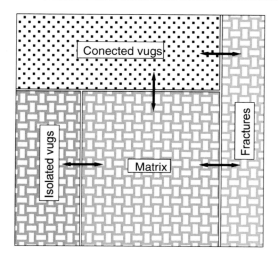

Figure 2.3. Schematical representation of triple-porosity systems.

Mavor and Cinco-Ley (1979) presented the solution in Laplace space for a well in a naturally fractured reservoir producing at a constant rate, including wellbore storage and skin effects.

$$P(\bar{s}) = \frac{K_o\left(\sqrt{\bar{s}f(\bar{s})}\right) + s\left(\sqrt{\bar{s}f(\bar{s})}\right)K_1\left(\sqrt{\bar{s}f(\bar{s})}\right)}{\bar{s}\left[\left(\sqrt{\bar{s}f(\bar{s})}\right)K_1\left(\sqrt{\bar{s}f(\bar{s})}\right) + \bar{s}C_D\left\{K_o\left(\sqrt{\bar{s}f(\bar{s})}\right) + s\left(\sqrt{\bar{s}f(\bar{s})}\right)K_1\left(\sqrt{\bar{s}f(\bar{s})}\right)\right\}\right]} \quad (2.4)$$

The $f(\bar{s})$ function for pseudosteady-state interporosity flow is given as:

$$f(\bar{s}) = \frac{\omega(1-\omega)\bar{s} + \lambda}{(1-\omega)\bar{s} + \lambda} \quad (2.5)$$

The derivative of Equation (2.4) is:

$$P_D{}'(\bar{s}) = \frac{K_o\left(\sqrt{\bar{s}f(\bar{s})}\right) + s\left(\sqrt{\bar{s}f(\bar{s})}\right)K_1\left(\sqrt{\bar{s}f(\bar{s})}\right)}{\left(\sqrt{\bar{s}f(\bar{s})}\right)K_1\left(\sqrt{\bar{s}f(\bar{s})}\right) + \bar{s}C_D\left\{K_o\left(\sqrt{\bar{s}f(\bar{s})}\right) + s\left(\sqrt{\bar{s}f(\bar{s})}\right)K_1\left(\sqrt{\bar{s}f(\bar{s})}\right)\right\}} \quad (2.6)$$

CONVENTIONAL ANALYSIS

Synthetic pressure buildup data is provided in the Horner plot of Figure 2.4. The early data on the first straight line (radial flow behavior) respond to heterogeneous systems, while the second straight line acts as the radial flow in homogeneous reservoirs. As for homogeneous systems (chapter 1), the permeability is found with Equation (2.10) and the skin factor is found with Equation (2.17) for buildup or (2.11) for drawdown tests using the extrapolation of the second line to the time of 1 hr. This same line is used to read the P^* for estimating the average reservoir pressure. The skin factor in heterogeneous formations is different than the one in homogeneous formations, which refer to a mechanical damage in the well. In this case (naturally fractured formations), the skin factor measures the effective connectivity between the wellbore and fractures.

Rewriting Equations (1.10), (1.11) and (1.17) for double-porosity reservoirs yields:

$$k_{fb} = \left| \frac{162.6 q \mu B}{mh} \right| \quad (2.7)$$

For the first semilog line drawdown:

$$s = 1.1513 \left[\frac{P_{1hr} - P_i}{m} \log \left(\frac{k_{fb}}{\phi_f c_f \mu r_w^2} \right) - 3.2275 \right] \quad (2.8)$$

For the second semilog line drawdown:

$$s = 1.1513 \left[\frac{P_{1hr} - P_i}{m} \log \left(\frac{k_{fb}}{(\phi c_t)_{f+m} \mu r_w^2} \right) - 3.2275 \right] \quad (2.9)$$

For the first semilog line buildup:

$$s = 1.1513 \left[\frac{P_{1hr} - P_{wf}}{m} - \log \left(\frac{k_{fb}}{\phi_f c_f \mu r_w^2} \right) + 3.2275 \right] \quad (2.10)$$

For the second semilog line buildup:

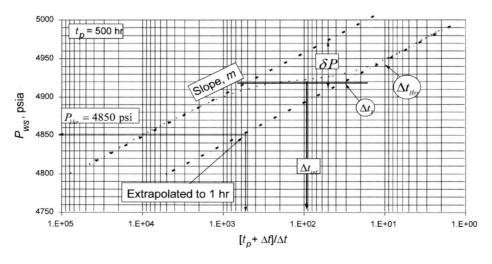

Figure 2.4. Synthetic buildup curve for a double-porosity system.

$$s = 1.1513\left[\frac{P_{1hr} - P_{wf}}{m}\log\left(\frac{k_{fb}}{(\phi c_t)_{f+m}\mu r_w^2}\right) - 3.2275\right] \tag{2.11}$$

Kazemi (1969) used the vertical separation between the two semilog lines, labeled as δP in Figure 2.4, to find the storativity ratio:

$$\omega = 10^{-\delta P/|m|} \tag{2.12}$$

A straight line can be drawn along the transition time given between the two semilog lines. The intersection of such lines with the semilog lines can be used to estimate λ. However, such formulations are of impractical use since normally the first straight line is masked by wellbore storage effects. Thus, Tiab and Escobar (2003) developed an expression to estimate from the inflection point, t_{inf} or Δt_{inf}, which is pointed out in Figure 2.4.

$$\lambda = \frac{3792(\phi c_t)_{m+f}\mu r_w^2}{k_{fb}\Delta t_{inf}} \tag{2.13}$$

Equations (11.47) to (11.49) have been applied here. Also, Stewart (2011) presented two practical expressions to find λ when wellbore storage masks the first straight line:

$$\lambda = \frac{4[t_D]_I(\phi c_t)_t\mu r_w^2}{k_{fb}t_I} \tag{2.14}$$

$$\lambda = \frac{4[t_D]_{bts}(\phi c_t)_t\mu r_w^2}{k_{fb}t_{bts}} \tag{2.15}$$

Suffix I in Equation (2.9) stands for intercept. Dimensionless time, $[t_D]_I$, is the time at which the second semilog line intercepts a near-zero slope drawn along the transition period, and t_{bts} is the beginning of the second semilog line, which is better identified in the derivative plot. These two points are indicated inside circles in Figure 2.4.

TDS TECHNIQUE

Engler and Tiab (1996) presented a set of equations and procedures for a very practical interpretation of transient pressure data in double-porosity systems. They started with the solution to the diffusivity equation for naturally fractured reservoirs, excluding wellbore storage and skin effects:

$$P_D = \frac{1}{2}\left[\ln t_D + 0.80908 + Ei\left(-\frac{\lambda t_D}{\omega(1-\omega)}\right) - Ei\left(-\frac{\lambda t_D}{1-\omega}\right)\right] + s \tag{2.16}$$

The pressure derivative function of the above equation is:

$$t_D * P_D' = \frac{1}{2}\left[1 - \exp\left(-\frac{\lambda t_D}{1-\omega}\right) + \exp\left(-\frac{\lambda t_D}{\omega(1-\omega)}\right)\right] \qquad (2.17)$$

and the dimensionless quantities given in oil-field units are defined as:

$$t_D = \frac{0.0002637 k_{fb} t}{(\phi c_t)_{f+m} \mu r_w^2} \qquad (2.18)$$

$$P_D = \frac{k_{fb} h (P_i - P)}{141.2 q \mu B} \qquad (2.19)$$

Analogous to the homogeneous reservoir case, the fracture bulk permeability is found from the radial pressure derivative during any of the radial flow regimes (1 or 2; they should be the same):

$$k_{fb} = \frac{70.6 q \mu B}{h (t * \Delta P')_{r1,r2}} \qquad (2.20)$$

The skin factor can be found with either of the below equations, depending upon the selected radial flow regime:

$$s = \frac{1}{2}\left[\left(\frac{\Delta P}{t * \Delta P'}\right)_{r1} - \ln\left(\frac{k_{fb}\, t_{r1}}{(\phi c_t)_{m+f} \mu r_w^2} \frac{1}{\omega}\right) + 7.43\right] \qquad (2.21)$$

$$s = \frac{1}{2}\left[\left(\frac{\Delta P}{t * \Delta P'}\right)_{r2} - \ln\left(\frac{k_{fb}\, t_{r2}}{(\phi c_t)_{m+f} \mu r_w^2}\right) + 7.43\right] \qquad (2.22)$$

Engler and Tiab (1996) developed several useful analytical equations and correlations for the determination of ω and λ using some of the characteristic points found on the pressure and pressure derivative against time log-log plot—see Figure 2.5—as follows:

$$\omega = 0.15866\left\{\frac{(t * \Delta P')_{min}}{(t * \Delta P')_{r1,r2}}\right\} + 0.54653\left\{\frac{(t * \Delta P')_{min}}{(t * \Delta P')_{r1,r2}}\right\}^2 \qquad (2.23)$$

$$\omega = 0.19211\left\{\frac{5 t_{min}}{t_{b2}}\right\} + 0.80678\left\{\frac{5 t_{min}}{t_{b2}}\right\}^2 \qquad (2.24)$$

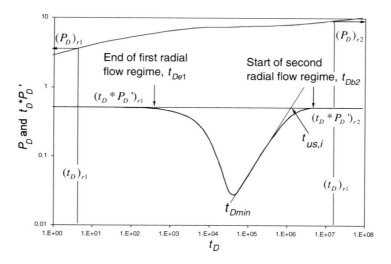

Figure 2.5. Characteristic features of a double-porosity system found on a dimensionless pressure and pressure derivative versus dimensionless time log-log plot.

$$\lambda = \frac{1}{\beta} \frac{1}{t_{usi}} \quad (2.25)$$

where β is defined by:

$$\beta = \frac{0.0002637 k_{fb}}{(\phi c_t)_{m+f} \mu r_w^2} \quad (2.26)$$

$$\lambda = \frac{\omega(1-\omega)}{50 \beta t_{e1}} = \frac{\omega \ln(1/\omega)}{\beta t_{min}} = \frac{1}{\beta t_{usi}} = \frac{5(1-\omega)}{\beta t_{b2}} \quad (2.27)$$

$$\lambda = \left[\frac{42.5 h (\phi c_t)_{f+m} r_w^2}{qB} \right] \frac{(t * \Delta P')_{min}}{t_{min}} \quad (2.28)$$

Tiab, Igbokoyi and Restrepo (2007) came up with other practical expressions:

$$\omega = \left(2.9114 + 4.5104 \frac{(t \times \Delta P')_{r1,r1}}{(t \times \Delta P')_{min}} - 6.5452 e^{0.7912 \frac{(t \times \Delta P')_{min}}{(t \times \Delta P')_{r1,r2}}} \right)^{-1} \quad (2.29)$$

$$\omega = \left(2.9114 - \frac{3.5688}{-\lambda t_{Dmin}} - \frac{6.5452}{e^{-\lambda t_{Dmin}}} \right)^{-1} \quad (2.30)$$

Double-Porosity, Single-Permeability Reservoirs

Table 2.1. Ranges for checking wellbore storage effects on minimum pressure derivative point, adapted from Tiab, Igbokoyi and Restrepo (2007)

λ	C_D
10^{-4}	$C_D > 10$
10^{-5}	$C_D > 100$
10^{-6}	$C_D > 10^3$
10^{-7}	$C_D > 10^4$
10^{-8}	$C_D > 10^5$

Equation (2.28) is valid for $0 < \omega < 0$. They advised that using $(\phi c_t)_f \approx (\phi c_t)_m$ would lead to a big overestimation of fracture porosity; thus, they introduced the following expression:

$$(\phi c_t)_{f+m} = (\phi c_t)_m \left(1 + \frac{\omega}{1-\omega} \right) \tag{2.31}$$

They also pointed out that once ω is estimated, the fracture porosity can be estimated if matrix porosity, ϕ_m, total matrix compressibility, c_{tm}, and total fracture compressibility, c_{tf}, are known, as follows:

$$\phi_f = \left(\frac{\omega}{1-\omega} \right) \frac{c_{tm}}{c_{tf}} \phi_m \tag{2.32}$$

Tiab et al. (2007) concluded that the fracture permeability can be estimated from:

$$k_f = \frac{k_{fb}^2}{k_m} \tag{2.33}$$

where k_m is the matrix permeability, which is measured from representative cores, and k is the mean permeability obtained from pressure transient tests.

They also used Saidi's (1987) equation:

$$c_{tf} = \frac{1 - \left(k_f / k_{fi} \right)^{1/3}}{\Delta P} \approx \frac{1 - \left(k_f / k_{fi} \right)}{3\Delta P} \tag{2.34}$$

Combining Equations (2.32) and (2.33), Tiab et al. (2007) arrived at:

$$c_{tf} = \frac{1 - \left(k_{fb} / k_i \right)^{2/3}}{\Delta P} \tag{2.35}$$

where k_i is the average permeability obtained from a transient test run when the reservoir pressure is at or near initial conditions, P_i, and k_{fb} is the average permeability obtained from a transient test at the current average reservoir pressure. For this special case, $\Delta P = P_i - \overline{P}$.

Tiab and Donaldson (2004) combined Equations (2.29) and (2.30) to find an equation for the estimation of fracture porosity from two well tests:

$$\phi_f = \phi_m c_{tm} \left(\frac{\omega}{1-\omega}\right) \frac{\Delta P}{\left(1-(k_f/k_{fi})^{1/3}\right)} \tag{2.36}$$

The fracture width or aperture may be estimated from the following correlation given by Bona et al. (2003):

$$w_f = \sqrt{\frac{k_f}{33\omega\phi_t}} \tag{2.37}$$

where fracture width, w_f, is given in microns (10^{-3} mm), permeability in md, porosity in fraction and storage capacity in fraction.

Tiab et al. (2007) also found wellbore storage affects the minimum point during the transition period (trough) if any of the conditions in Table 2.1 are met. The dimensionless wellbore storage coefficient is defined by:

$$C_D = \frac{0.894C}{(\phi c_t)_{f+m} hr_w^2} \tag{2.38}$$

In the case in which the minimum point is affected by wellbore storage, this must be corrected using the following expressions, also presented by Tiab et al. (2007):

$$(t*\Delta P')_{min} = (t*\Delta P')_{r1,r2} + \frac{(t*\Delta P')_{minO} - (t*\Delta P')_{r1,r2}\left[1+2D_1D_2\right]}{1+D_2\left[\ln\left(\frac{C}{(\phi c_t)_{f+m} hr_w^2}\right)+2s-0.8801\right]} \tag{2.39}$$

where:

$$D_1 = \left[\ln\left(\frac{qBt_{minO}}{(t*\Delta P')_r (\phi c_t)_{f+m} hr_w^2}\right)+2s-4.17\right] \tag{2.40}$$

$$D_2 = \frac{48.02C}{qB}\left(\frac{(t*\Delta P')_{r1,r2}}{t_{minO}}\right) \tag{2.41}$$

Double-Porosity, Single-Permeability Reservoirs

Engler and Tiab (1996) also found the minimum point during the transition period affected by wellbore storage is the ratio between the time at the maximum point (during wellbore storage) and the time at the trough and is less than ten ($t_x/t_{min}<10$).

Example 2.1

Tiab et al. (2007) reported a field example of pressure tests conducted in a double-porosity reservoir. Pressure tests in the first few wells located in a naturally fractured reservoir yielded a similar average permeability of the system of 82.5 md. An interference test also yielded the same average reservoir permeability, which implied fractures are uniformly distributed. The total storativity, $(\phi c_t)_{m+f} = 1.0 \times 10^{-5}$ psi^{-1}, was obtained from this interference test. Only the porosity, permeability and compressibility of the matrix could be determined from the recovered cores.

The pressure drop from the initial reservoir pressure to the current average reservoir pressure is 300 psi. Digitized pressure and pressure derivative data are provided in Table 2.2. The characteristics of the rock, fluid and well are given below:

$q = 3000$ STB/D	$h = 25$ ft	$\phi_m = 10\%$
$r_w = 0.4$ ft	$\mu = 1$ cp	$B = 1.25$ rb/STB
$c_{tm} = 1.35 \times 10^{-5}$ psi^{-1}	$k_m = 0.10$ md	

For this example, one must (1) estimate formation permeability, storage capacity ratio, ω, and the interporosity flow parameter, λ, by conventional semilog analysis and the *TDS* technique and (2) estimate the three fracture properties: permeability, porosity and width.

Solution by Conventional Analysis

The following information was obtained from the semilog plot given in Figure 2.6:

$$\delta P = 130 \text{ psi} \qquad m = 325 \text{ psi/cycle} \qquad \Delta t_{inf} = 2.5 \text{ hrs}$$

The average permeability of the fracture system estimated from the slope of the semilog straight line using Equation (2.7) yields:

$$k_{fb} = \left| \frac{162.6 q \mu B}{mh} \right| = \frac{162.6(3000)(1.25)(1)}{(325)(25)} = 75 \text{ md}$$

The fluid storage coefficient is estimated using Equation (2.12):

$$\omega = \omega = 10^{-\delta P/|m|} = 10^{(-130/325)} = 0.40$$

The interporosity fluid parameter is given by Equation (2.13):

Freddy Humberto Escobar

Table 2.2. Pressure and pressure derivative data for example 2.1

t, hr	ΔP, psi	$t^*\Delta P$, psi	t, hr	ΔP, psi	$t^*\Delta P$, psi
0.175	173.390	134.417	1.248	432.287	86.414
0.259	228.744	149.274	1.414	448.779	78.399
0.342	269.708	156.134	2.428	472.930	72.200
0.426	297.284	160.881	3.386	502.126	78.399
0.506	335.122	152.665	4.371	533.125	89.042
0.591	358.485	144.869	5.413	541.170	97.413
0.677	377.777	112.307	6.355	557.624	110.638
0.758	386.360	101.130	7.416	578.897	123.789
0.917	398.107	98.145	9.461	619.256	144.869
1.095	410.212	88.377	12.360	652.581	140.594
			14.338	677.477	138.504

$$\lambda = \frac{3792(\phi c_t)_{m+f}\,\mu r_w^2}{k_{fb}\,\Delta t_{inf}} = \frac{3792(1\times10^{-5})(1)(0.4^2)}{(75)(2.5)}\left[0.40\ln\left(\frac{1}{0.40}\right)\right] = 1.19\times10^{-5}$$

Solution by TDS Technique

The pressure and pressure derivative plot is given in Figure 2.7, from which the following characteristic points were obtained:

$$t_{min} = 2.427 \text{ hrs} \qquad (t^*\Delta P')_{r2} = 139 \text{ psi} \qquad (t^*\Delta P')_{min} = 70.5 \text{ psi}$$

The fracture bulk permeability is determined from Equation (2.20):

$$k_{fb} = \frac{70.6q\mu B}{h(t^*\Delta P')_{r1,r2}} = \frac{70.6(3000)(1)(1.25)}{(25)(139)} = 76.2 \text{ md}$$

The coordinates of the trough are used to estimate the interporosity fluid parameter using Equation (2.28):

$$\lambda = \left[\frac{42.5h(\phi c_t)_{f+m}r_w^2}{qB}\right]\frac{(t^*\Delta P')_{min}}{t_{min}} = \left(\frac{(42.5)(25)(1\times10^{-5})(0.4)^2}{(3000)(1.25)}\right)\frac{(70.5)}{2.427} = 1.31\times10^{-5}$$

The storage coefficient ω is calculated from Equation (2.29):

$$\omega = \left(2.9114 + 4.5104\frac{139}{70.5} - 6.5452e^{0.7912\frac{70.5}{139}}\right)^{-1} = 0.49$$

Both the *TDS* and conventional techniques yield approximately similar values for the reservoir parameters; thus, Tiab et al. (2007) suggested this match is because both parallel straight lines are well defined, suggesting that the trough is not influenced by wellbore storage and, therefore, that it is not necessary to apply Equation (2.37).

Estimation of Fracture Permeability

The current fracture permeability is calculated from Equation (2.33):

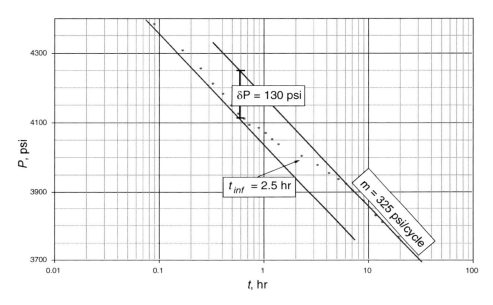

Figure 2.6. Semilog plot for example 2.1.

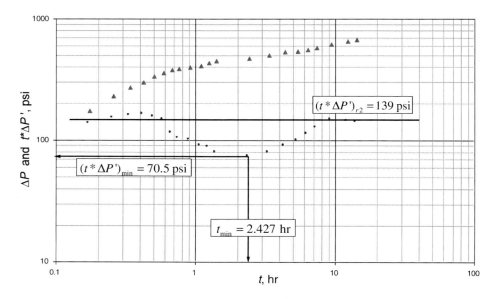

Figure 2.7. Pressure and pressure derivative plot for example 2.1.

$$k_f = \frac{k_{fb}^2}{k_m} = \frac{76.2^2}{0.10} = 58064 \text{ md}$$

Notice that k_{fb} in the above equation refers to the value obtained from well test analysis. The fracture permeability at initial reservoir pressure is:

$$k_{fi} = \frac{k_{fbi}^2}{k_m} = \frac{82.5^2}{0.10} = 68062 \text{ md}$$

Estimation of Fracture Porosity
Fracture compressibility can be estimated from Equation (2.35):

$$c_{tf} = \frac{1-(k/k_i)^{2/3}}{\Delta P} = \frac{1-(58064/68062)^{1/3}}{300} = 1.72 \times 10^{-4} \text{ psi}^{-1}$$

The compressibility ratio is:

$$\frac{c_{tf}}{c_{tm}} = \frac{1.72 \times 10^{-4}}{1.35 \times 10^{-5}} = 12.7$$

The fracture porosity from Equation (2.32) is:

$$\phi_f = \left(\frac{\omega}{1-\omega}\right)\frac{c_{tm}}{c_{tf}}\phi_m = \left(\frac{0.40}{1-0.40}\right)\frac{0.1}{12.7} = 0.0052 = 0.52 \text{ \%}$$

Tiab et al. (2007) also concluded that if they had assumed $c_{tf} = c_{tm}$ (as normally assumed), they would have found an erroneous value of $\phi_f = 6.6\%$, a tenfold overestimation of the fracture pore space.

Estimation of Fracture Width
The total porosity of this naturally fractured reservoir is:

$$\phi_t = \phi_m + \phi_f = 0.10 + 0.0052 = 0.1052$$

The current fracture width is estimated from Equation (2.37):

$$w_f = \sqrt{\frac{k_f}{33\omega\phi_t}} = \sqrt{\frac{58064}{33(0.40)(0.1052)}} = 205 \text{ microns} = 0.205 \text{ mm}$$

AVERAGE RESERVOIR PRESSURE

The estimation of average reservoir pressure in vertical wells in homogeneous reservoirs by the *TDS* technique was introduced by Chacon, Djebrouni and Tiab (2004). Their formulation also includes hydraulically fractured wells. New equations for estimating the shape factor, C_A, were also presented by Chacon et al. (2004).

According to Chaco et al. (2004), the average reservoir pressure and shape factor can be obtained by reading a pressure and pressure derivative point on the late pseudosteady-state period. It is recommended to take the latest point in which the pseudosteady state is completely developed in the whole system:

$$\overline{P} = P_i - 70.6 \frac{q\mu B}{kh} \left[\left(\frac{(t*\Delta P')_{pss}}{(\Delta P_w)_{pss} - (t*\Delta P')_{pss}} \right) \ln \left(\frac{2.2458\,A}{C_A r_w^2} \right) \right] \tag{2.42}$$

$$C_A = \frac{2.2458\,A}{r_w^2} e^{\left[-\frac{kt_{pss}}{301.77\phi\mu c_t\,A} \left(\frac{(\Delta P)_{pss}}{(t*\Delta P')_{pss}} - 1 \right) \right]} \tag{2.43}$$

where $(\Delta P)_{pss}$ and $(t*\Delta P')_{pss}$ are the values of the pressure and pressure derivative, respectively, at any arbitrary time during the pseudosteady state, t_{pps}. For cases in which a steady state is observed, these values are read during steady-state acting.

According to Djebrouni (2003), reservoir area can be estimated using an arbitrary point during the late pseudosteady-state pressure derivative:

$$A = \frac{0.2337qBt_{pss}}{\phi c_t h\,(t*\Delta P')_{pss}} \tag{2.44}$$

However, due to noise in the data, it is better to have a more representative average value; thus, it is better to extrapolate the pseudosteady-state line to a time of 1 hr and read the pressure derivative value, $(t*\Delta P')_{p1}$, there.

$$A = \frac{0.2337qB}{\phi c_t h\,(t*\Delta P')_{p1}} \tag{2.45}$$

The average reservoir pressure in rectangular drainage areas drained by a fractured vertical well having a uniform flux, biradial flow or linear flow is estimated, respectively, from:

$$\overline{P} = P_i - \frac{q\mu B}{kh} \left\{ \frac{0.2337kt_{pss}}{\phi\mu c_t\,A} \left(\frac{(\Delta P)_{pss}}{(t*\Delta P')_{pss}} \right) - 70.6 \ln \left[\left(\frac{x_e}{x_f} \right)^2 \left(\frac{2.2458}{C_A} \right) \right] \right\} \tag{2.46}$$

$$\overline{P} = P_i - 5.64 \left[\frac{q\mu B}{kh} \left(\frac{x_e}{x_f} \right)^{0.72} \left(\frac{k}{\phi\mu c_t A} \right)^{0.36} \right] t_{BRPi}^{0.36} \tag{2.47}$$

$$\overline{P} = P_i - 4.06 \left[\frac{qB\sqrt{\mu}}{h \sqrt{\phi c_t k}} \left(\frac{x_e}{x_f} \right) \right] \sqrt{t_{LPi}} \tag{2.48}$$

The shape factor in Equation (2.46) can be determined by:

$$C_A = 2.2458 \left(\frac{x_e}{x_f} \right)^2 \left\{ \exp \left[\frac{kt_{pss}}{301.77\phi\mu c_t A} \left(1 - \frac{(\Delta P)_{pss}}{(t*\Delta P')_{pss}} \right) \right] \right\} \tag{2.49}$$

Customary conventional methods for the estimation of average reservoir pressure in homogeneous reservoirs are also used indiscriminately on naturally fractured formations, which may not be very accurate. To overcome this situation, Molina et al. (2005) presented the formulation for estimating average reservoir pressure in double-porosity, naturally occurring formations:

$$\overline{P} = P_{wf} + \Delta P_{pss} + \frac{141.2q\mu B}{k_{fb}h} \left(\frac{(t*\Delta P')_{pss}}{\Delta P_{pss} - (t*\Delta P')_{pss}} \right) \left(\ln \frac{r_e}{r_w} - \frac{3}{4} + \frac{0.1987 C_A r_w^2 (1-\omega)^2}{\lambda A} \right) \tag{2.50}$$

The shape factor can be approximated by Equation (2.43). Molina et al. (2005) also provided a formulation for homogeneous reservoirs:

$$\overline{P} = P_i - \frac{70.6 q_n \mu B}{kh} \left[\left(\frac{(t*\Delta P_q')_{pss}}{(\Delta P_q)_{pss} - (t*\Delta P_q')_{pss}} \right) \ln \left(\frac{2.2458A}{C_A r_w^2} \right) \right] \tag{2.51}$$

The above expressions for the determination of average reservoir pressure are, of course, applied to buildup tests only. Since shutting in the well will incur lost income, Escobar, Ibagon and Montealegre (2007) formulated a methodology for estimating the average reservoir pressure in multirate tests. The expressions are given as follows:

$$\overline{P} = P_i - \frac{141.2 q_n \mu B}{k_{fb}h} \left[\frac{(t*\Delta P_q')_{pss}}{(\Delta P_q)_{pss} - (t*\Delta P_q')_{pss}} \left(\ln \frac{r_e}{r_w} - 0.75 + \frac{2r_w^2(1-\omega)^2}{\lambda r_e^2} \right) \right] \tag{2.52}$$

$$C_A = \frac{2.2458A}{r_w^2} \frac{1}{e^{\left[\frac{k_{fb}(t_{eq})_{pss}}{301.77\phi\mu c_t A} \left(\frac{(\Delta P_q)_{pss}}{(t*\Delta P_q')_{pss}} - 1 \right) \right]}} \tag{2.53}$$

It is worth noting that Equations (2.50) and (2.52) can be applied to homogeneous reservoirs if $\lambda = 1$ and ω is set to zero.

Care must be taken when using Equations (2.51) through (2.53). With the application of time superposition—and using the solution of the diffusivity equation given by Equation (1.8), implying radial flow regime—one can easily obtain the following expression (Earlougher 1977) for radial:

$$\frac{P_i - P_{wf}(t)}{q_N} = \frac{162.6\mu B}{kh}\left\{\sum_{j=1}^{N}\left(\frac{q_j - q_{j-1}}{q_N}\right)\log\left(t - t_{j-1}\right) + \log\frac{k}{\phi\mu c_t r_w^2} - 3.2275 + 0.8686s\right\} \quad (2.55)$$

As expressed above, this equation works under the radial flow regime; thus, the superposition function during the radial flow regime, X_{n_rad}, is:

$$X_{n\_rad} = \sum_{j=1}^{n}\frac{\left(q_j - q_{j-1}\right)}{q_n}\log\left(t - t_{j-1}\right) \quad (2.55)$$

Then, the equivalent time, t_{eq}, in Equation (2.52) is:

$$t_{eq\_rad} = 10^{X_{n\_rad}} \quad (2.56)$$

However, the given Equations (2.51) through (2.53) are meant for a late pseudosteady-state period.

Escobar, Alzate and Moreno (2013, 2014) concluded that using radial flow-equivalent time leads to less accurate results, especially when estimating both the shape factor and the fracture length of vertical hydraulically fractured wells. Thus, Escobar et al. (2013, 2014) used superposition functions for a pseudosteady-state period:

$$X_{n\_pss} = \sum_{j=1}^{n}\frac{\left(q_j - q_{j-1}\right)}{q_n}\left(t - t_{j-1}\right) \quad (2.57)$$

and the pseudosteady-state–equivalent time is:

$$t_{eq\_pss} = X_{n\_pss} \quad (2.58)$$

So, for estimating average reservoir pressure in multirate tests it is recommended to use Equations (2.57) and (2.58) instead of Equations (2.55) and (2.56).

Table 2.3. Flow rate schedule for example 2.2

t, hr	q, BPD	t, hr	q, BPD
0.5	300	3700	350
5.5	430	4000	425
64	330	40000	360
130	380	8000	315
300	310	20000	225
800	270	60000	306

Example 2.2

Escobar et al. (2013, 2014) presented a multirate test of an elongated reservoir simulated with the information below and the flow rates given in Table 2.4. The normalized pressure and pressure derivative log-log plot along with some characteristic points are given in both Figure 2.8 and Table 2.4. Estimating the average reservoir pressure was done using the radial-equivalent time function and the pseudosteady-state–equivalent function.

$q = 3000$ STB/D	$h = 30$ ft	$\phi_m = 10\%$
$r_w = 0.3$ ft	$\mu = 3$ cp	$B = 1.3$ rb/STB
$c_t = 1.9 \times 10^{-5}$ psi^{-1}	$k = 200$ md	$P_i = 3000$ psi
$C = 0.005$ psi/bbl	$X_E = 3000$ ft	$Y_E = 30000$ ft

Solution

Even though this example is for homogeneous systems, it is presented to indicate how the *TDS* methodology works. The following information was taken from Figure 2.8:

$(t_{eq\_rad})_{pss} = 20713.3$ hr $(t_{eq\_rad}*\Delta P_q')_{pss} = 1.09$ psi/BPD
$(t_{eq\_pss})_{pss} = 20239.5$ hr $(t_{eq\_pss}*\Delta P_q')_{pss} = 1.034$ psi/BPD
$(\Delta P_q)_{pss} = 1.767$ psi/BPD

Using the radial superposition time, Escobar et al. (2007) found the values of shape factor and average reservoir pressure, estimated with the use of Equations (2.51) and (2.53), applied to a homogenous system. Their results are reported in Table 2.5 along with the results of Escobar et al. (2013, 2014).

Table 2.4. Pressure, time and derivative data for example 2.2

t, hr	P_{wf}, psi	t_{eq_rad}, hr	t_{eq_pss}, hr	ΔP_q, psi/BPD	$t_{eq\_rad}*\Delta P_q'$, psi/BPD	$t_{eq\_pss}*\Delta P_q'$, psi/BPD
0	2972.16	0	0		0.00	0.00
0.01	2951.14	0.01	0.01	0.09	0.08	0.08
0.02	2934.68	0.02	0.02	0.16	0.12	0.12
0.04	2910.94	0.04	0.04	0.26	0.16	0.16

Double-Porosity, Single-Permeability Reservoirs

t, hr	P_{wf}, psi	t_{eq_rad}, hr	t_{eq_pss}, hr	ΔP_q, psi/BPD	$t_{eq\_rad}{*}\Delta P_q{'}$, psi/BPD	$t_{eq\_pss}{*}\Delta P_q{'}$, psi/BPD
0.06	2895.00	0.06	0.06	0.33	0.16	0.16
0.08	2883.81	0.08	0.08	0.37	0.15	0.15
0.101	2874.47	0.101	0.101	0.40	0.14	0.14
0.127	2866.19	0.127	0.127	0.43	0.12	0.12
0.160	2858.99	0.160	0.160	0.46	0.10	0.10
0.202	2852.77	0.202	0.202	0.48	0.09	0.09
0.285	2844.87	0.285	0.285	0.51	-0.77	0.07
0.402	2838.13	0.402	0.402	0.53	-0.77	-0.86
0.500	2870.51	0.500	0.500	0.55	-0.78	-0.81
0.565	2847.85	0.342	0.444	0.34	0.08	0.17
0.639	2837.93	0.485	0.553	0.37	0.05	0.07
0.800	2829.56	0.748	0.789	0.39	0.04	0.05
1.146	2820.88	1.275	1.298	0.41	0.04	0.04
1.893	2811.32	2.381	2.393	0.43	0.03	0.04
3.500	2801.09	4.747	4.752	0.45	0.04	0.04
5.452	2798.04	5.844	5.849	0.47	0.07	0.35
6.139	2782.50	13.06	8.27	0.68	0.43	-0.01
6.646	2786.86	12.28	10.62	0.65	0.47	0.02
7.393	2787.62	14.96	14.08	0.64	0.04	0.04
9.000	2786.13	21.99	21.55	0.64	0.04	0.04
12.463	2781.93	37.84	37.62	0.65	0.04	0.04
19.925	2775.17	61.76	61.64	0.67	-0.42	-0.47
50.034	2763.89	56.32	68.66	0.71	-0.42	-0.45
72.044	2766.17	75.58	82.64	0.61	0.04	0.05
79.49	2762.35	123.56	126.83	0.62	0.05	0.05
90.44	2758.42	172.53	174.63	0.63	0.06	0.07
114.03	2752.36	425.82	242.06	0.64	0.36	0.42
164.87	2745.81	346.23	265.60	0.66	0.53	0.03
206.46	2732.25	366.80	330.47	0.87	0.49	0.08
220.44	2732.54	459.48	439.98	0.86	0.12	0.11
244.03	2730.57	973.81	628.87	0.86	-0.69	0.60
294.87	2724.76	853.78	679.58	0.88	-0.52	0.21
404.39	2712.91	866.96	754.19	0.90	-0.54	0.14
500.00	2701.07	907.47	819.33	0.93	-0.60	0.15
564.63	2699.13	1372.04	1338.11	1.11	-1.13	-1.27
800.00	2681.84	1021.27	1295.92	1.15	0.10	0.22
1223.17	2665.59	1592.24	1737.86	1.23	0.17	0.21
1330.00	2685.80	2399.99	2484.01	0.89	0.22	0.24
1439.25	2676.59	3772.13	3820.40	0.91	-0.66	0.29
1600.00	2664.56	4610.21	4648.47	0.94	-0.72	-0.87
1946.33	2643.39	3071.92	4245.96	0.98	-0.57	-0.88
2692.48	2607.12	3716.12	4592.30	1.06	0.17	-0.87
564.63	2699.13	1372.04	1338.11	1.11	-1.13	-1.27

Table 2.4. Pressure, time and derivative data for example 2.2. (continued)

t, hr	P_{wf}, psi	t_{eq_rad}, hr	t_{eq_pss}, hr	ΔP_q, psi/BPD	$t_{eq\_rad}*\Delta P_q{}'$, psi/BPD	$t_{eq\_pss}*\Delta P_q{}'$, psi/BPD
4028.88	2570.42	4735.94	5338.44	1.18	0.23	0.32
4856.94	2563.48	6273.18	6674.84	1.24	0.30	0.35
5139.25	2587.46	7327.10	7652.90	0.96	0.35	0.77
5440.34	2576.27	14372.97	9680.65	0.98	1.61	1.11
6392.48	2540.01	12757.01	10773.13	1.04	1.48	0.44
8413.88	2503.40	13832.16	12794.53	1.15	1.57	0.61
9300.00	2444.14	15556.19	14849.53	1.53	0.88	0.74
10392.48	2423.30	17441.40	16904.53	1.57	0.94	0.85
13783.88	2363.83	20713.26	20329.53	1.73	1.09	1.03
16523.88	2314.28	23382.03	23069.53	1.87	1.22	1.17
19263.88	2264.29	26073.11	25809.53	2.01	1.35	1.31
22003.88	2214.19	28777.44	28549.53	2.15	1.48	1.46
24743.88	2164.02	32169.46	31974.53	2.29	1.65	1.63
27483.88	2113.86	37612.81	37454.53	2.43	1.93	1.91
30223.88	2051.08	43067.76	42934.53	2.57	2.20	2.19
37073.88	1925.58	49696.80	49585.04	2.91	13.56	13.50

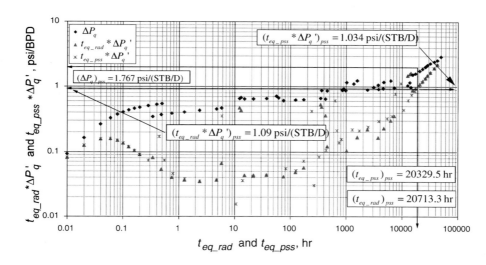

Figure 2.8. Pressure and pressure derivative plot for example 2.2.

For the case of pseudosteady-state superposition time, shape factor and average pressure were estimated with Equations (2.53) and (2.51):

$$\overline{P} = P_i - \frac{70.6(225)(3)(1.3)}{(200)(30)} \left[\frac{1.09}{1.767-1.09} \ln\left(\frac{2.5458(9\times10^7)}{18.4(0.3^2)} \right) \right] = 2566.1 \text{ psi}$$

Double-Porosity, Single-Permeability Reservoirs

Table 2.5. Summary of results for example 2.2

Parameter	Equivalent time function	
	Radial	Pseudosteady
C_A	135.71	18.4
$\overline{P}$	2557.9	2566.1

$$C_A = \frac{2.2458(9 \times 10^7)}{0.3^2} \frac{1}{e^{\left[\frac{(200)(20239.5)}{301.77(0.1)(3)(1.9\times10^{-5})(9\times10^7)}\left(\frac{1.767}{1.034}-1\right)\right]}} = 18.4$$

A summary of results is given in Table 2.5. Even though the C_A values are different, the average pressure values are practically the same.

Nomenclature

A	Drainage area, ft^2
B	Volume factor, bbl/STB
C	Wellbore storage, bbl/psi
C_A	Reservoir shape factor
c_t	Total or system compressibility, 1/psi
h	Formation thickness, ft
h_f	Fracture thickness, ft
h_m	Matrix thickness, ft
k	Permeability, md
k_{fb}	Fracture-bulk permeability, md
k_m	Matrix permeability, md
m	Slope of P-vs-log t plot
n	Number of normal fracture planes
P	Pressure, psi
$\overline{P}$	Average reservoir pressure, psi
$P(\overline{s})$	Pressure in Laplace domain
P_D	Dimensionless pressure
P_i	Initial reservoir pressure, psia
P_{wf}	Well flowing pressure, psi
P_{ws}	Well shut-in or static pressure, psi
P_{1hr}	Intercept of the semilog plot, psi
q	Flow rate, bbl/D
r_D	Dimensionless radius
r_e	Drainage radius, ft
r_w	Well radius, ft
s	Skin factor
$\overline{s}$	Laplace parameter
s_t	Total skin factor
t	Time, hr

$t_D{*}P_D{'}$	Dimensionless pressure derivative
t_p	Production (Horner) time before shutting in a well, hr
$t{*}\Delta P'$	Pressure derivative function, psi
t_D	Dimensionless time based on well radius
t_{DA}	Dimensionless time based on reservoir area
X_E, x_e	Reservoir length, ft
x_f	Fracture half-length, ft
X_N	Superposition time
X_{N_rad}	Superposition time with log function
X_{N_pss}	Superposition time with Cartesian function
Y_E	Reservoir width, ft
w_f	Fracture width, microns

Greek

Δ	Change, drop
δP	Pressure separation between the two semilog lines, psi
Δt	Shut-in time, hr
κ	Capacity flow ratio
ϕ	Porosity, fraction
λ	Interporosity flow parameter
μ	Viscosity, cp
ω	Storativity ratio

Suffixes

bts	Beginning of second radial flow
$b2$	Beginning of second radial flow
$BRPi$	Biradial-pseudosteadystate intercept
D	Dimensionless
DL	Dual linear
DA	Dimensionless ith respect to area
$e1$	End of first radial flow
eq	Equivalent
eq_rad	Equivalent with radial superposition function
eq_pss	Equivalent with pseudosteady-state superposition function
f	Fracture
fi	Fracture at initial conditions
fb	Fracture bulk
$f{+}m$	Fracture plus matrix
I, i	Intersection or initial conditions
inf	Inflection
L	Linear or hemilinear
LPi	Intercept of linear and pseudosteady-state lines

m	Matrix
min	Minimum
$minO$	Minimum observed
o	Oil
pss	Pseudosteady
$P1$	Pseudostate period at time of 1 hr
SS	Steady
US,i	Intercept of radial and pseudosteady-state lines during the transition period
$r1$	Radial flow of the first semilog line
$r2$	Radial flow of the second semilog line
tf	Total fracture
tm	Total matrix
w	Well, water
x	Maximum point (peak) during wellbore storage

REFERENCES

Barenblatt, G. I. & Zheltov, Y. P. (1960). "Fundamental Equations of Filtration of Homogeneous Liquids in Fissured Rocks." *Soviet Physics*, Doklady, *5*, 522.

Bona, N., Radaelli, F., Ortenzi, A., De Poli, A., Peduzzi, C. & Giorgioni, M. (2003), August 1). Integrated Core Analysis for Fractured Reservoirs: Quantification of the Storage and Flow Capacity of Matrix, Vugs, and Fractures. *Society of Petroleum Engineers.* doi:10.2118/85636-PA.

Camacho-Velazquez, R., Vasquez-Cruz, M., Castrejon-Aivar, R. & Arana-Ortiz, V. (2002, January 1). "Pressure Transient and Decline Curve Behaviors in Naturally Fractured Vuggy Carbonate Reservoirs." *Society of Petroleum Engineers.* doi:10.2118/77689-MS.

Chacon, A., Djebrouni, A. & Tiab, D. (2004, January 1). "Determining the Average Reservoir Pressure from Vertical and Horizontal Well Test Analysis Using the Tiab's Direct Synthesis Technique." *Society of Petroleum Engineers.* doi:10.2118/88619-MS.

Djebrouni, A. (2003). "*Average Reservoir Pressure Determination Using Tiab's Direct Synthesis Technique*" M. S. Thesis, The University of Oklahoma, Norman, OK, U.S.A.

Engler, T. & Tiab, D. (1996). "Analysis of Pressure and Pressure Derivative without Type Curve Matching, 4. Naturally Fractured Reservoir*s*." *Journal of Petroleum Science and Engineering*, *15*. p. 127-138.

Earlougher, R. C., Jr. (1977). "*Advances in Well Test Analysis*." Monograph Series Vol. *5*, *Society of Petroleum Engineers*, Dallas, TX.

Escobar, F. H., Ibagón, O. E. & Montealegre-M. M. (2007). "Average Reservoir Pressure Determination for Homogeneous and Naturally Fractured Formations from Multi-Rate Testing with the TDS Technique." *Journal of Petroleum Science and Engineering.* Vol. *59*, p. 204-212, 2007.

Escobar, F. H., Alzate, H. D. & Moreno, L. (2013). "Effect of Extending the Radial Superposition Function to Other Flow Regimes." *Journal of Engineering and Applied Sciences.* Vol. *8*. Nro. 8. P. 625-634. Aug. 2013.

Escobar, F. H., Alzate, H. D. & Moreno, L. (2014, May 21). "Effect of Extending the Radial Superposition Function to Other Flow Regimes." *Society of Petroleum Engineers.* doi:10.2118/169473-MS

Kazemi, H. (1969, December 1). "Pressure Transient Analysis of Naturally Fractured Reservoirs with Uniform Fracture Distribution." *Society of Petroleum Engineers.* doi:10.2118/2156-A.

Molina, M. D., Escobar, F. H., Montealegre-M, M. & Restrepo, D. P. (2005). "Application of the TDS Technique for Determining the Average Reservoir Pressure for Vertical Wells in Naturally Fractured Reservoirs." *CT&F – Ciencia, Tecnología y Futuro.* Vol. 2, No. 6. P. 45-55. Dec.

Saidi, M. A. 1987. "*Reservoir Engineering of Fractured Reservoirs.*" Total Edition Press. 289p.

Stewart, G. (2011). "*Well test Design & Analysis.*" PennWell Corporation. 1544p.

Mavor, M. J. & Cinco-Ley, H. (1979, January 1). "Transient Pressure Behavior Of Naturally Fractured Reservoirs." *Society of Petroleum Engineers.* doi:10.2118/7977-MS.

Tiab, D. & Escobar, F. H. (2003) "*Determinación del Parámetro de Flujo Interporoso a Partir de un Gráfico Semilogarítmico.*" X Congreso Colombiano del petróleo (Colombian Petroleum Symposium). Oct. 14-17, 2003. ISBN 958-33-8394-5. Bogotá (Colombia).

Tiab, D., Igbokoyi, A. & Restrepo, D. P. (2007). "*Fracture Porosity From Pressure Transient Data.*" Paper IPTC 11164 presented at the International Petroleum Technology Conference held in Dubai, U.A. E., 4–6 Dec.

Tiab, D. & E. C. Donaldson. (2004). "*Petrophysics.*" 2nd Edition, Elsevier, 889p.

Warren, J. E. & Root, P. J. (1963). "The Behavior of Naturally Fractured Reservoirs." *SPE Journal.* p. 245-255.

Wijesinghe, A. M. & Culham, W. E. (1984, January 1). "Single-Well Pressure Testing Solutions for Naturally Fractured Reservoirs With Arbitrary Fracture Connectivity." *Society of Petroleum Engineers.* doi:10.2118/13055-MS

Wong, D. W., Harrington, A. G. & Cinco-Ley, H. (1986). "*Application of the Pressure-Derivative Function in the Pressure-Transient Testing of Fractured Wells.*" SPEFE, Oct.: 470-480, 1986.

Chapter 3

TRANSITION PERIOD OFF FROM RADIAL FLOW REGIME

BACKGROUND

The transition period, which takes place once fracture fluid is depleted inside the fractures, normally occurs during radial flow regimes since most customary values of the interporosity flow parameter are found between 10^{-5} and 10^{-6}. Actually, interporosity flow parameters can reach such values as 10, indicating the transition period may show up before the radial flow regime develops. That is the case in some wells drilled in naturally fractured formations that are subjected to hydraulic fracturing. Furthermore, λ may reach such small values as 10^{-9} or even lower, indicating the transition period could be seen after radial flow regime has vanished. Such may be the case of a naturally fractured elongated reservoir, as reported by Marhaendrajana, Blasingame and Rushing (2004), where the interporosity flow parameter goes to the order of 10^{-8}.

Arab (2003) and Escobar, Sanchez and Cantillo (2008) found the transition period can show up during the late pseudosteady-state period under constant bottom-hole pressure testing conditions. Moreover, Escobar, Rojas and Cantillo (2011a); Escobar, Rojas and Bonilla (2012a); and Escobar, Rojas and Cantillo (2011a) also found the appearance of the transition period during late pseudosteady-period in elongated systems tested under transient-rate conditions.

This chapter, then, will provide conventional and *TDS* techniques for interpreting pressure tests in double-porosity formations when radial flow takes place during linear, bilinear or biradial flow regimes caused by hydraulic fracturing or when the transition period takes place during linear or hemilinear flow regimes in long reservoirs.

TRANSITION PERIOD OCCURS BEFORE RADIAL FLOW REGIME— CONVENTIONAL ANALYSIS

Tiab and Bettam (2007) introduced a technique to interpret pressure and pressure derivative tests in heterogeneous formations drained by a hydraulically fractured vertical well. An important number of pressure tests are conducted in vertically fractured wells, which may have a heterogeneous nature with very high mass transfer capacity between fracture network and matrix or hydraulic fracture. In this case, the transition period takes place before the radial flow develops. Once the flux in the hydraulic fractured is depleted, the naturally occurring fractures feed the hydraulic fracture, allowing the development of the transition period.

Escobar, Montealegre and Martínez (2009) and Escobar, Martinez and Cantillo (2010a) used the conventional analysis method for interpretation of tests in the aforementioned situation based upon the governing equations developed by Tiab and Bettam (2007), whose main assumptions are: a slightly compressible and constant viscosity fluid flows throughout a constant thickness reservoir with constant matrix and fracture permeability and porosity; and the well fully penetrates de producing formation. Flow from the natural fracture network to either a hydraulic fracture or a matrix occurs under pseudosteady-state conditions. Wellbore storage, geomechanical skin factors and gravity effects are not considered.

The dimensionless quantities defined by Tiab and Bettam (2007) are:

$$P_D = \frac{k_{fb} h \Delta P}{141.2 q \mu B} \tag{3.1}$$

$$t_D * P_D' = \frac{k_{fb} h (t * \Delta P')}{141.2 q \mu B} \tag{3.2}$$

$$t_D = \frac{0.0002637 k_{fb} t}{\mu (\phi c_t)_{f+m} r_w^2} \tag{3.3}$$

$$t_{DA} = \frac{0.0002637 k_{fb} t}{\mu (\phi c_t)_{f+m} A} \tag{3.4}$$

$$t_{Dxf} = \frac{0.0002637 k_{fb} t}{\mu (\phi c_t)_{f+m} x_f^2} \tag{3.5}$$

$$C_{fD} = \frac{k_f w_f}{k_{fb} x_f} \tag{3.6}$$

$$\omega = \frac{\phi_f c_f}{\phi_f c_f + \phi_m c_m} \tag{3.7}$$

$$\lambda = \frac{4n(n+2)k_m r_w^2}{4 k_{fb} h_m^2} \tag{3.8}$$

Equations (3.7) and (3.8) and their variables have already been discussed in Chapter 1.

According to Tiab and Bettam (2007) the bilinear flow regime of a finite-conductivity hydraulic fracture in a heterogeneous formation is governed by:

$$P_D = \frac{2.451}{\sqrt{C_{fD}}} \left(\frac{t_{Dx_f}}{\omega} \right)^{1/4} \tag{3.9}$$

Substituting the dimensionless quantities, Equations (3.1), (3.5) and (3.6), into Equation (3.9) produces:

$$\Delta P = \frac{44.102 q \mu B}{k_{fb} h \sqrt{C_{fD}}} \left(\frac{k_{fb}}{\mu (\phi c_t)_{f+m} x_f^2 \omega} \right)^{0.25} t^{0.25} \tag{3.10}$$

For pressure buildup analysis, application of time superposition is required; therefore, Equation (3.10) becomes:

$$\Delta P_{ws} = \frac{44.102 q \mu B}{k_{fb} h \sqrt{C_{fD}}} \left(\frac{k_{fb}}{\mu (\phi c_t)_{f+m} x_f^2 \omega} \right)^{0.25} \left(\sqrt[4]{t_p + \Delta t} - \sqrt[4]{\Delta t} \right) \tag{3.11}$$

The above two expressions imply a Cartesian plot of ΔP versus either $t^{0.25}$ (for drawdown tests) or $[(t_p+\Delta t)^{0.25} - \Delta t^{0.25}]$ (for buildup tests) will yield a straight line with a slope, m_{BL}, which is good for estimating the fracture conductivity:

$$k_f w_f = \frac{1944.96}{\sqrt{\omega \mu (\phi c_t)_{f+m} k_{fb}}} \left(\frac{q B \mu}{h m_{BL}} \right)^2 \tag{3.12}$$

Figure 3.1 is a plot of $P_D C_{fD}^{0.5} \lambda_f^{0.25}$ versus $\lambda\, t_{Dxf}/\omega$. It is observed there that during the pseudosteady-state transition period $P_D \lambda_f^{0.5}$ yields a horizontal line defined by Tiab and Bettam (2007) as:

$$P_{Dpss} C_{fD}^{0.5} \lambda_f^{0.25} = \frac{\pi}{\sqrt{2}} \tag{3.13}$$

An expression to find the interporosity flow parameter is found once Equations (3.1) and (3.6) are plugged into the above expression:

$$\sqrt{\frac{k_f w_f}{k_{fb} x_f}} \lambda_f^{0.25} = 313.667 \frac{q \mu B}{k_{fb} h (\Delta P)_{pss}} \tag{3.14}$$

It is observed for the first bilinear flow regime, Figure 3.1, that the curves for the same interporosity flow parameter coincide for different values of the dimensionless capacity ratio. Escobar, Martinez and Montealegre (2009) found a correlation valid for $\omega \geq 0.005$ with a correlation coefficient of 0.99995:

$$\omega = \frac{0.0097452700024 \lambda_f k_{fb}}{\mu(\phi c_t)_{f+m} x_f^2} \left(\frac{k_{fb} h \Delta P_{1hr} C_{fD}^{0.5} \lambda_f^{0.25}}{141.2 q \mu B} \right)^{-3.96430663} \quad (3.15)$$

As presented by Tiab (1994) for the case of an infinite-conductivity fractured well in a homogeneous system, the linear flow regime becomes shorter as the x_e/x_f ratio increases. When this ratio is higher than 16, the linear flow regime is no longer observed and a biradial flow regime dominates the early time data. Based on these criteria, Tiab (1994) presented the governing equation for biradial flow regime:

$$P_D = 2.14 \left(\frac{x_e}{x_f} \right)^{0.72} \left(\frac{t_{DA}}{\xi} \right) t_{DA}^{0.36} \quad (3.16)$$

Notice that ξ was originally excluded in Tiab's (YEAR) model since it was only presented for homogeneous formations. After, Escobar, Ghisays-Ruiz and Bonilla (2014a) added the ξ symbol. When $\xi = 1$, Equation (3.16) accounts for homogeneous reservoirs. In the case of naturally fractured formations, $\xi = \omega$. The above expression, although correct, presents a major drawback in cases where the pressure test is not long enough to reach the reservoir boundaries. In such cases, the drainage area is unknown and Equation (3.16) may have no applicability if the user is not very familiar with the *TDS* technique. The equation does have applicability using intercept points between radial and biradial straight-line flow regimes, t_{RBRi}.

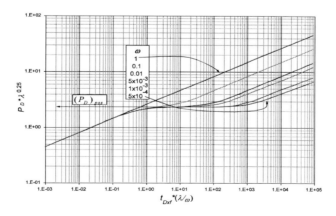

Figure 3.1. Effect of the dimensionless storativity ratio on the dimensionless pressure behavior for a finite-conductivity fracture, adapted from Tiab and Bettam (2007).

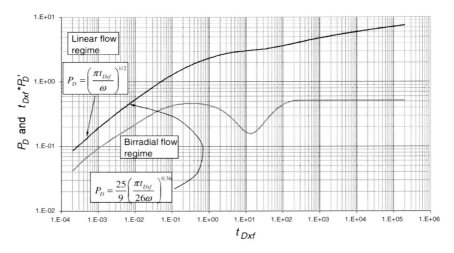

Figure 3.2. Dimensionless pressure and pressure derivative behavior for an infinite-conductivity fractured vertical well in a heterogeneous reservoir, $\lambda = 1 \times 10^{-8}$ and $\omega = 0.1$, adapted from Escobar et al. (2014a).

With the purpose of making procedures easier for *TDS* technique users, Escobar et al. (2014a) developed a new mathematical model for biradial flow regime that excludes the reservoir area. Figure 3.2 shows the dimensionless pressure and pressure derivative versus dimensionless time based on fracture length. The linear flow regime is observed at an early time point in the test. After a short transition, the biradial flow regime is observed. The new expressions for pressure and pressure derivative that Escobar et al. (2014a) came up with are given below:

$$P_D = \frac{25}{9}\left(\frac{\pi t_{Dxf}}{26\xi}\right)^{0.36} \quad (3.17)$$

$$t_D * P_D' = \left(\frac{\pi t_{Dxf}}{26\xi}\right)^{0.36} \quad (3.18)$$

After replacing in Equation (3.17) the dimensionless quantities given by Equations (3.1) and (3.5), it yields:

$$\Delta P = \frac{9.4286 q \mu B}{kh}\left(\frac{k}{\omega\mu(\phi c_t)_{m+f} x_f^2}\right)^{0.36} t^{0.36} \quad (3.19)$$

or

$$\Delta P = m_{ell} t^{0.36} \quad (3.20)$$

which implies a Cartesian plot of ΔP versus $t^{0.36}$ (for drawdown) or ΔP versus $[(t_p+\Delta t)^{0.36} - \Delta t^{0.36}]$ (for buildup) provides a straight line whose slope, m_{ell}, provides the half-fracture length:

$$x_f = \sqrt{9.4286 \frac{q\mu B}{khm_{ell}} \left(\frac{k}{\omega\mu(\phi c_t)_{m+f}}\right)^{0.36}} \qquad (3.21)$$

Notice Escobar et al. (2014a) did not provide the conventional technique for the determination of the storativity ratio. Also, according to Tiab and Bettam (2007), the linear flow regime during an early time is governed by:

$$P_D = \sqrt{\frac{\pi t_{Dxf}}{\omega}} \qquad (3.22)$$

Substituting in Equation (3.21) the dimensionless quantities, Equations (3.1) and (3.5), yields:

$$\Delta P_{wf} = 4.064 \left(\frac{qB}{x_f h}\right) \sqrt{\frac{\mu}{\omega(\phi c_t)_{f+m} k_{fb}}} t^{0.5} \qquad (3.23)$$

Application of time superposition in Equation (3.23) leads to:

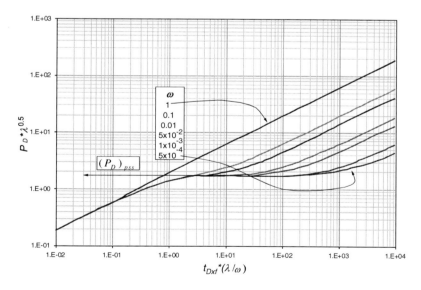

Figure 3.3. Effect of the dimensionless storativity ratio on the dimensionless pressure behavior for an infinite-conductivity fracture, adapted from Tiab and Bettam (2007).

$$\Delta P_{ws} = 4.064 \left(\frac{qB}{x_f h} \right) \sqrt{\frac{\mu}{\omega(\phi c_t)_{f+m} k_{fb}}} \left(\sqrt{t_p + \Delta t} - \sqrt{\Delta t} \right) \qquad (3.24)$$

Equations (3.22) and (3.23) imply a Cartesian plot of ΔP versus either $t^{0.5}$ or the famous tandem square root, $[(t_p+\Delta t)^{0.5} - \Delta t^{0.5}]$, will yield a straight line with a slope, m_L, which allows obtaining the half-fracture length:

$$x_f = 4.064 \left(\frac{qB}{m_L h} \right) \sqrt{\frac{\mu}{(\phi c_t)_{f+m} k_{fb} \omega}} \qquad (3.25)$$

Figure 3.3 is a plot of $P_D \lambda_f^{0.5}$ versus $\lambda_f t_{Dxf}/\omega$. As pointed out by Tiab and Bettam (2007), it is observed that during the pseudosteady-state transition period $P_D \lambda_f^{0.5}$ is a horizontal line:

$$P_{Dpss} \lambda_f^{0.5} = \frac{\pi}{2} \qquad (3.26)$$

Substituting Equation (3.1) for the pressure dimensionless term and solving for λ_f results in:

$$\lambda_f^{0.5} = \frac{221.8 q \mu B}{k_{fb} h (\Delta P)_{pss}} \qquad (3.27)$$

Notice that, for the first linear flow, the lines for the same interporosity flow parameter coincide for different values of the storativity coefficient ratio. An empirical expression using the pressure drop at the time of 1 hr, $\Delta P1_{hr}$, whose correlation coefficient is 0.99998, is provided:

$$\omega = \frac{0.00083696 \lambda_f k_{fb}}{\phi \mu (c_t)_{f+m} x_f^2} \left(\frac{k_{fb} h \Delta P_{1hr} \lambda_f^{0.5}}{141.2 q \mu B} \right)^{-1.984944590} \qquad (3.28)$$

The interporosity flow parameter can be approximated by the equation provided by Tiab and Escobar (2003):

$$\lambda = \frac{3792(\phi c_t)_{f+m} \mu r_w^2}{k_{fb} \Delta t_{inf}} \left[\omega \ln \left(\frac{1}{\omega} \right) \right] \qquad (3.29)$$

The half-fracture length can be found from an equation presented by Tiab (2001). Notice this equation can also be used to obtain the conductivity when the half-fracture length is known:

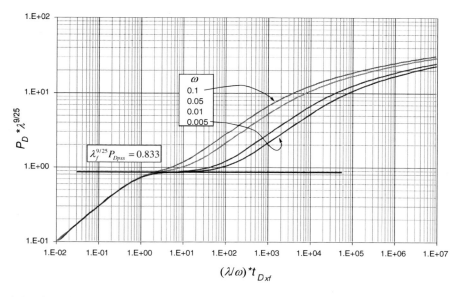

Figure 3.4. Dimensionless pressure times the interporosity flow parameter to the power 0.36 versus dimensionless time multiplied by λ/ω, $\lambda_f = 100$; adapted from Escobar et al. (2015).

$$x_f = \frac{1.92173}{\dfrac{e^S}{r_w} - \dfrac{3.31739k}{k_f w_f}} \qquad (3.30)$$

The above expression as well as the *TDS* technique for finite-conductivity vertically fractured wells in homogeneous reservoirs was included in the work by Tiab et al. (1999).

For the case of infinite-conductivity fractures, Escobar et al. (2015) presented the following expression, which has a similar application to Equation (3.27):

$$\lambda_f = 564069.34 \left[\frac{q\mu B}{kh\Delta P_{pss}} \right]^{25/9} \qquad (3.31)$$

As seen in Figure 3.4, once the transition period vanishes, a line with a slope of 0.4 develops. Draw such line and draw a parallel line through the points just before the development of the transition period. Read two points in both lines at the same time value and take their ratio, which is used in the expression below. The storativity ratio is found from such separation:

$$\omega = 0.3625 e^{-0.9877 \frac{\Delta P_{high}}{\Delta P_{low}}} \qquad (3.32)$$

TRANSITION PERIOD OCCURS BEFORE RADIAL FLOW REGIME—
TDS TECHNIQUE

Bilinear Flow Regime

Making use of Equation (3.9), after replacing the dimensionless fracture conductivity given by Equation (3.6), Tiab and Bettam (2007) arrived at the following expression:

$$k_f w_f = \frac{1947.46}{\sqrt{\omega\mu(\phi c_t)_{f+m} k_{fb}}} \left(\frac{q\mu B}{h\Delta P_{BL1}} \right)^2 \tag{3.33}$$

which uses the value of the pressure drop, ΔP_{BL1}, of the bilinear flow regime at the time of 1 hr, extrapolated if needed, as well as the pressure derivative of Equation (3.9):

$$t_{Dx_f} * P_D' = \frac{0.6127}{\sqrt{C_{fD}}} \left(\frac{t_{Dx_f}}{\omega} \right)^{1/4} \tag{3.34}$$

Tiab and Bettam (2007) also arrived at an equation to estimate fracture conductivity by reading the pressure derivative value, $(t^*\Delta P')_{BL1}$, of the bilinear flow regime line at the time of 1 hr:

$$k_f w_f = \frac{171.74}{\sqrt{\omega\mu(\phi c_t)_{f+m} k_{fb}}} \left(\frac{q\mu B}{h(t^*\Delta P')_{BL1}} \right)^2 \tag{3.35}$$

Biradial Flow Regime

Escobar et al. (2014a) used Equation (3.18) to solve for the half-fracture length based upon reading the value of the pressure derivative, ΔP_{BR1}, of the biradial flow straight line at a time of 1 hr:

$$x_f = 22.5632 \left(\frac{qB}{h\Delta P_{BR1}} \right)^{1.3889} \sqrt{\frac{1}{\xi(\phi c_t)_{f+m}} \left(\frac{\mu}{k} \right)^{1.778}} \tag{3.36}$$

As presented by Escobar et al. (2014a), the pressure derivative of Equation (3.16) is:

$$t_D * P_D' = \left(\frac{\pi t_{Dxf}}{26\xi} \right)^{0.36} \tag{3.37}$$

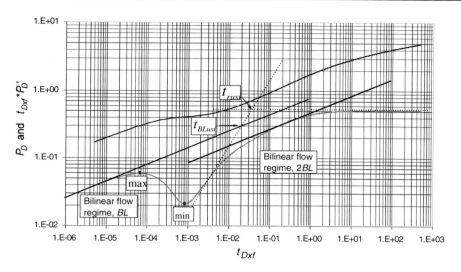

Figure 3.5. Pressure and pressure derivative behavior for a finite-conductivity–fractured well draining a double-porosity reservoir, $\lambda = 1 \times 10^{-4}$, $x_f = 400$ ft, $k_{fb} = 0.1$ md and $\omega = 0.05$; adapted from Tiab and Bettam (2007).

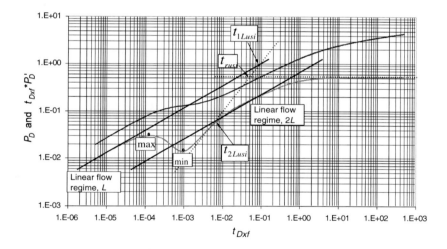

Figure 3.6. Pressure and pressure derivative behavior for an infinite-conductivity–fractured well draining a double-porosity reservoir, $\lambda = 1 \times 10^{-4}$, $x_f = 400$ ft, $k_{fb} = 0.1$ md and $\omega = 0.05$; adapted from Tiab and Bettam (2007).

Replacing the dimensionless quantities given by Equations (3.2) and (3.5) and solving for the half-fracture length yields:

$$x_f = 5.4595 \left(\frac{qB}{h(t*\Delta P')_{BR1}} \right)^{1.3889} \sqrt{\frac{1}{\xi(\phi c_t)_{f+m}} \left(\frac{\mu}{k_{fb}} \right)^{1.778}} \qquad (3.38)$$

Linear Flow Regime

Also, for the linear flow regime, Tiab and Bettam (YEAR) used Equation (3.23) to obtain an expression for the determination of the half-fracture length from reading the pressure drop, ΔP_{L1}, of the linear flow regime at the time of 1 hr, extrapolated if needed.

$$x_f = \frac{4.064qB}{(\Delta P)_{L1}h\sqrt{\omega}}\sqrt{\frac{\mu}{(\phi c_t)_t\, k_{fb}}} \tag{3.39}$$

By the same token, they used the pressure derivative during linear flow, $(t*\Delta P')_{L1}$, at a time of 1 hr:

$$x_f = \frac{2.032qB}{(t*\Delta P')_{L1}h\sqrt{\omega}}\sqrt{\frac{\mu}{(\phi c_t)_{f+m}k_{fb}}} \tag{3.40}$$

Naturally Fractured Reservoir Parameters

As expressed by Tiab and Bettam (2007), when the transition period takes place during the fracture acting time, depending on the case, two bilinear or linear flow straight lines are shown, labeled as 1 or 2, as depicted in Figures 3.5 and 3.6. In both cases the first line ($BL1$ or $L1$) corresponds to the heterogeneous behavior and the second one ($2BL1$ or $2L1$) to the total system dominated flow. If the two lines are present, Tiab and Bettam (2007) found a very useful equation by dividing Equation (3.36) by Equation (3.35):

$$\omega = \left[\frac{(t*\Delta P')_{2BL1}}{(t*\Delta P')_{BL1}}\right]^4 \tag{3.41}$$

They also found the interporosity flow parameter can be found from the time intersection of the radial flow regime and the unit-slope line developed during the transition period, t_{rusi}, and the intercept of the unit-slope line with the bilinear flow regime line:

$$\lambda_f = 51410\,(k_{fb}x_f)^2(k_f w_f)^{2/3}\left[\frac{(\phi c_t)_{f+m}\mu}{k^3 t_{USi}}\right]^{4/3} \tag{3.42}$$

$$\lambda_f = \frac{4323.1}{\omega^{1/3}}\left[\frac{(\phi c_t)_{f+m}\mu(kx_f)^2}{k^3 t_{BLUSi}}\right] \tag{3.43}$$

Tiab and Bettam (2007) also developed useful equations that use the maximum and minimum point coordinates of the pressure derivative:

$$\omega = 0.08868 \left\{ \frac{(t*\Delta P')_{min}}{(t*\Delta P')_{max}} \right\} + 0.1707 \left\{ \frac{(t*\Delta P')_{min}}{(t*\Delta P')_{max}} \right\}^2 \qquad (3.44)$$

$$\omega = \exp\left(-0.229 \frac{t_{min}}{t_{max}} \right) \qquad (3.45)$$

$$\lambda_f = \frac{5231423.93}{C_{fD}^2} \left(\frac{47.825 \, qB\mu}{k_{fb}h \, (t*\Delta P')_{max}} \right)^4 \qquad (3.46)$$

$$\omega = \frac{kt_{max}}{1896.1\lambda\mu(\phi c_t)_{f+m} x_f^2} \qquad (3.47)$$

Analogous relationships were also developed by Tiab and Bettam (2007) for the linear flow regime:

$$\omega = \left[\frac{(t*\Delta P')_{2L1}}{(t*\Delta P')_{1L1}} \right]^2 \qquad (3.48)$$

$$\lambda_f = \frac{4828.364(\phi c_t)_{f+m} \mu x_f^2}{\omega k_{fb} t_{1LUSi}} \qquad (3.49)$$

$$\lambda_f = \frac{4828.364(\phi c_t)_{f+m} \mu x_f^2}{k_{fb} t_{2LUSi}} \qquad (3.50)$$

$$\omega = \frac{(t*\Delta P')_{2LUSi}}{(t*\Delta P')_{1LUSi}} \qquad (3.51)$$

$$\lambda_f = 2916 \left(\frac{qB\mu}{k_{fb}h(t*\Delta P')_{max}} \right)^2 \qquad (3.52)$$

$$\omega = 0.0666 \left\{ \frac{(t*\Delta P')_{min}}{(t*\Delta P')_{max}} \right\} + 0.0963 \left\{ \frac{(t*\Delta P')_{min}}{(t*\Delta P')_{max}} \right\}^2 \qquad (3.53)$$

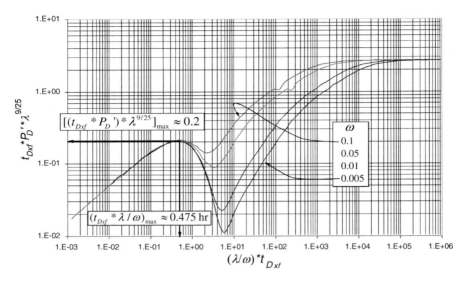

Figure 3.7. Correlation of the maximum and minimum pressure derivative coordinates during an elliptical flow period, $\lambda_f = 100$; adapted from Escobar et al. (2015).

For infinite-conductivity fractures, Escobar et al. (2015) presented expressions that work in the same fashion. For instance, the interporosity flow parameter is found from the derivative of the minimum point:

$$\lambda_f \approx 26.364 \left[\frac{q\mu B}{kh(t*\Delta P')_{min}} \right]^{25/9} \quad (3.54)$$

Referring to Figure 3.7, the found from the maximum and minimum points:

$$\lambda_f \approx 10719.63 \left[\frac{q\mu B}{kh(t*\Delta P')_{max}} \right]^{25/9} \quad (3.55)$$

$$\omega = \frac{\lambda_f k t_{max}}{1801.29 \phi \mu c_t x_f^2} \quad (3.56)$$

$$\lambda_f = \frac{1801.29 \omega \phi \mu c_t x_f^2}{k t_{max}} \quad (3.57)$$

The ratio between the times at which the maximum and minimum take place was correlated to find the storativity ratio:

$$\omega = 0.0002\left(\frac{t_{min}}{t_{max}}\right)^2 - 0.0155\left(\frac{t_{min}}{t_{max}}\right) + 0.2487 \tag{3.58}$$

An expression analogous to Equation (3.48) is given as:

$$\omega = \left[\frac{(t*\Delta P')_{2BR1}}{(t*\Delta P')_{BR1}}\right]^{2.7778} \tag{3.59}$$

Equation (3.59) can be used if biradial flow occurs before and after the transition. However, there are some cases in which linear flow develops before the transition; for this case, we use instead:

$$\omega = 0.13853\left(\frac{h}{qB}\right)^{0.778}\left[\frac{(t*\Delta P')_{2BR1}^{1.3889}}{(t*\Delta P')_{L1}}\right]^2\left(\frac{k}{\mu}\right)^{0.778} \tag{3.60}$$

Example 3.1

An example presented by Tiab and Bettam (2007) used the following information:

$q = 1000$ STB/D	$h = 100$ ft	$\phi_t = 20\%$
$r_w = 0.25$ ft	$\mu = 0.65$ cp	$B = 1.05$ rb/STB
$(c_t)_{f+m} = 4\times10^{-6}$ psi^{-1}	$k_{fb} = 40$ md	$P_i = 3000$ psi
$x_f = 200$ ft	$\lambda_f = 3.2$	$\omega = 0.05$
$\lambda = 5\times10^{-6}$	$C_{fD} = 1$	

Pressure data is given in Table 3.1. This reservoir is characterized using both conventional analysis and the *TDS* technique.

Solution by Conventional Analysis

In this example the bilinear flow regime occurs before the transition period. From Figure 3.8, values of $\Delta P_{1hr} = 101.5$ psi, $t_{min} = 0.095$ hr, $t_{max} = 0.0073$ hr and $\Delta P_{pss} = 37.5$ psi are obtained. A value of $m_{BL} = 96.96$ psi/hr is obtained from Figure 3.9. The interporosity flow parameter between a fracture network and a hydraulic fracture can be found using Equation (3.14):

$$\lambda_f = \left(\frac{313.667q\mu B}{k_{fb}h(\Delta P)_{pss}}\right)^4\left(\frac{k_{fb}x_f}{k_f w_f}\right)^2 = \left(\frac{313.667(1000)(0.65)(1.05)}{(40)(100)(37.5)}\right)^4(1)^2 = 4.15$$

The storativity ratio is found from Equation (3.15):

$$\omega = \frac{0.009745270024(4.15)(40)}{(0.65)(0.2)(4\times10^{-6})200^2}\left(\frac{(400)(100)(101.5)\sqrt{1}(4.15)^{0.25}}{141.2(1000)(0.65)(1.05)}\right)^{-3.96430663} = 0.0634$$

The fracture conductivity can be found with Equation (3.12):

$$k_f w_f = \frac{1944.96}{\sqrt{(0.0634)(0.65)(0.2)(4\times10^{-6})(40)}}\left(\frac{1000(1.05)(0.65)}{100(96.96)}\right)^2 = 8319.9 \text{ md-ft}$$

The interporosity flow parameter between the fracture network and matrix is estimated with Equation (3.29):

$$\lambda = \frac{3792(0.2)(4\times10^{-6})(0.65)(0.25^2)}{(40)(0.0095)}\left[0.0634\ln\left(\frac{1}{0.0634}\right)\right] = 4.86\times10^{-6}$$

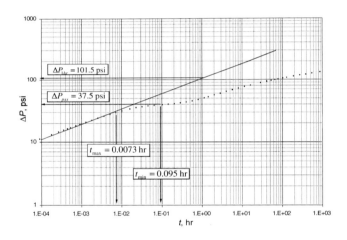

Figure 3.8. Log-log plot of pressure drop vs. time for example 3.1.

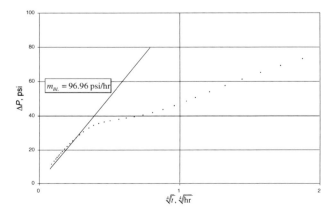

Figure 3.9. Cartesian plot of pressure drop vs. $t^{0.25}$ for example 3.1.

Table 3.1. Pressure data for example 3.1

t, hr	ΔP, psi	t*ΔP, psi	t, hr	ΔP, psi	t*ΔP, psi	t, hr	ΔP, psi	t*ΔP, psi
0.0001	10.482	2.500	0.02537	33.370	3.872	6.372	64.205	11.380
0.0002	12.374	3.027	0.03583	34.591	3.181	9.000	68.192	11.596
0.0003	13.627	3.293	0.05061	35.542	2.507	12.7128	72.233	11.546
0.0004	14.585	3.475	0.07149	36.260	2.024	17.9574	76.311	11.767
0.0005	15.368	3.596	0.113	37.038	1.819	31.9574	83.169	12.036
0.0006	16.036	3.751	0.160	37.662	2.114	45.9574	87.513	12.317
0.0008	17.140	3.913	0.226	38.456	2.721	59.9574	90.699	12.053
0.00101	18.078	4.064	0.319	39.513	3.566	73.9574	93.217	12.001
0.00143	19.544	4.310	0.451	40.907	4.610	87.9574	95.298	12.095
0.00201	21.095	4.588	0.637	42.700	5.795	115.957	98.618	11.885
0.00285	22.725	4.797	0.900	44.933	7.091	143.957	101.218	11.862
0.00402	24.423	4.965	1.271	47.610	8.274	211.957	105.871	11.930
0.00568	26.171	5.066	1.600	49.625	8.949	305.957	110.288	12.065
0.00802	27.944	5.116	2.261	52.938	9.834	433.957	114.495	11.962
0.01133	29.699	4.975	3.193	56.525	10.571	613.957	118.673	12.356
0.01796	31.907	4.398	4.511	60.302	11.055	971.957	124.205	12.356

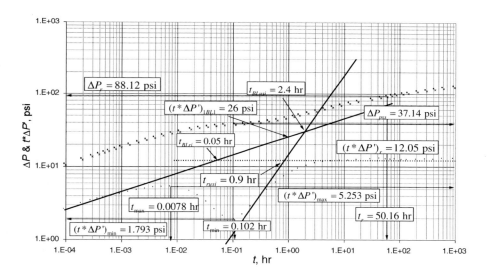

Figure 3.10. Pressure and pressure derivative log-log plot for example 3.1.

Solution by TDS Technique

The following information was obtained from Figure 3.10:

t_r = 5.16 hr $(t*\Delta P')_r$ = 12.05 psi ΔP_r = 88.12 psi
t_{min} = 0.102 hr ΔP_{pss} = 37.14 psi t_{max} = 0.0078 hr
$(t*\Delta P')_{max}$ = 5.253 psi t_{usi} = 0.9 hr t_{BLri} = 0.05 hr
t_{BLusi} = 2.4 hr $(t*\Delta P')_{1BL1}$ = 26 psi

Permeability is estimated with Equation (1.18):

$$k = \frac{70.6q\,B\,\mu}{h\,\left(t*\Delta P'\right)_r} = \frac{(70.6)(1000)(1.05)(0.65)}{(100)(12.05)} = 39.98 \text{ md}$$

The storativity ratio coefficient is obtained from Equations (3.44) and (3.45):

$$\omega = 0.08868\left\{\frac{0.102}{5.253}\right\} + 0.1707\left\{\frac{0.102}{5.253}\right\}^2 \approx 0.05$$

and

$$\omega = \exp\left(-0.229\frac{t_{min}}{t_{max}}\right) = \exp\left(-0.229\frac{1.793}{0.0078}\right) = 0.05$$

The fracture conductivity is calculated from Equation (3.33):

$$k_f w_f = \frac{121.74}{\sqrt{(0.05)(0.65)(0.2)(4\times10^{-6})}}\left(\frac{(1000)(0.65)(1.05)}{(100)(26)}\right)^2 = 8227.1 \text{ md-ft}$$

The skin factor can be estimated with Equation (1.21):

$$s = 0.5\left[\frac{88.12}{12.05} - \ln\left(\frac{(39.98)(50.16)}{(0.2)(4\times10^{-6})(0.65)(0.25)^2}\right) + 7.43\right] = -5.05$$

The half-fracture length can be found with Equation (3.30):

$$x_f = \frac{1.92173}{\dfrac{e^{-5.05}}{0.25} - \dfrac{3.31739\times39.98}{8227.1}} = 203 \text{ ft}$$

Table 3.2. Comparison of results from example 3.1

Parameter	TDS (Tiab & Bettam 2007)	Equation	Conventional (Escobar et al. 2009)	Equation
λ_f	2.6	3.42	4.15	3.14
λ_f	4.2	3.43		
ω	0.05	3.44	0.0634	3.15
ω	0.05	3.45		
$k_f w_f$, md-ft	8227.1	3.33	8319.9	3.12

Thus, the dimensionless fracture conductivity is calculated with Equation (3.6):

$$C_{fD} = \frac{k_f w_f}{k_{fb} x_f} = \frac{8227.1}{39.89 \times 203} = 1.015$$

The interporosity flow parameter is calculated by Equations (3.42) and (3.43). The following results were found:

$$\lambda_f = 51410(39.98 \times 203)^2 (8227.1)^{2/3} \left[\frac{(0.2)(4 \times 10^{-6})}{39.98^3 (0.9)} \right]^{4/3} = 2.6$$

$$\lambda_f = \frac{4323.1}{0.05^{1/3}} \left[\frac{(0.2)(4 \times 10^{-6})(0.65)(203)^2}{39.98(2.4)} \right] = 4.2$$

Equation (3.29) also belongs to the *TDS* technique, but the time of inflection is changed by the time at the minimum during the transition period. Results from both techniques are reported in Table 3.2. Two important observations are drawn: (1) The *TDS* technique allows verifying the estimated parameters; and (2) the *TDS* technique allows estimating all the parameters without knowing the others. In the worked examples, the parameters were estimated using the input data for the simulation, which is not quite practical.

Table 3.3. Pressure data for example 3.2

t, hr	ΔP, psi	$t*\Delta P'$, psi	t, hr	ΔP, psi	$t*\Delta P'$, psi	t, hr	ΔP, psi	$t*\Delta P'$, psi
1.280E-06	0.412	0.198	3.036E-04	3.248	0.183	0.072	8.677	3.324
1.707E-06	0.473	0.244	4.049E-04	3.294	0.115	0.096	9.683	3.721
2.277E-06	0.549	0.274	5.400E-04	3.324	0.085	0.128	10.812	4.148
3.036E-06	0.640	0.305	7.204E-04	3.355	0.092	0.171	12.078	4.590
4.049E-06	0.732	0.351	9.601E-04	3.385	0.125	0.228	13.465	5.048
5.400E-06	0.839	0.412	1.280E-03	3.431	0.168	0.303	14.975	5.490
7.204E-06	0.961	0.457	1.707E-03	3.477	0.229	0.405	16.622	5.902
9.601E-06	1.113	0.518	2.277E-03	3.553	0.305	0.540	18.361	6.252
1.280E-05	1.266	0.579	3.036E-03	3.660	0.396	0.720	20.206	6.542
1.707E-05	1.449	0.656	4.049E-03	3.782	0.518	0.960	22.127	6.786
2.277E-05	1.632	0.717	0.0054	3.950	0.656	1.280	24.110	6.984
3.036E-05	1.845	0.762	0.007	4.163	0.839	1.707	26.138	7.137
4.049E-05	2.074	0.793	0.010	4.438	1.052	2.277	28.212	7.244
5.400E-05	2.303	0.793	0.013	4.773	1.296	3.036	30.316	7.335
7.204E-05	2.531	0.762	0.017	5.185	1.586	4.049	32.436	7.411
9.601E-05	2.745	0.686	0.023	5.688	1.906	5.400	34.571	7.457
1.280E-04	2.928	0.579	0.030	6.283	2.242	7.201	36.736	7.503
1.707E-04	3.065	0.442	0.041	6.984	2.577	9.603	38.886	7.533
2.277E-04	3.172	0.305	0.054	7.777	2.943			

Example 3.2

The naturally fractured reservoir parameters must be found using a synthetic example provided by Escobar et al. (2015) generated with the following input information:

q = 30 STB/D h = 100 ft ϕ_t = 5%
μ = 1.5 cp B = 1.2 rb/STB $(c_t)_{f+m}$ = 1×10^{-4} psi^{-1}
k_{fb} = 5 md x_f = 10 ft λ_f = 50
ω = 0.01

The pressure test data are given in Table 3.3.

Solution by Conventional Analysis

In Figure 3.11, a horizontal line was drawn through the transition points. The following value of pressure drop was observed:

$$\Delta P_{pss} = 3.2 \text{ psi}$$

Using the above value in Equation (3.31) yields:

$$\lambda_f = 564069.34 \left[\frac{(50)(1.5)(1.2)}{(5)(100)(3.2)} \right]^{25/9} = 46.05$$

Then, a 0.4-slope line is drawn after the transition period. Another parallel line goes just after the initiation of the transition period so the amplitude can be estimated. Two pressure drop values are read at the same time from both straight lines, so:

ΔP_{high} = 22 psi
ΔP_{low} = 6 psi

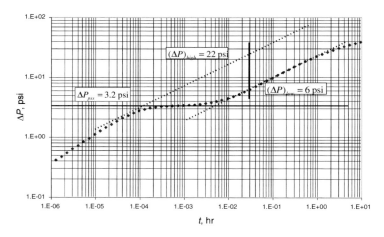

Figure 3.11. Pressure drop versus time log-log plot for example 3.2; adapted from Escobar et al. (2015).

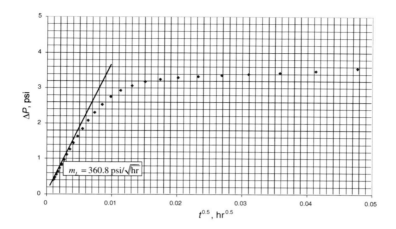

Figure 3.12. Cartesian plot of pressure versus the square root of time for example 3.2.

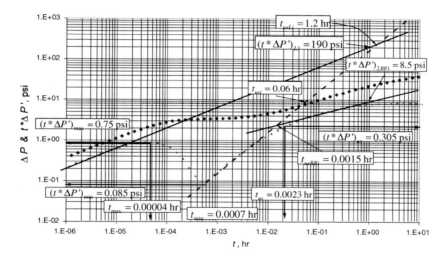

Figure 3.13. Pressure and pressure derivative versus time log-log plot for example 3.2.

Equation (3.32) leads to the estimation of the storativity ratio:

$$\omega = 0.3625\,e^{-0.9877\frac{22}{6}} = 0.0097 \approx 0.01$$

The half-fracture length is found from the slope of a Cartesian plot (m_L= 360.8 psi/hr$^{0.5}$) of the pressure versus the square root of time—see Figure 3.12—using Equation (3.25):

$$x_f = 4.064\left(\frac{30(1.2)}{(100)(360.8)}\right)\sqrt{\frac{1.5}{(0.05)(0.0001)(5)(0.01)}} = 9.93 \text{ ft}$$

Solution by TDS Technique

The following information was read obtained the pressure and pressure derivative log-log plot provided in Figure 3.13:

$t_{max} = 0.00004$ hr	$(t^*\Delta P')_{2BR1} = 8.5$ psi	$(t^*\Delta P')_{max} = 0.75$ psi
$(t^*\Delta P')_{L1} = 190$ psi	$(t^*\Delta P')_{min} = 0.085$ psi	$t_{us} = 0.0023$ hr
$(t^*\Delta P')_{us} = 0.305$ psi	$t_{usi} = 0.06$ hr	$t_{usBRi} = 0.015$ hr
$t_{usLi} = 1.2$ hr	$(t^*\Delta P')_{2BR1} = 8.5$ psi	$t_{min} = 0.0007$ hr

using Equation (3.40):

$$x_f = 2.032\left(\frac{30(1.2)}{(190)(100)\sqrt{0.01}}\right)\sqrt{\frac{1.5}{0.05(0.0001)(5)}} = 9.43\,\text{ft}$$

The half-fracture length is also found with Equation (3.38):

$$x_f = 5.4595\left(\frac{30(1.2)}{100(8.5)}\right)^{1.3889}\sqrt{\frac{1}{0.05(0.0001)}\left(\frac{1.5}{5}\right)^{1.778}} = 10.37\,\text{ft}$$

The interporosity flow parameter is estimated with Equations (3.54), (3.55) and (3.57):

$$\lambda_f \approx 26.364\left[\frac{30(1.5)(1.2)}{5(100)(0.085)}\right]^{25/9} = 51.28$$

$$\lambda_f \approx 10719.63\left[\frac{30(1.5)(1.2)}{5(100)(0.75)}\right]^{25/9} = 49.24$$

$$\lambda_f = \frac{1801.29(0.1)(0.05)(1.5)(0.0001)(10^2)}{5(0.00004)} = 45.03$$

The storativity ratio is estimated with Equations (3.58), (3.60), (29) and (31):

$$\omega = 0.0002\left(\frac{0.0007}{0.00004}\right)^2 - 0.0155\left(\frac{0.0007}{0.00004}\right) + 0.2487 = 0.039$$

$$\omega = 0.13853\left(\frac{100}{30(1.2)}\right)^{0.778}\left[\frac{(8.5)^{1.3889}}{(190)}\right]^2\left(\frac{5}{1.5}\right)^{0.778} = 0.0083$$

TRANSITION PERIOD OCCURS AFTER RADIAL FLOW REGIME—CONVENTIONAL ANALYSIS

This situation takes places in elongated systems in which a linear flow regime develops. The first case takes place when the well is centered and only a dual-linear flow develops. For such a case, Escobar et al. (2009) found a correlation to estimate the storativity ratio using the pressure drop, ΔP_{1hr}, read at a time of 1 hr on the radial flow regime of the semilog plot:

$$\omega = \text{anti} \log\left[-3.22454 + 1.00005\log\left(\frac{k_{fb}}{\mu(\phi c_t)_{f+m}r_w^2}\right) - \frac{k_{fb}h\Delta P_{1hr}}{162.447q\mu B} - s_r\right] \quad (3.61)$$

For this case Equation (3.29) is still valid for determining the interporosity flow parameter. Escobar et al. (2010b) found the governing equation for the case:

$$P_D = \frac{2\sqrt{\pi t_D}}{W_D\sqrt{\omega}} + s_{DL} \quad (3.62)$$

Once the dimensionless parameters are substituted into Equation (3.62), it yields:

$$\Delta P_{wf} = \frac{8.1282}{Y_E}\frac{qB}{h}\left(\frac{\mu}{(\phi c_t)_{f+m}k_{fb}\omega}\right)^{0.5}\sqrt{t} + \frac{141.2q\mu B}{k_f h}s_{DL} \quad (3.63)$$

Equation (3.63) indicates a Cartesian plot of ΔP versus either $t^{0.5}$ or $[(t_p+\Delta t)^{0.5} - \Delta t^{0.5}]$ (for buildup cases) will yield a straight line during dual-linear flow behavior whose slope, m_{DLF}, and intercept, b_{DLF}, are used to obtain reservoir width, Y_E, once the storativity coefficient ratio is determined and dual-linear skin factor, s_{DL}.

$$Y_E = \frac{8.1282qB}{m_{DLF}h}\left[\frac{\mu}{k_{fb}(\phi c_t)_{f+m}\omega}\right]^{0.5} \quad (3.64)$$

$$s_{DL} = \frac{k_{fb}hb_{DLF}}{141.2q\mu B} \quad (3.65)$$

If the linear or hemilinear flow regime takes place after the transition period, then the equations given in chapter 1 for homogeneous systems will apply. In the contrary case, the governing equation presented by Escobar et al. (2009) is:

$$P_D = \frac{10\pi\sqrt{t_D}}{6W_D\sqrt{\omega}} + s_L \tag{3.66}$$

Once the dimensionless quantities are replaced in Equation (3.66), the following expression is obtained:

$$\Delta P_{wf} = \frac{12.006}{Y_E} \frac{qB}{h} \left(\frac{\mu}{(\phi c_t)_{f+m} k_{fb}\omega} \right)^{0.5} \sqrt{t} + \frac{141.2q\mu B}{k_{fb}h} s_L \tag{3.67}$$

which implies a Cartesian plot of ΔP versus either $t^{0.5}$ or $[(t_p+\Delta t)^{0.5} - \Delta t^{0.5}]$ (for buildup cases) will yield a straight line during linear flow behavior whose slope, m_{LF}, and intercept, b_{LF}, are used to obtain reservoir width, Y_E, once the storativity coefficient ratio is determined and single-linear skin factor, s_L.

$$Y_E = \frac{12.006qB}{m_{LF}h} \left[\frac{\mu}{k_{fb}(\phi c_t)_{f+m}\omega} \right]^{0.5} \tag{3.68}$$

$$s_L = \frac{k_{fb}hb_{LF}}{141.2q\mu B} \tag{3.69}$$

TRANSITION PERIOD OCCURS AFTER RADIAL FLOW REGIME— *TDS* TECHNIQUE

The formulation for this case was developed by Escobar, Hernandez and Saavedra (2010b). A typical pressure behavior for this case is reported in Figure 3.14. When the transition period takes place during a dual-linear flow regime, the authors took derivative to Equation (3.62):

$$(t_D * P_D')_{DL} = \frac{\sqrt{\pi t}}{W_D\sqrt{\omega}} \tag{3.70}$$

Escobar, Hernandez and Hernandez (2007a) defined the dimensionless reservoir width, W_D, as:

$$W_D = \frac{Y_E}{r_w} \tag{3.71}$$

Substituting the dimensionless quantities given by Equations (3.2), (3.3) and (3.62) into Equation (3.70) and solving for reservoir width yields:

$$Y_E = \frac{4.064\, qB}{h\,(t*\Delta P')_{DL1}} \sqrt{\frac{\mu}{k_{fb}(\phi c_t)_{f+m}\omega}} \qquad (3.72)$$

The equation for skin factor was obtained from the division of Equation (3.62) by Equation (3.70):

$$s_{DL} = \left(\frac{\Delta P_{DL}}{(t*\Delta P')_{DL}} - 2\right)\frac{1}{19.601\, Y_E}\sqrt{\frac{k_{fb} t_{DL}}{(\phi c_t)_{f+m}\mu\omega}} \qquad (3.73)$$

where ΔP_{DL} and $(t*\Delta P')_{DL}$ are the pressure and pressure derivative points, respectively, read at any arbitrary time during dual-linear flow regime, t_{DL}; ω can be obtained from any arbitrary point of the dual-linear flow regime on the pressure derivative curve. As treated in chapter 1, a given geometric skin factor contains the former skin factors, for instance, $s_{DL} = s_{DL}' + s$.

If the reservoir width is known—for instance, from the intercept of radial and dual linear— then the storativity ratio can be solved from Equation (3.72) using any arbitrary point read on the dual-linear flow line:

$$\omega = \frac{16.5186\,\mu\, t_{DL}}{(\phi c_t)_{f+m} k_{fb}}\left[\frac{qB}{Y_E h (t*\Delta P')_{DL}}\right]^2 \qquad (3.74)$$

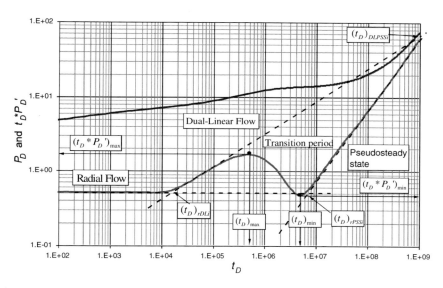

Figure 3.14. Dimensionless pressure and pressure derivative behavior versus dimensionless time for a long double-porosity reservoir with a $\omega = 0.05$, y $\lambda = 2\times 10^{-8}$ well centered in the reservoir; adapted from Escobar et al. (2010b).

Table 3.4. Constants for Equation (3.76)

Constant	Value	Constant	Value
a	-0.001453744345531936	g	-0.00884465599095826
b	-0.4034531139231534	h	-6557972683918.12
c	0.004969243560711644	i	-371412686858.0194
d	9384837.434697306	j	-10246537.3284442
e	-25245.96831604798	k	2190827.21495808
f	0.03947187690326912	x	$(t_D*P_D')_{min}$

Escobar et al. (2010b) used the maximum point observed when the dual-linear flow vanishes for the estimation of the interporosity flow parameter:

$$\lambda = \left[\frac{0.0320524\, q\mu B}{k_{fb} h\, (t*\Delta P')_{max}} \right]^2 \tag{3.75}$$

They also correlated the minimum point at the trough for the estimation of the storativity ratio:

$$\omega = \frac{a + cx + e\lambda + gx^2 + i\lambda^2 + k\lambda x}{1 + bx + d\lambda + fx^2 + h\lambda^2 + j\lambda x} \tag{3.76}$$

whose constants are given in Table 3.4. This correlation—Equation (3.76)—is recommended since it has an error of 0.12%. The intersection of the late pseudosteady-state line with the dual-linear flow regime pressure derivative line, allowed obtaining an expression to estimate reservoir area:

$$A = 0.05829 \sqrt{\frac{Y_E^2 \omega k_{fb} t_{DLPSSi}}{(\phi c_t)_{f+m}\, \mu}} \tag{3.77}$$

The intercept of the radial flow line with the dual-linear flow line leads to the determination of reservoir width:

$$Y_E = 0.0575652 \sqrt{\frac{k_{fb}\, t_{rDLi}}{\mu(\phi c_t)_{f+m}\, \omega}} \tag{3.78}$$

If the linear flow regime takes place after the transition period, as schematically described in Figure 3.15, then the reservoir behaves homogenously and the equations provided in chapter 1 are used for the purpose of interpretation. Figure 3.16 reports the opposite case, when the transition period takes place after the hemilinear or single-linear flow regime. Equation (3.66) governs such a case (Escobar et al. 2010b), whose pressure derivative is:

$$(t_D * P_D')_L = \frac{5\pi \sqrt{t_D}}{6w_D \sqrt{\omega}} \tag{3.79}$$

Again, once the dimensionless parameters are substituted into Equation (3.79), an expression for finding reservoir width is developed:

$$Y_E = \frac{6qB}{h(t * \Delta P')_L} \sqrt{\frac{t_L \mu}{k_{fb}(\phi c_t)_{f+m} \omega}} \tag{3.80}$$

Dividing Equation (3.66) by Equation (3.79) and solving for the geometrical skin factor yields:

$$s_L = \left(\frac{\Delta P_L}{(t * \Delta P')_L} - 2 \right) \frac{1}{23.522 Y_E} \sqrt{\frac{k_{fb} t_L}{(\phi c_t)_{f+m} \mu \omega}} \tag{3.81}$$

It is worth reminding that the geometrical skin factors involve the skin from the former flow regime. This means the dual-linear skin factor, s_{DL}, includes the mechanical skin factor, and s_L includes the dual-linear skin factor.

Also, the intersection points between the hemilinear, radial and pseudosteady-state extrapolated lines leads to the estimation of reservoir width and area, respectively:

$$Y_E = 0.08503 \sqrt{\frac{k_{fb} t_{RLi}}{\mu (\phi c_t)_{f+m} \omega}} \tag{3.82}$$

$$A = \sqrt{\frac{k_{fb} t_{LPPSi} \omega Y_E^2}{658.366 (\phi c_t)_{f+m} \mu}} \tag{3.83}$$

As shown in Figure 3.16, the maximum point presented once the single-linear flow has been interrupted by the transition period proves to be useful for the estimation of the interporosity flow parameter. Escobar et al. (2010b) came up with the following expression:

Table 3.5. Constants for equation (3.85)

Constant	Value	Constant	Value
a	-503.3662694927747	f	-121.8994586826316
b	67719.22178724448	g	525.9038673208586
c	1023359.583295485	h	-229.1239182054168
d	-2121.757955433415	i	-139.4598412994527
e	20.68488201371427	x	$(t_D * P_D')_{min}$

$$\lambda = \left[\frac{0.06213\, q\mu B}{k_{fb} h\, (t*\Delta P')_{max}} \right]^2 \tag{3.84}$$

Escobar et al. (2010b) also provided a correlation with an error less than 0.085% to estimate ω:

$$\ln \omega = a + b\sqrt{\lambda} \ln \lambda + c\sqrt{\lambda} + d/\ln \lambda + ex + f\sqrt{x} \ln x + g\sqrt{x} + h \ln x + i/\sqrt{x} \tag{3.85}$$

The coefficients of Equation (3.85) are given in Table 3.5. When the transition period takes place during the dual-linear flow regime (see Figure 3.15), the fissures are fed by the matrix under the pseudosteady-state flow regime. The expression governing this is given by:

$$\ln(t_D * P_D')_{US} = -16.591198 + 0.97816 \ln t_{D,US} + 3621.0204\sqrt{\lambda} \tag{3.86}$$

The former expression has a standard error of 0.707% and a correlation coefficient of 0.99997 and should be used for dimensionless time values between 300000 and 3.69×10^7. The intercept of Equation (3.86) with the radial flow dimensionless pressure derivative line ($t_D * P_D' = 0.5$) provides the following expression:

$$\lambda = \left\{ \frac{-0.69315 + 16.591198 - 0.97816 \ln t_{D,USi}}{3621.0204} \right\}^2 \tag{3.87}$$

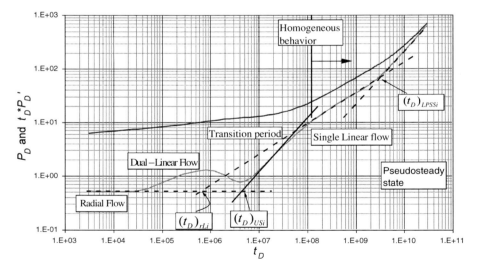

Figure 3.15. Dimensionless pressure and pressure derivative behavior for an elongated naturally fractured reservoir for a $\omega = 0.05$, $\lambda = 1 \times 10^{-8}$ and $X_E = 29000$ ft well off-centered in the reservoir; dual-linear flow is interrupted by the transition; adapted from Escobar et al. (2010b).

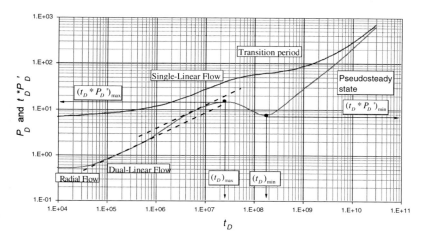

Figure 3.16. Dimensionless pressure and pressure derivative behavior for an elongated naturally fractured reservoir for a $\omega = 0.05$, $\lambda = 1\times10^{-9}$ and $X_E = 29000$ ft well off-centered in the reservoir; single-linear flow is interrupted by the transition; adapted from Escobar et al. (2010b).

When the transition occurs after the single-linear flow regime has been observed, the governing expression for the unit-slope transition line is similar to Equation (3.86), with different coefficients. For this case, the correlation is:

$$\lambda = \left\{ \frac{-0.69315 + 17.252793 - 0.9518448 \ln t_{D,USi}}{23920.204652} \right\}^2 \tag{3.87}$$

For homogeneous cases, reservoir width can also be found from the intersection of the dual-linear and linear flow lines with the radial lines:

$$Y_E = 0.05756 \sqrt{\frac{kt_{RDL_i}}{\phi \mu c_t}} \tag{3.88}$$

$$Y_E = 0.1020 \sqrt{\frac{kt_{RL_i}}{\phi \mu c_t}} \tag{3.89}$$

And also for the homogeneous case, as reported by Escobar, Hernandez and Hernandez (2007a), drainage area can be found from the intersection of the dual-linear and linear flows with the pseudosteady-state lines:

$$A = \sqrt{\frac{kt_{DLPSS_i} Y_E^2}{301.77 \phi \mu c_t}} \tag{3.90}$$

Table 3.6. Pressure and pressure derivative data versus time for example 3.3

t, hr	ΔP, psi	$t*\Delta P$, psi	t, hr	ΔP, psi	$t*\Delta P$, psi	t, hr	ΔP, psi	$t*\Delta P$, psi
0.0002	2.66	1.47	0.0544	16.71	4.62	0.40	25.19	5.61
0.0010	5.27	3.86	0.0552	16.66	4.70	0.43	25.72	6.04
0.0019	8.45	5.46	0.0560	16.7	4.76	0.48	26.39	6.46
0.0027	10.03	4.86	0.0569	16.83	4.54	0.54	27.09	6.91
0.0035	12.35	3.71	0.0585	16.8	5.04	0.62	28.11	7.23
0.0044	12.29	6.09	0.0652	17.96	5.27	0.70	29	7.52
0.0052	13.54	2.87	0.0727	18.13	5.01	0.78	29.81	7.79
0.0100	11.98	1.99	0.0744	18.32	5.12	0.87	30.8	7.92
0.0144	12.31	1.99	0.08	18.45	5.05	0.99	31.71	8.19
0.0152	12.36	1.67	0.09	21.31	5.19	1.11	32.72	8.47
0.0160	12.35	2.33	0.09	19.49	5.14	1.25	33.75	8.54
0.0185	14	2.64	0.10	20.42	5.32	1.40	34.86	8.84
0.0210	13.68	2.90	0.10	20.56	5.02	1.60	35.97	8.96
0.0227	13.75	3.11	0.11	20.91	4.52	1.84	37.18	9.13
0.0252	14.07	3.44	0.13	21.54	4.14	2.14	38.57	9.18
0.0277	14.43	3.76	0.14	21.8	3.48	2.46	39.93	9.40
0.0302	14.8	2.90	0.15	22.12	3.29	2.79	41.12	9.41
0.0310	14.78	2.83	0.16	22.27	2.74	3.17	42.4	9.65
0.0319	14.89	3.46	0.17	22.33	2.95	3.60	43.44	9.93
0.0335	15.01	2.98	0.18	22.52	2.75	4.19	44.98	10.03
0.0352	15.14	3.19	0.20	22.73	2.89	4.81	46.32	10.26
0.0369	15.25	4.23	0.23	22.97	3.07	5.53	47.95	10.08
0.0385	15.46	3.44	0.25	23.03	3.21	6.44	49.54	10.44
0.0410	15.74	3.94	0.27	23.17	3.60	7.07	50.61	10.77
0.0435	16.06	3.84	0.29	23.55	3.88	7.67	51.56	11.14
0.0452	16.18	3.93	0.32	23.88	4.39	8.23	52.18	11.47
0.0460	16.31	3.97	0.35	24.35	4.70	8.92	52.74	11.97
0.0510	16.55	4.45	0.37	24.83	5.14	9.46	53.33	12.40

$$A = \sqrt{\frac{kt_{LPSS_i} Y_E^2}{948.047\phi\mu c_t}} \tag{3.91}$$

Example 3.3

Escobar et al. (2010b) presented an example taken from a pressure test run in a South American well. Reservoir, fluid and well parameters are provided below, and pressure data are given in Table 3.6. The reservoir permeability of 2700 md was obtained from a previous test. From this, the reservoir width, reservoir area, skin factor, interporosity flow parameter and the dimensionless storativity coefficient can be found.

$q = 457$ STB/D $\qquad h = 84$ ft $\qquad \phi_t = 7.34\%$

$r_w = 0.5$ ft $\qquad \mu = 9.4$ cp $\qquad B = 1.49$ rb/STB

$(c_t)_{f+m} = 9.899 \times 10^{-6}$ psi^{-1} $\qquad P_i = 2500$ psi $\qquad P_{wf} = 2376.08$ psi

The reservoir width, geometric skin factor, interporosity flow parameter and the dimensionless storativity coefficient can be found by both conventional analysis and the *TDS* technique.

Solution by Conventional Analysis

In this example, the single-linear flow regime occurs after the transition period. Since the permeability is known, the semilog slope can be found from the classical conventional equation, giving a value of 4.58 psi/cycle. Then, $\Delta P_{1hr} = 21.16$ psi is obtained using any point on the radial flow regime straight line from Figure 3.17. The skin factor is then computed with Equation (2.11) to be -2.29. The storativity ratio can be estimated with Equation (3.61):

$$\omega = 10^{-3.22454+1.00005\log\left(\frac{2700}{(9.4)(0.0734)(9.899\times10^{-6})(0.5^2)}\right)-\frac{(2700)(84)(21.16)}{162.447(457)(9.4)(1.49)}+2.29} = 0.035$$

From the semilog plot of Figure 3.18, the time at which inflection takes place, t_{inf}, is 0.2 hr. The interporosity flow parameter is found with Equation (3.29):

$$\lambda = \frac{3792(0.0734)(9.899\times10^{-6})(9.7)(0.5^2)}{(2700)(0.2)}\left[0.035\ln\left(\frac{1}{0.035}\right)\right] = 1.4\times10^{-6}$$

Values of $m_{DLF} = 35.87$ psi/hr, $b_{DLF} = 8.47$, $m_{LF} = 18.55$ psi/hr and $b_{LF} = 13.53$ psi are taken from Figure 3.18. The reservoir width is from each linear flow, Equations (3.64) and (3.68), respectively, as follows:

$$Y_E = \frac{8.1282(457)(1.49)}{(35.87)(84)}\left[\frac{9.7}{(2700)(0.0734)(9.899\times10^{-6})(0.035)}\right]^{0.5} = 679.65 \text{ ft}$$

$$Y_E = \frac{12.006(457)(1.49)}{(18.55)(84)}\left[\frac{9.7}{(2700)(0.0734)(9.899\times10^{-6})(0.035)}\right]^{0.5} = 435.92 \text{ ft}$$

The geometrical skin factors were found to be $s+s_{DL} = 6.69$ and $s_{DL}'+s_L = 6.76$ with Equations (3.65) and (3.69), respectively.

Solution by TDS Technique

As seen in the plot, the dual-linear flow is very noisy. The following data were read from Figure 3.19:

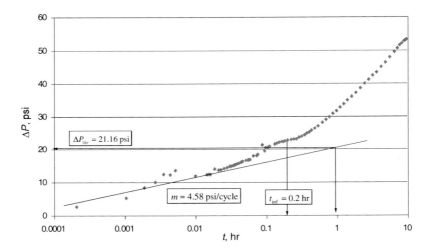

Figure 3.17. Semilog plot of pressure drop vs. time for the example 3.3; adapted from Escobar et al. (2010b).

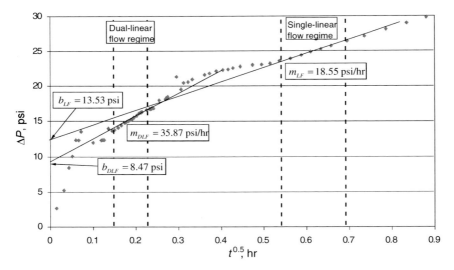

Figure 3.18. Cartesian plot of pressure drop vs. $t^{0.5}$ for the example 3.3; adapted from Escobar et al. (2010b).

$(t*\Delta P')_r = 1.99$ psi
$t_{rDLi} = 0.01$ hr
$t_{DL} = 0.03$ hr
$t_L = 0.626$ hr
$(t*\Delta P')_{min} = 2.93$ psi
$t_{LPSSi} = 42$ hr
$t_{max} = 0.0877$ psi

$\Delta P'_r = 11.98$ psi
$(t*\Delta P')_{DL} = 3.2$ psi
$(t*\Delta P')_{max} = 5.2$ psi
$(t*\Delta P')_L = 7.17$ psi
$t_{min} = 0.2$ psi
$t_{rPSSi} = 1.4$ hr

$t_r = 0.01$ hr
$\Delta P_{DL} = 16.06$ psi
$\Delta P_L = 28.2$ psi
$t_{rLi} = 0.018$ hr
$t_{DLPSSi} = 165$ hr
$t_{USi} = 0.13$ hr

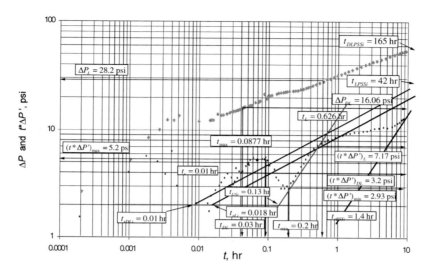

Figure 3.19. Pressure and pressure derivative plot for example 3.3; adapted from Escobar et al. (2010b).

Equations (3.2) and (3.3) allow converting the coordinates of the minimum to be $(t_D*P_D')_{min} = 0.74$ and $(t_D)_{min} = 92768.12$. Using these values in Equation (3.76), we obtain a storativity ratio value of 0.052. Using the value of maximum pressure derivative, the interporosity flow parameter can be found with Equation (3.75):

$$\lambda = \left[\frac{0.0320524(457)(9.4)(1.49)}{2700(84)(5.2)} \right]^2 = 3.8 \times 10^{-8}$$

Reservoir width is found with Equations (3.72) and (3.78), respectively.

$$Y_E = \frac{4.064(457)(1.49)}{87(3.2)} \sqrt{\frac{(9.4)(0.03)}{2700(0.0734)(9.899 \times 10^{-6})(0.052)}} = 522.62 \text{ ft}$$

$$Y_E = 0.0575652 \sqrt{\frac{2700(0.01)}{(9.4)(0.0734)(9.899 \times 10^{-6})(0.052)}} = 502.027 \text{ ft}$$

Notice the value of the pressure derivative at 1 hr was not used in Equation (3.72). Instead, a point on the dual-linear flow was employed, as expressed in Equation (1.48). Well drainage area is determined using the point of intersection between the dual-linear and pseudosteady-state straight lines using Equation (3.77), as follows:

$$A = 0.05828558 \sqrt{\frac{502.027^2 (0.052)(2700)(165)}{(9.4)(0.0734)(9.899 \times 10^{-6})}} = 1835227.6 \text{ ft}^2$$

The area can be verified with Equation (1.29):

Transition Period off from Radial Flow Regime

Table 3.7. Summary of results for example 3.3

Parameter	Conventional analysis	Equation	*TDS* Technique	Equation
			Value	
ω	0.035	3.61	0.052	3.76
ω			0.0713	3.72
ω			0.078	3.74
λ	1.4×10^{-6}	3.29	3.8×10^{-8}	3.75
Y_E, ft	679.65	3.64	522.62	3.72
Y_E, ft	435.92	3.68	502.02	3.78
Y_E, ft			446.4	1.47
A, ft^2			1835227.6	3.77
A, ft^2			1834004.93	1.29
A, ft^2			1847210.2	3.91
s	-2.29	2.11	-2.84	2.22
s_{DL}	6.69	3.65	12.2	3.73
s_L	6.76	3.69	4.32	1.49

$$A = \frac{2700(1.4)}{301.77(9.4)(0.0734)(9.899\times10^{-6})} = 1834004.93 \text{ ft}^2$$

Since the hemilinear flow is better defined, the reservoir width can be determined with Equation (1.47):

$$Y_E = \frac{7.2034\,(457)(1.49)}{84(7.17)} \sqrt{\frac{0.627(9.4)}{2700(0.0734)(9.9\times10^{-6})}} = 446.4 \text{ ft}$$

With this reservoir width value, the storativity ratio is verified with Equations (3.72) and (3.74), respectively:

$$\sqrt{\omega} = \frac{4.064(457)(1.49)}{87(3.2)(446.4)} \sqrt{\frac{(9.4)(0.03)}{2700(0.0734)(9.899\times10^{-6})}}; \omega = 0.0713$$

$$\omega = \frac{16.5186(9.4)(0.03)}{2700(0.0734)(9.889\times10^{-6})} \left[\frac{457(1.5)}{441.4(84)(3.2)}\right]^2 = 0.078$$

Area is again found with Equation (3.91):

$$A = \sqrt{\frac{2700(42)(441.4^2)}{948.047(9.4)(0.0734)(9.899\times10^{-6})}} = 1847210.2 \text{ ft}^2$$

The mechanical skin factor can be found with Equation (2.22):

$$s = 0.5 \left(\frac{11.98}{1.99} - \ln \left(\frac{2700\,(0.01)}{9.4\,(0.0734)(9.899 \times 10^{-6})(0.5^2)(0.078)} \right) + 7.43 \right) = -2.84$$

The geometrical skin factors are found with Equations (3.73) and (1.49):

$$s_{DL} = \left(\frac{16.06}{3.2} - 2 \right) \frac{1}{19.601(441.4)} \sqrt{\frac{2700(0.03)}{9.4\,(0.0734)(9.899 \times 10^{-6})(0.078)}} = 8.6$$

$$s_L = \left(\frac{28.2}{7.17} - 2 \right) \frac{1}{34.743(441.6)} \sqrt{\frac{2700(0.626)}{9.4\,(0.0734)(9.899 \times 10^{-6})}} = 3.96$$

A summary of the results is given in Table 3.7.

GAS FLOW

The dimensionless pseudopressure and pseudopressure derivative are defined by:

$$m(P)_D = \frac{kh[m(P_i) - m(P)]}{1422.52qT} \tag{3.92}$$

$$t_D * m(P)'_D = \frac{kh[t * \Delta m(P)']}{1422.52qT} \tag{3.93}$$

Agarwal (1949) introduced the pseudotime function to account for the time dependence of gas viscosity and total system compressibility:

$$t_a = \int_{to}^{t} \frac{dt}{\mu(t)c_t(t)} \tag{3.94}$$

Pseudotime is better defined as a function of pressure as a new function given in hr psi/cp:

$$t_a(P) = \int_{Po}^{P} \frac{(dt/dP)}{\mu(p)c_t(P)} dP \tag{3.95}$$

Notice that μ and c_t are now pressure-dependent properties, rewriting Equation (3.3):

$$t_D = \frac{0.0002637 k_{fb} t}{\phi(\mu c_t)_i r_w^2} \tag{3.96}$$

Including the pseudotime function, t_a (P), in Equation (3.94), the dimensionless pseudotime is given by:

$$t_{Da} = \left(\frac{0.0002637 k}{\phi r_w^2}\right) t_a(P) \tag{3.97}$$

Notice the viscosity-compressibility product is not seen in Equation (3.97) since they are included in the pseudotime function. However, if we multiply and then divide by $(\mu c_t)_i$, a similar equation to the general dimensionless time expression, Equation (3.96), is obtained:

$$t_{Da} = \left(\frac{0.0002637 k}{\phi(\mu c_t)_i r_w^2}\right) \left[(\mu c_t)_i \times t_a(P)\right] \tag{3.98}$$

For gas flow, instead of substituting in the governing functions of Equations (3.1) and (3.2), they are replaced by Equations (3.92) and (3.93). The product $q\mu B$ is replaced by qT in all the equations. If viscosity is not in the product, then dividing by viscosity is expected for example, recall Equation (3.12):

$$k_f w_f = \frac{1944.96}{\sqrt{\omega \mu (\phi c_t)_{f+m} k_{fb}}} \left(\frac{qB\mu}{hm_{BL}}\right)^2 \tag{3.12}$$

Table 3.8. Constants for gas flow equations

Equation	New Constant	Equation	New Constant
3.12	197405.14	3.46	481.81 (inside)
3.14	3160.05	3.54	16133.9
3.21	95.05	3.55	6560125.65
3.25	44.944	3.61	
3.27	2234.49	3.64	81.888
3.31	345195361.5	3.26, 3.69	1422.52
3.33	197405.14	3.68	120.952
3.35	12337.82	3.74	1676.397
3.36	558.65	3.75	0.00318154
3.38	135.173	3.80	60.476
3.39, 3.72	40.944	3.84	0.006167
3.40	20.472		

Take into account the porosity and compressibility refer to matrix and fractures. Thus, for gas flow, Equation (3.12) becomes:

$$k_f w_f = \frac{197405.14}{\sqrt{\omega\mu(\phi c_t)_{f+m} k_{fb}}} \left(\frac{qT}{hm_{BL}} \right)^2 \qquad (3.99)$$

Based upon the above considerations, the constants for the equations applied to gas flow are reported in Table 3.8.

The correlation given in Equation (3.15) will convert to gas flow:

$$\omega = \frac{0.0097452700024\lambda_f k_{fb}}{\mu(\phi c_t)_{f+m} x_f^2} \left(\frac{k_{fb} h\Delta m(P)_{1hr} C_{fD}^{0.5} \lambda_f^{0.25}}{1422.52qT} \right)^{-3.96430663} \qquad (3.100)$$

The correlation given in Equation (3.28) will convert to gas flow:

$$\omega = \frac{0.0008369600544\lambda_f k_{fb}}{\phi\mu(c_t)_{f+m} x_f^2} \left(\frac{k_{fb} h\Delta m(P)_{1hr} \lambda_f^{0.5}}{1422.52qT} \right)^{-1.984944590} \qquad (3.101)$$

Although the estimation of permeability works well when working with rigorous time, there are many applications in which pseudotime is recommended. For instance, the determination of reservoir area provides a 4% error, compared to pseudotime, if estimated with actual time, as reported by Escobar, Lopez and Cantillo (2007b). Escobar, Muñoz and Cerquera (2011c, 2012c) found a better performance of pseudotime in horizontal wells in naturally occurring formations, and finally, Escobar, Martinez and Mendez (2012d) more accurately determined the hydraulic fracture parameters in a vertical well when pseudotime was used. Also, Escobar, Zhao and Zhang (2014d) successfully used the pseudotime function in a gas composite reservoir, and Escobar, Castro and Mosquera (2014e) also applied this function in a transient rate analysis of fractured vertical gas wells.

The equations given in Table 3.8 can be easily converted to pseudotime when this function is estimated by Equation (3.97) by dropping the viscosity-total compressibility product (ϕc_t) and changing the actual time, t, to the pseudotime function, $t_a(P)$. For example, expressing Equation (3.88) as a function of pseudotime yields:

$$Y_E = 0.05756\sqrt{\frac{kt_a(P)_{RDL_i}}{\phi}} \qquad (3.102)$$

Or, as another example, expressing Equation (1.31), given in Table 1.3, yields:

$$A = \frac{kt_a(P)_{rpssi}}{283.66\phi} \qquad (3.103)$$

Nomenclature

A	Drainage area, ft^2
B	Volume factor, bbl/STB
C_{fD}	Dimensionless fracture conductivity
c_t	Total or system compressibility, 1/psi
h	Formation thickness, ft
k	Permeability, md
k_{fb}	Fracture-bulk permeability, md
$k_f w_f$	Fracture conductivity, md-ft
m	Slope of P-vs-log t plot
m_{BL}	Slope of P-vs- $t^{0.25}$ plot
m_{ell}	Slope of P-vs- $t^{0.36}$ plot
m_L	Slope of P-vs- $t^{0.5}$ plot
$m(P)$	Pseudopressure function, psi^2/cp
$m(P)_D$	Dimensionless pseudopressure function
n	Number of normal fracture planes
P	Pressure, psi
P_D	Dimensionless pressure
P_i	Initial reservoir pressure, psia
P_{wf}	Well flowing pressure, psi
P_{ws}	Well shut-in or static pressure, psi
P_{1hr}	Intercept of the semilog plot, psi
q	Flow rate, bbl/D; for gas wells, Mscf/D
r_w	Well radius, ft
s	Skin factor
T	Reservoir temperature, °R
$t_a(P)$	Pseudotime function, hr psi/cp
$t_D * P_D'$	Dimensionless pressure derivative
t_p	Production (Horner) time before shutting in a well, hr
$t * \Delta P'$	Pressure derivative function, psi
$t * \Delta m(P)'$	Pseudopressure derivative function based on actual time, psi^2/cp
$t_D * m(P)'_D$	Dimensionless pseudopressure derivative function based on actual time
$t_a(P) * \Delta m(P)$	Pseudopressure derivative function based on pseudotime, hr psi^3/cp^2
$t_{Da} * m(P)'_D$	Dimensionless pseudopressure derivative function based on pseudotime
t_{Da}	Dimensionless pseudotime function
t_D	Dimensionless time based on well radius
t_{DA}	Dimensionless time based on reservoir area
t_{Dxf}	Dimensionless time based on half-fracture length
X_E, x_e	Reservoir length, ft
x_f	Fracture half-length, ft
Y_E	Reservoir width, ft
W_D	Dimensionless reservoir width
w_f	Fracture width, microns

Greek

Δ	Change, drop
Δt	Shut-in time, hr
ϕ	Porosity, fraction
λ	Interporosity flow parameter between matrix-fracture network
λ_f	Interporosity flow parameter between hydraulic fracture-fracture network
ξ	Dummy variable for establishing whether reservoir is fractured or not
μ	Viscosity, cp
ω	Storativity ratio

Suffixes

$2BR$	Biradial flow in the homogeneous region (second biradial)
$2BR$	Second biradial flow at the time of 1 hr
$1hr$	Time of 1 hour
$1BLUSi$	Intercept of first bilinear and transition pseudosteady-state lines
$1LUSi$	Intercept of first linear and transition pseudosteady-state lines
$2BL1$	Second bilinear at 1 hr
$2BLUSi$	Intercept of second bilinear and transition pseudosteady-state lines
$2LUSi$	Intercept of second linear and transition pseudosteady-state lines
$2L1$	Second linear at 1 hr
BL	Bilinear
$BL1$	Bilinear at 1 hr
$BLUSi$	Intercept of bilinear and transition unit-slope lines
BR	Biradial
$BR1$	Biradial at 1 hr
D	Dimensionless
DL	Dual linear
$DL1$	Dual linear at 1 hr
$DLPSSi$	Intercept of dual linear and late pseudosteady-state lines
DA	Dimensionless ith respect to area
ell	Elliptical
f	Fracture
fb	Fracture bulk
$f+m$	Fracture plus matrix
I, i	Intersection or initial conditions
inf	Inflection
L	Linear, single linear or hemilinear
$L1$	Linear at 1 hr
$LPSSi$	Intercept of linear and pseudosteady-state lines
m	Matrix
max	Maximum

min	Minimum
pss	Pseudosteady
US,i	Intercept of radial and pseudosteady-state lines during the transition period
r	Radial flow
rDLi	Intercept of radial and dual-linear lines
rLi	Intercept of radial and linear lines
w	Well, water
wf	Well flowing
ws	Well shut in
x	Maximum point (peak) during wellbore storage

REFERENCES

Agarwal, R. G. (1979, January 1). "Real Gas Pseudo-Time - A New Function For Pressure Buildup Analysis Of MHF Gas Wells." *Society of Petroleum Engineers.* doi:10.2118/8279-MS.

Arab, N. (2003). *"Application of Tiab's Direct Synthesis Technique to Constant Bottom Hole Pressure Test."* The University of Oklahoma. M.Sc. Thesis.

Marhaendrajana, T., Blasingame, T. A. & Rushing, J. A. (2004, January 1). "Use of Production Data Inversion to Evaluate Performance of Naturally Fractured Reservoirs." *Society of Petroleum Engineers.* doi:10.2118/90013-MS.

Escobar, F. H., Hernandez, Y. A. & Hernandez, C. M. (2007a). "Pressure transient analysis for long homogeneous reservoirs using TDS technique." *Journal of Petroleum Science and Engineering*, Volume *58*, Issue 1-2, pages 68-82.

Escobar, F. H., Lopez, A. M. & Cantillo, J. H. (2007b). "Effect of the Pseudotime Function on Gas Reservoir Drainage Area Determination." *CT&F – Ciencia, Tecnología y Futuro.* Vol. *3*, No. 3. p. 113-124. Dec.

Escobar, F. H., Sanchez, J. A. & Cantillo, J. H. (2008). "Rate Transient Analysis for Homogeneous and Heterogeneous Gas Reservoirs using The TDS Technique." *CT&F – Ciencia, Tecnología y Futuro.* Vol. *4*, No. 4. p. 45-59. Dec. 2008.

Escobar, F. H., Martínez, J. A. & Montealegre-M.,M. (2009). "Conventional Pressure Analysis for Naturally Fractured Reservoirs with Transition Period before and After the Radial Flow Regime." *CT&F – Ciencia, Tecnología y Futuro.* Vol. *3*, No. 5. p. 85-106. ISSN 0122-5383. Dec.

Escobar, F. H., Martinez, J. & Cantillo, J. H. (2010a, January 1). "Pressure-Transient Analysis for Naturally Fractured Reservoirs With Transition Period Before and After the Radial-Flow Regime." *Society of Petroleum Engineers.* doi:10.2118/138490-MS.

Escobar, F. H., Hernandez, D. P. & Saavedra, J. A. (2010b). *"Pressure and Pressure Derivative Analysis For Long Naturally Fractured Reservoirs Using The TDS Technique."* Dyna, Year 77, Nro. 163, p. 102-114. Sept. 2010.

Escobar, F. H., Rojas M. M., Cantillo, J. H. (2011a). *"Rate-Transient Analysis for Long Homogeneous Reservoirs by the TDS Technique."* XIV Congreso Colombiano del Petróleo. ISBN 958-33-8394-5. Nov. 22-25.

Escobar, F. H., Muñoz, Y. E.M and Cerquera, W. M., (2011b). "Pressure and Pressure Derivate Analysis vs. Pseudotime for a Horizontal Gas Well in a Naturally Fractured

Reservoir using the TDS Technique." *Entornos Journal*. ISSN 0124-7905. No. *24*. P.39-54. Sep.

Escobar, F. H., Rojas, M. M. & Bonilla, L. F. (2012a). "Transient-Rate Analysis for Long Homogeneous and Naturally Fractured Reservoir by The TDS Technique." *Journal of Engineering and Applied Sciences*. Vol. *7*. Nro. 3. P. 353-370. Mar. 2012.

Escobar, F. H., Rojas, M. M. & Cantillo, J. H. (2012b). *"Straight-Line Conventional Transient Rate Analysis for Long Homogeneous and Heterogeneous Reservoirs."* *Dyna*. year 79, Nro. *172*, 153-163. April.

Escobar, F. H., Muñoz, Y. E.M y Cerquera, W. M. (2012c). "Pseudotime Function Effect on Reservoir Width Determination in Homogeneous and Naturally Fractured Gas Reservoir Drained by Horizontal Wells." *Entornos Journal*. No. *24*. P. 221-231. April.

Escobar, F. H., Martinez, L. Y., Méndez, L. J. & Bonilla, L. F. (2012d). "Pseudotime Application to Hydraulically Fractured Vertical Gas Wells and Heterogeneous Gas Reservoirs Using the TDS Technique." *Journal of Engineering and Applied Sciences*. Vol. *7*. Nro. 3. P. 260-271. Mar.

Escobar, F. H., Ghisays-Ruiz, A. & Bonilla, L. F. (2014a). "New Model for Elliptical Flow Regime in Hydraulically-Fractured Vertical Wells in Homogeneous and Naturally-Fractured Systems." *Journal of Engineering and Applied Sciences*. Vol. *9*. Nro. 9. p. 1629-1636.

Escobar, F. H., Zhao, Y. L. & Zhang, L. H. (2014b). "Interpretation of Pressure Tests in Hydraulically-Fractured Wells in Bi-Zonal Gas Reservoirs." *Ingeniería e Investigación*. Vol. *34*. Nro. 4. P. 76-84. 2014.

Escobar, F. H., Castro, J. R. & Mosquera, J. S. (2014c) "Rate-Transient Analysis for Hydraulically Fractured Vertical Oil and Gas Wells." *Journal of Engineering and Applied Sciences*. Vol. *9*. Nro. 5. P. 739-749. May.

Escobar, F. H., Zhao, Y. L. & Zhang, L. H. (2014d). "Interpretation of Pressure Tests in Hydraulically-Fractured Wells in Bi-Zonal Gas Reservoirs." *Ingeniería e Investigación*. Vol. *34*. Nro. 4. P. 76-84. 2014.

Escobar, F. H., Castro, J. R. & Mosquera, J. S. (2014e) "Rate-Transient Analysis for Hydraulically Fractured Vertical Oil and Gas Wells." *Journal of Engineering and Applied Sciences*. Vol. *9*. Nro. 5. P. 739-749. May.

Escobar, F. H., Zhao, Y. L. & Fahes, M. (2105). "Characterization of the naturally fractured reservoir parameters in infinite-conductivity hydraulically-fractured vertical wells by transient pressure analysis." *Journal of Engineering and Applied Sciences*. Vol. *10*. Nro. 12. p. 5352-5362. July 2015.

Tiab, D., Azzougen, A., Escobar, F. H. & Berumen, S. (1999, January 1). "Analysis of Pressure Derivative Data of Finite-Conductivity Fractures by the 'Direct Synthesis' Technique." *Society of Petroleum Engineers*. doi:10.2118/52201-MS.

Tiab, D. & Bettam, Y. (2007, January 1). "Practical Interpretation of Pressure Tests of Hydraulically Fractured Wells in a Naturally Fractured Reservoir." *Society of Petroleum Engineers*. doi:10.2118/107013-MS

Tiab, D. (1994). "Analysis of Pressure and Pressure Derivative without Type Curve Matching: Vertically Fractured Wells in Closed Systems." *Journal of Petroleum Science and Engineering*. *11*. p. 323-333.

Tiab, D. (2001). *"Advances in Pressure Transient Analysis,"* UPTEC Training Manual, Norman, OK (May).

Tiab, D. & Escobar, F. H. (2003) *"Determinación del Parámetro de Flujo Interporoso a Partir de un Gráfico Semilogarítmico."* X Congreso Colombiano del petróleo (Colombian Petroleum Symposium). Oct. 14-17, 2003. ISBN 958-33-8394-5. Bogotá (Colombia).

Chapter 4

DOUBLE-POROSITY AND DOUBLE-PERMEABILITY RESERVOIRS

BACKGROUND

Double-porosity models have been widely used for characterizing naturally fractured reservoirs. As commented in chapter 2, for the double-porosity case, the well is only fed by the fractures. However, there is a need to use more complex mathematical models such as those of double-permeability to better describe fractured reservoirs in which the well is fed by both fractures and matrix. It has been noticed that the behavior of such reservoirs does not have the classical unit slope during the transition period in double-porosity systems, and the homogeneous behavior is also different since the second radial flow regime has a lower pressure derivative value because the matrix permeability contributes to the flow capacity. In this chapter, the work of Escobar, Vega and Diaz (2012) is presented for the characterization of double-permeability systems. They used a model previously presented in the literature to understand the pressure and pressure derivative behavior of such systems for the development of a practical and easy methodology for their characterization.

MATHEMATICAL FORMULATION

The dimensionless quantities introduced by Bremer, Hubert and Saul (1985) are:

$$P_{1D,2D} = \frac{(k_1 h_1 + k_2 h_2)}{\alpha_p q \mu} (P_i - P_{1,2}) \tag{4.1}$$

$$t_{DV} = \frac{\alpha_t (k_1 h_1 + k_2 h_2)t}{\left[(\phi c_t h)_1 + (\phi c_t h)_2\right] \mu r_w^2} \tag{4.2}$$

$$r_D = \frac{r}{r_w} \tag{4.3}$$

$$C_{DV} = \frac{\alpha_c C}{\left[(\phi c_t h)_1 + (\phi c_t h)_2\right] r_w^2} \tag{4.4}$$

$$\lambda = \frac{r_w^2}{(k_1 h_1 + k_2 h_2)} \frac{k_v}{\Delta h} \tag{4.5}$$

$$\omega = \frac{(\phi c_t h)_1}{(\phi c_t h)_1 + (\phi c_t h)_2} \tag{4.6}$$

Besides the λ and ω parameters introduced by Warren and Root (1963), a third one, κ, which represents the matrix-fracture flow capacity ratio, is introduced for the double-permeability model:

$$\kappa = \frac{k_1 h_1}{k_1 h_1 + k_2 h_2} \tag{4.7}$$

Bremer et al. (1986) presented pressure solutions in the Laplace space by simulating a two-layer system. The Laplace dimensionless pressure solutions for layers 1 and 2 are represented by Equations (4.8) and (4.12). These solutions were used by Escobar et al. (2012) to develop the interpretation technique.

$$\bar{P}_{1D}(s) = [1/z - C_{DV} z \bar{P}_{wD}(z)] \times \left\{ \frac{1}{\kappa(a_2 - a_1)} \left[\frac{a_2 K_0(\sigma_2 r_D)}{\sigma_2 K_1(\sigma_2)} - \frac{a_1 K_0(\sigma_1 r_D)}{\sigma_1 K_1(\sigma_1)} \right] \right\} \tag{4.8}$$

where:

$$a_{1,2} = \frac{(1-\kappa)}{\lambda} \left[\frac{(1-\omega)z + \lambda}{1-\kappa} - \sigma_{1,2}^2 \right] \tag{4.9}$$

$$\sigma_{1,2}^2 = \frac{1}{2} \left[\frac{(1-\omega)z + \lambda}{1-\kappa} + \frac{\omega z + \lambda}{\kappa} \right] \pm \frac{1}{2} \left\{ \left[\frac{(1-\omega)z + \lambda}{1-\kappa} - \frac{\omega z + \lambda}{\kappa} \right]^2 + \frac{4\lambda^2}{\kappa(1-\kappa)} \right\}^{1/2} \tag{4.10}$$

$$K_1^0(z) = K_0(z) / \left[z K_1(z) \right] \tag{4.11}$$

$$\bar{P}_{2D}(s) = [1/z - C_{DV} z \bar{P}_{wD}(z)] \times \left\{ \frac{1}{\kappa(a_2 - a_1)} \left[\frac{K_0(\sigma_2 r_D)}{\sigma_2 K_1(\sigma_2)} - \frac{K_0(\sigma_1 r_D)}{\sigma_1 K_1(\sigma_1)} \right] \right\} \tag{4.12}$$

$$\bar{P}_{wD}(z) = C_{DV}z^2 + \frac{\kappa z}{\dfrac{1}{a_2 - a_1}\left[a_2 K_1^0(\sigma_2) - a_1 K_1^0(\sigma_1)\right] + s} \quad (4.13)$$

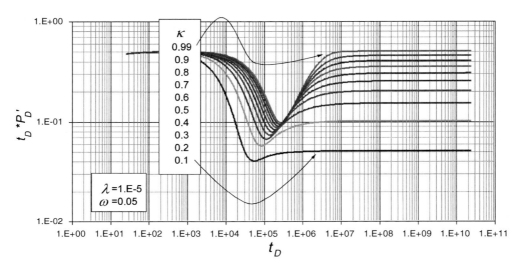

Figure 4.1. Pressure derivative behavior for $\omega = 0.05$, $\lambda = 1\times 10^{-5}$ and $0.1 \leq \kappa \leq 0.99$; adapted from Escobar et al. (2012).

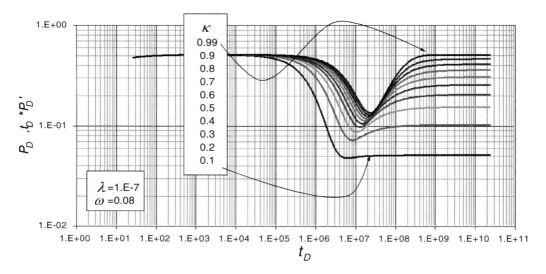

Figure 4.2. Pressure derivative behavior for $\omega = 0.08$, $\lambda = 1\times 10^{-7}$ and $0.1 \leq \kappa \leq 0.99$; adapted from Escobar et al. (2012).

TDS METHODOLOGY

The nature of the interpretation technique follows the philosophy of the *TDS* technique proposed originally by Tiab (1993). In order to study the pressure and pressure derivative behavior, several simulations for different combinations of ω, λ and κ were run. For instance, notice in Figures 4.1 and 4.2 that as κ becomes smaller the matrix permeability increases and the homogeneous radial flow pressure derivative decreases, indicating the total permeability has increased. In Figure 3.4, a comparison between the double-porosity and double-permeability models is given. A double-porosity system is a special case of a double-permeability system in which $\kappa = 1$. For such cases, the transitional behavior takes longer for the double-permeability case since the mass transfer is delayed due to the complexity of the system.

As observed in Figures 4.1 and 4.2, the matrix-fracture flow capacity ratio is a function of the value of the derivative after the transition. Then, using regression analysis makes it possible to find an expression to estimate κ from the ratio of the pressure derivative values during the radial flow regime:

$$\ln(\kappa) = 0.00073002267 + 0.999971542 * \ln\left(\frac{(t*\Delta P')_{r2}}{(t*\Delta P')_{r1}}\right) - \frac{0.00030874828}{\omega^{0.5}} \quad (4.14)$$

Another correlation using the same ratio and the value of the matrix-fracture flow capacity ratio was obtained for estimating the dimensionless storage coefficient, ω:

$$\ln(\omega) = 13.46 - 81.83(\kappa) + 80.26(\kappa)^{0.5} * \ln(\kappa) + 16.78\ln(\kappa) + 39.06\exp(-\kappa) + 48.17\left\{\frac{(t*\Delta P')_{min}}{(t*\Delta P')_{r2}}\right\} \quad (4.15)$$

$$-22.88\left\{\frac{(t*\Delta P')_{min}}{(t*\Delta P')_{r2}}\right\}^{1.5} + 11.46\left\{\frac{(t*\Delta P')_{min}}{(t*\Delta P')_{r2}}\right\}^{0.5} + 46.86\exp\left(-\frac{(t*\Delta P')_{min}}{(t*\Delta P')_{r2}}\right)$$

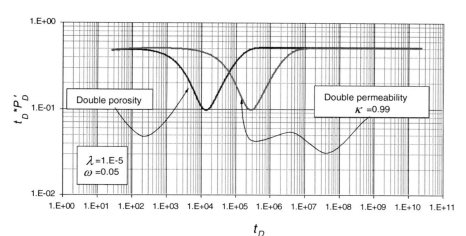

Figure 4.3 Comparison of double-porosity and double-permeability models for $\omega = 0.05$, $\lambda = 1\times10^{-5}$ and $\kappa = 0.99$; adapted from Escobar et al. (2012).

Table 4.1. Equation of κ as a function of ω and the ratio of the radial pressure derivatives

ω	Equation
0.005	$\kappa = 0.998739(t*\Delta P')_{r_2} / (t*\Delta P')_{r_1} - 0.0002501$
0.05	$\kappa = 0.999891(t*\Delta P')_{r_2} / (t*\Delta P')_{r_1} - 0.0000115$
0.08	$\kappa = 1.000066(t*\Delta P')_{r_2} / (t*\Delta P')_{r_1} + 0.0000056$
0.2	$\kappa = 1.000024(t*\Delta P')_{r_2} / (t*\Delta P')_{r_1} - 0.00000001$
0.4	$\kappa = 1.0000078(t*\Delta P')_{r_2} / (t*\Delta P')_{r_1} - 0.0000015$
0.8	$\kappa = 1.0000306(t*\Delta P')_{r_2} / (t*\Delta P')_{r_1} - 0.0000332$
0.99	$\kappa = 1.0000235(t*\Delta P')_{r_2} / (t*\Delta P')_{r_1} - 0.0000355$

Some other expressions for κ as a function of ω are presented in Table 4.1. These correlations use the ratio of the first radial and second radial (homogeneous-acting behavior) pressure derivatives. Since the influence of ω is very similar in double-porosity and double-permeability situations, the relationships presented by Engler and Tiab (1996), which use the ratio between the radial flow and minimum pressure derivatives and beginning and ending of the radial flow regimes, Equations (2.23) and (2.24), can be used. The following equation was found for the determination of κ:

$$\lambda = \frac{\kappa + \xi}{t_{D\min}} \ln\left(\frac{1}{\omega}\right) \qquad (4.16)$$

where the constant, ξ, is found graphically in Figure 4.4.

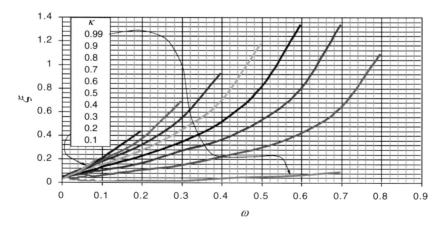

Figure 4.4. Graphical correlation for estimating the value of λ. $0.1 \leq \kappa \leq 0.99$; adapted from Escobar et al. (2012).

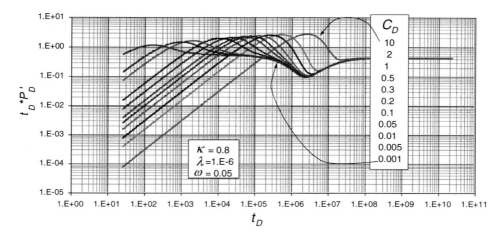

Figure 4.5. Effect of wellbore storage coefficient on the trough of the pressure derivative; adapted from Escobar et al. (2012).

Table 4.2. Conditions for effects of wellbore storage on the minimum pressure derivative

κ	C_D			
	$\lambda=10^{-4}$	$\lambda=10^{-5}$	$\lambda=10^{-6}$	$\lambda=10^{-7}$
0.1	3.89E+01	3.11E+02	1.94E+03	1.94E+04
0.2	3.89E+01	3.11E+02	1.94E+03	1.94E+04
0.3	3.89E+01	3.11E+02	3.11E+03	2.33E+04
0.4	7.77E+01	7.77E+01	7.77E+03	3.85E+04
0.5	7.77E+01	7.77E+01	7.77E+03	7.77E+04
0.6	7.77E+01	7.77E+01	7.77E+03	7.77E+04
0.7	7.77E+01	7.77E+01	7.77E+03	7.77E+04
0.8	7.77E+01	7.77E+01	7.77E+03	7.77E+04
0.9	1.55E+02	1.55E+03	7.77E+03	7.77E+04
0.99	1.55E+02	7.77E+02	7.77E+03	7.77E+04

INFLUENCE OF WELLBORE STORAGE

As seen in Figure 4.5, the minimum value of the pressure derivative is affected by wellbore storage when $C_D \geq 0.2$ for $\kappa = 0.8$, $\lambda = 1 \times 10^{-6}$ and $\omega = 0.05$. Escobar et al. (2012) also concluded the minimum pressure derivative (trough) is not affected by wellbore storage effects for any value of ω only if the wellbore storage coefficient is not greater than the values provided in Table 4.2 for a given value of λ.

If wellbore storage affects the trough, then Equations (2.39), (2.40) and (2.41) apply. Care must be taken with the derivative during the radial flow regime after the transition, which, for double-porosity systems, has the same values on both sides of the transition period.

Example 4.1

Figure 4.6 presents the pressure and pressure derivative data for a simulated test run with the information given below. It is requested to validate the equations for estimating the double-permeability reservoir parameters.

q = 800 STB/D $\qquad$ h = 120 ft $\qquad$ ϕ_m = 22%
r_w = 0.5 ft $\qquad$ μ = 20 cp $\qquad$ B = 1.45 rb/STB
c_t = 1×10^{-6} psi^{-1} $\qquad$ k_m = 160 md $\qquad$ P_i = 5000 psi
λ = 1×10^{-5} $\qquad$ ω = 0.4 $\qquad$ κ = 0.7

Solution

The following information was obtained from Figure 4.6:

$(t*\Delta P')_{min}$ = 12.315 psi $\qquad$ t_{min} = 7.049 hr $\qquad$ $(t*\Delta P')_{r2}$ = 57.623 psi

Equation (4.14) is used to find the matrix-fracture flow capacity ratio:

$$\ln(\kappa) = 0.00073002267 + 0.999971542 * \ln\left(\frac{57.63}{80.038}\right) - \frac{0.00030874828}{0.4^{0.5}}$$

Then, κ = 0.7174.
Equation (4.15) is used to find the storativity ratio:

$$\ln(\omega) = 13.46 - 81.83(0.7) + 80.26(0.7)^{0.5} * \ln(0.7) + 16.78\ln(0.7) + 39.06\exp(-\kappa) + 48.17\left\{\frac{21.3154}{57.623}\right\}$$

$$-22.88\left\{\frac{21.3154}{57.623}\right\}^{1.5} + 11.46\left\{\frac{21.3154}{57.623}\right\}^{0.5} + 46.86\exp\left(-\frac{21.3154}{57.623}\right)$$

Then, ω = 0.4017.
The value for time at the minimum point can be converted into dimensionless form using Equation (1.2):

$$t_{Dmin} = \frac{0.0002637kt}{\phi\mu c_t r_w^2} = \frac{0.0002637(160)(7.05)}{(0.22)(20)(1\times10^{-6})(0.5^2)} = 106383.78$$

Entering the values of ω and κ from Figure 4.4 provides a value of ξ = 0.52. Replacing this value in Equation (4.16) obtains:

$$\lambda = \frac{0.7 + 0.52}{106383.78}\ln\left(\frac{1}{0.4}\right) = 1.046\times10^{-5}$$

Notice that for practical purposes the expressions given in chapter 2 can be applied to this example.

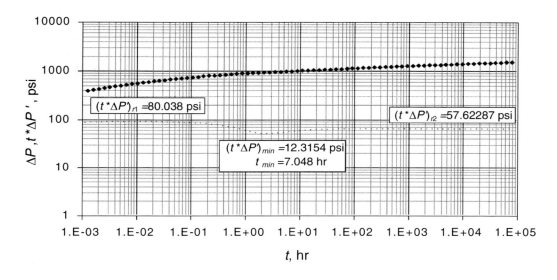

Figure 4.6. Pressure and pressure derivative log-log plot for example 4.1.

Nomenclature

B	Volume factor, bbl/STB
c_t	Total or system compressibility, 1/psi
h	Formation thickness, ft
k	Permeability, md
k_{fb}	Fracture-bulk permeability, md
P	Pressure, psi
P_D	Dimensionless pressure
P_i	Initial reservoir pressure, psia
P_{wf}	Well flowing pressure, psi
P_{ws}	Well shut-in or static pressure, psi
q	Flow rate, bbl/D; for gas wells, Mscf/D
r_w	Well radius, ft
s	Skin factor
t_p	Production (Horner) time before shutting in a well, hr
$t*\Delta P'$	Pressure derivative function, psi
t_D	Dimensionless time based on well radius

Greek

Δ	Change, drop
Δt	Shut-in time, hr
ϕ	Porosity, fraction
λ	Interporosity flow parameter between matrix-fracture network

Double-Porosity and Double-Permeability Reservoirs 103

ξ Dummy variable for finding κ

μ Viscosity, cp

ω Storativity ratio

Suffixes

D Dimensionless

f Fracture

fb Fracture bulk

$f+m$ Fracture plus matrix

I, i Intersection or initial conditions

m Matrix

max Maximum

min Minimum

r Radial flow

w Well

wf Well flowing

ws Well shut in

REFERENCES

Bremer, R. E., Hubert, W. & Saul, V. (1985, June 1). "Analytical Model for Vertical Interference Tests Across Low-Permeability Zones." *Society of Petroleum Engineers.* doi:10.2118/11965-PA.

Engler, T. & Tiab, D. (1996). "Analysis of Pressure and Pressure Derivative without Type Curve Matching, 4. Naturally Fractured Reservoirs." *Journal of Petroleum Science and Engineering, 15.* p. 127-138.

Escobar, F. H., Vega, J. & Diaz, M. R. (2012). "Pressure and Pressure Derivative Analysis for Double-Permeability Systems without Type-Curve Matching." *Journal of Engineering and Applied Sciences.* Vol. 7. Nro. 10. Oct.

Tiab, D. (1993a, January 1). "Analysis of Pressure and Pressure Derivatives Without Type-Curve Matching: I-Skin and Wellbore Storage." *Society of Petroleum Engineers.* This paper was prepared for presentation at the Production Operations Symposium held in Oklahoma City, OK, U.S.A., March 21-23. doi:10.2118/25426-MS.

Warren J. & Root P. (1963). "The Behavior of Naturally Fractured Reservoirs.*" SPE Journal,* September.

Chapter 5

TRIPLE-POROSITY. SINGLE-PERMEABILITY RESERVOIRS

BACKGROUND

Recent studies have shown the presence of cavities or vugs in naturally fractured carbonate reservoirs affects the well-pressure behavior since they possess a complex porous system identified as a triple-porosity, naturally fractured reservoir. Several cases of this are present in Mexico. As a consequence, strange anomalies are observed in the slope of the semilog plot during the transition period. The behavior of the dimensionless pressure versus dimensionless time has an alteration of the normal slope, reflected as an additional depression in the curve. This abnormality in the slope changes is caused by the presence of an additional pore system with different petrophysical properties in the reservoir due to the presence of fractures, vugs and matrix or large fractures, small fractures and matrix, which are especially present in naturally fractured carbonate reservoirs.

From the 1960s to now, there have been various formulations to conceptualize naturally fractured formations and to establish the fluid flow modeling in this type of rock. Initially, dual-porosity models were proposed; the most used model in the oil industry to present a better field-scale application is the one developed by Warren and Root (1963). It is classified as a dual-medium formulation (double-porosity model) in which the fractures form a network of channels, providing fluid flow parallel to the main permeability axis, and the matrix subsystem is constituted of discrete homogeneous and isotropic blocks, providing the capacity of storage. These dual-medium formulations were also applied to characterize triple-porosity systems, where the presence of vugs was calculated as part of the fractures system or as part of the matrix system, simplifying the calculations. However, this assumption did not correctly describe the fluid mechanics behavior in the reservoir because vugs and matrices do not have the same effect or interaction with the fracture network. Taking into account the limitations of these models, different authors have reformulated the theoretical principles to try to establish a model that captures the reality of the process flow taking place in triple-porosity reservoirs.

Abdassah and Ershaghi (1986) presented a triple-porosity model and unique permeability. They considered a model for unsteady flow between the system of fractures, with two types of matrix blocks, and only the primary flow through the fracture system. Also, they considered the existence of parallel flow between the fracture system, with homogeneous properties, and its interaction with two separate groups of matrix blocks having different permeabilities and porosities.

Wu, Liu and Bodvarson (2004) proposed a conceptual model of triple porosity and triple permeability. They conceptualized the fracture-matrix system formed by a matrix and two types of fractures—large and small fractures—and extended the concept of dual permeability by adding a connection (with small fractures) between the large fractures and matrix blocks.

Camacho et al. (2005) presented a study to model secondary porosities, mainly in naturally fractured vuggy carbonate reservoirs. This model utilizes the approximation of pseudosteady interporosity flow, which means the fluid transfer among the matrix, vugs and fractures is directly proportional to the difference in the average pressure in volume with the macroscopic matrix, fractures and vugs. They proposed solutions for two different cases: the first one, when no primary flow occurs through vugs, which is an extension of the model of Warren and Root (1963), and the second one, where the dissolution process has created an interconnected system of vugs. The later will be treated in chapter 6. In both cases, there exists an interaction between the matrix, vugs and fracture system. Based on the analytical solution introduced by Camacho et al. (2005) and by referring to the studies and analysis techniques presented by Escobar et al. (2004) and Mirshekari et al. (2007), a methodology was developed by Escobar, Rojas and Rojas (2014) to interpret the pressure behavior and the pressure derivative of the transient flow period and the period of flow dominated by bordering in naturally fractured vuggy carbonate reservoirs; in that way, the dimensionless storativity coefficients and the interporosity flow parameters for matrix fracture, matrix vugs and vug fractures can be estimated.

MATHEMATICAL MODELING

Camacho et al. (2005) developed a complex model for the well pressure behavior of a triple-porosity and single-permeability reservoir:

$$\overline{P}_D(u) = \frac{K_0\left[\sqrt{g(u)}\right] + s\sqrt{g(u)}K_1\left[\sqrt{g(u)}\right]}{u\left(\sqrt{g(u)}K_1\left[\sqrt{g(u)}\right] + C_D u\left\{K_0\left[\sqrt{g(u)}\right] + s\sqrt{g(u)}K_1\left[\sqrt{g(u)}\right]\right\}\right)} \quad (5.1)$$

where:

$$g(u) = \lambda_{mf}(1 - [\lambda_{mv}b_1\lambda_{mf}b_3 + u(\lambda_{mv}b_2 + \lambda_{mf}b_4) + u^2\lambda_{mf}b_5 / \{b_3(\lambda_{mv} + \lambda_{mf}) +$$

$$u[(\lambda_{mv} + \lambda_{mf})b_4 + (1 - \omega_f - \omega_v)b_3] + u^2[(\lambda_{mv} + \lambda_{mf})b_5 + (1 - \omega_f - \omega_v)b_4] + \quad (5.2)$$

$$u^3(1 - \omega_f - \omega_v)b_5\}]) + \lambda_{mf}\left(1 - \frac{b_1 + b_2 u}{b_3 + b_4 u + b_5 u^2}\right) + \omega_f u$$

where:

$$b_1 = \lambda_{vf}(\lambda_{mv} + \lambda_{mf}) + \lambda_{mf}\lambda_{mv} \tag{5.3}$$

$$b_2 = \lambda_{vf}(1 - \omega_f - \omega_v) \tag{5.4}$$

$$b_3 = \lambda_{mv}(\lambda_{vf} + \lambda_{mf}) + \lambda_{mf}\lambda_{vf} \tag{5.5}$$

$$b_4 = \omega_v(\lambda_{mv} + \lambda_{mf}) + (1 - \omega_f - \omega_v)(\lambda_{mv} + \lambda_{vf}) \tag{5.6}$$

$$b_5 = (1 - \omega_f - \omega_v)\omega_v \tag{5.7}$$

Camacho et al. (2005) defined the following dimensionless quantities as:

$$P_{Dj} = \frac{2\pi k_f h(P_i - P_j)}{q\mu B} \tag{5.8}$$

where j = fractures or vugs. The pressure derivative is given by Equation (3.2). The definition by Camacho et al. (2005) for dimensionless time is:

$$t_D = \frac{k_f t}{\left[\left(\phi_f c_f + \phi_m c_m + \phi_v c_v\right)\mu r_w^2\right]} \tag{5.9}$$

They also defined the interporosity flow parameters as:

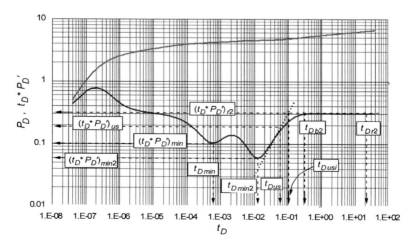

Figure 5.1. Points or fingerprints characteristic of triple-porosity reservoirs; adapted from Escobar et al. (2014).

$$\lambda_{mf} = \frac{\sigma_{mf} k_m r_w^2}{k_f} \qquad (5.10)$$

$$\lambda_{mv} = \frac{\sigma_{mv} k_m r_w^2}{k_f} \qquad (5.11)$$

$$\lambda_{vf} = \frac{\sigma_{vf} k_{vf} r_w^2}{k_f + k_v} \qquad (5.12)$$

where $k_{vf} = k_v$ if $P_v > P_f$ and $k_{vf} = k_f$ in the contrary case. σ is the shape factor between the media i and j. The dimensionless storativity coefficients are given as:

$$\omega_f = \frac{\phi_f c_f}{\phi_f c_f + \phi_m c_m + \phi_v c_v} \qquad (5.13)$$

$$\omega_v = \frac{\phi_v c_v}{\phi_f c_f + \phi_m c_m + \phi_v c_v} \qquad (5.14)$$

Using the model presented by Equation (5.1), Escobar, Rojas and Rojas (2014) studied the dimensionless pressure and pressure derivative behavior to find appropriate "fingerprints" so they could apply the *TDS* technique (Tiab 1993) to develop expressions useful for a practical characterization of pressure tests run in triple-porosity, single permeability reservoirs. Such fingerprints are schematically given in Figure 5.1.

PARAMETERS DETERMINATION

The estimation of the five naturally fractured parameters was presented individually by Escobar et al. (2014). Each one will be treated separately here.

Matrix-Fracture Interporosity Flow Parameter, λ_{mf}

Figure 5.2 shows the effect of the matrix-fracture interporosity flow parameter, λ_{mf}, on the behavior of the dimensionless pressure derivative versus dimensionless time for reservoirs with triple porosity (naturally fractured vuggy reservoirs) with constant values of $\lambda_{vf} = 1 \times 10^{-7}$, $\lambda_{mv} = 1 \times 10^{-10}$, $\omega_f = 1 \times 10^{-4}$ and $\omega_v = 1 \times 10^{-5}$. To avoid confusion with so many curves, the pressure behavior for the mentioned conditions is presented in Figure 5.3.

Basically, this parameter (λ_{mf}) affects the occurrence of the transition period. When the value of λ_{mf} decreases, the presence of the second minimum point, the unit-slope behavior just before the second radial flow (the radial flow formed in the homogenous systems once the transition period is no longer felt) and the same second radial flow occur later. In addition to this, the first transition zone is also affected by this interporosity flow parameter; however, the change is not significant and makes it impractical to use in a correlation.

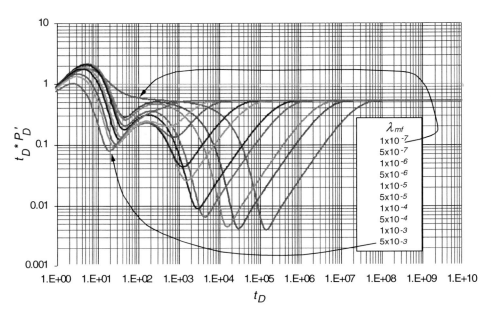

Figure 5.2. Effect of λ_{mf} on the dimensionless pressure derivative, $\lambda_{vf}=1\times10^{-7}$, $\lambda_{mv}=1\times10^{-10}$, $\omega_f=1\times10^{-4}$ and $\omega_v=1\times10^{-5}$; adapted from Escobar et al. (2014).

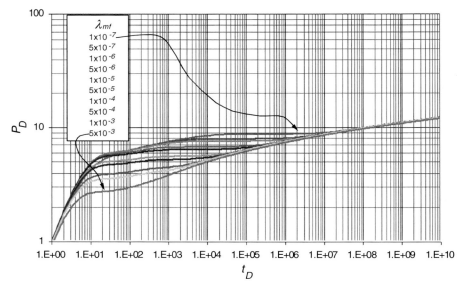

Figure 5.3. Effect of λ_{mf} on the dimensionless pressure, $\lambda_{vf}=1\times10^{-7}$, $\lambda_{mv}=1\times10^{-10}$, $\omega_f=1\times10^{-4}$ and $\omega_v=1\times10^{-5}$; adapted from Escobar et al. (2014).

Table 5.1. Constants for Equation (4)

Rank λ_{mf}	1×10^{-4} to 1×10^{-3}	1×10^{-5} to 1×10^{-3}	1×10^{-6} to 1×10^{-3}	1×10^{-7} to 1×10^{-3}	1×10^{-8} to 1×10^{-3}
λ_{vf}	1×10^{-4}	1×10^{-5}	1×10^{-6}	1×10^{-7}	1×10^{-8}
λ_{mv}	1×10^{-7}	1×10^{-8}	1×10^{-9}	1×10^{-10}	1×10^{-11}
ω_f	1×10^{-1}	1×10^{-2}	1×10^{-3}	1×10^{-4}	1×10^{-5}
ω_v	1×10^{-2}	1×10^{-3}	1×10^{-4}	1×10^{-5}	1×10^{-6}
A	18.750525	-0.01991929	1.4395861	0.61733522	0.87955429
B	-175.1714	-1.1282632	-33.795176	-14.552994	-21.273956
C	548.95252	9.4026458	311.82424	134.54613	201.21104
D	-559.03426	0	-1436.2235	-635.27347	-955.35918
E	0	0	3438.6978	1667.9363	2413.0321

Figure 5.3 shows the effects on pressure behavior when λ_{mf} decreases. Slope changes are better appreciated in the pressure curve as a consequence of the typical transitions generated by the influence of different porous media affecting fluid flow in naturally fractured vuggy reservoirs.

Considering these observations and the points or characteristic fingerprints, the expressions generated to calculate the matrix-fracture interporosity flow parameter are:

1) Using the second minimum point, $P_{Dmin2}/(t_D*P_D')_{min2}$ and t_{Dmin2}:

$$\lambda_{mf} = \frac{A}{t_{Dmin2}} + \frac{B}{t_{Dmin2}\ln\left(\frac{P_{Dmin2}}{(t_D*P_D')_{min2}}\right)} + \frac{C}{t_{Dmin2}\left(\ln\left(\frac{P_{Dmin2}}{(t_D*P_D')_{min2}}\right)\right)^2} + \frac{D}{t_{Dmin2}\left(\ln\left(\frac{P_{Dmin2}}{(t_D*P_D')_{min2}}\right)\right)^3}$$

$$+ \frac{E}{t_{Dmin2}\left(\ln\left(\frac{P_{Dmin2}}{(t_D*P_D')_{min2}}\right)\right)^4} + \frac{F}{t_{Dmin2}\left(\ln\left(\frac{P_{Dmin2}}{(t_D*P_D')_{min2}}\right)\right)^5} \quad (5.15)$$

The constants used in Equation (5.15) are given in Table 5.1.

2) Using the beginning of the second radial flow, t_{Db2}:

$$\lambda_{mf}^{-1} = A + B \times t_{Db2}^{1.5} + \frac{C \times t_{Db2}}{\ln(t_{Db2})} \quad (5.16)$$

Constants A, B and C are provided in Table 5.2.

Table 5.2. Constants for Equation (5.16)

Constant	Value
A	-122.95071
B	7.3564664×10^{-6}
C	0.9062836

Table 5.3. Constants for Equation (5.17)

Constant	Value
A	$-4.7766068 \times 10^{-8}$
B	12.367255
C	-622407
D	9.0548479×10^{10}
E	$-2.7050964 \times 10^{15}$
F	2.1249078×10^{19}

Equation (5.16) is applicable to $1 \times 10^{-6} < \lambda_{mf} < 1 \times 10^{-3}$, $1 \times 10^{-8} < \lambda_{vf} < 1 \times 10^{-4}$, $1 \times 10^{-11} < \lambda_{mv} < 1 \times 10^{-7}$, $1 \times 10^{-5} < \omega_f < 1 \times 10^{-1}$ and $1 \times 10^{-6} < \omega_v < 1 \times 10^{-2}$. Another expression developed using the beginning of the second radial flow is:

$$\lambda_{mf}^{-1} = A + \frac{B}{t_{Db2}} + \frac{C}{t_{Db2}^2} + \frac{D}{t_{Db2}^3} + \frac{E}{t_{Db2}^4} + \frac{F}{t_{Db2}^5} \tag{5.17}$$

with constants A through F given in Table 5.3. The above equation is applicable for the range of $1 \times 10^{-6} < \lambda_{mf} < 1 \times 10^{-3}$, $1 \times 10^{-8} < \lambda_{vf} < 1 \times 10^{-4}$, $1 \times 10^{-11} < \lambda_{mv} < 1 \times 10^{-7}$, $1 \times 10^{-5} < \omega_f < 1 \times 10^{-1}$ and $1 \times 10^{-6} < \omega_v < 1 \times 10^{-2}$.

3) Using a time point, t_{Dus}, read on the unit-slope line developed during the transition period before the start of the second radial flow regime, the pressure derivative is read at a time, t_{Dus}, which is located one log cycle after the second minimum point, t_{Dmin2}. The equation given for $\lambda_{vf} = 1 \times 10^{-6}$, $\lambda_{mv} = 1 \times 10^{-9}$, $\omega_f = 1 \times 10^{-3}$, $\omega_v = 1 \times 10^{-4}$ and $1 \times 10^{-6} < \lambda_{mf} < 1 \times 10^{-4}$ would be:

$$\ln(\lambda_{mf}) = A + B \times t_{Dus}^{0.5} + \frac{C}{t_{Dus}^{0.5}} + \frac{D \times \ln(t_{Dus})}{t_{Dus}} + E \times (t_D * P_D')_{us} \times \ln((t_D * P_D')_{us})$$
$$+ F \times \left(\ln \left((t_D * P_D')_{us} \right) \right)^2 + G \times (t_D * P_D')_{us}^{0.5} + H \times \ln(t_D * P_D')_{us} + \frac{I}{\ln(t_D * P_D')_{us}} \tag{5.18}$$

Table 5.4. Constants for Equations (5.18) and (5.19)

Constant	Value	Value
A	1543.0125	-7517.7246
B	-0.0031036453	-0.043074373
C	348.57011	39.852572
D	-1018.2157	0.48824847
E	-5442.053	8247.468
F	32.722676	121.74982
G	-11958.569	2708.9279
H	494.01527	2249.5118
I	-4460.9516	19002.275

for which constants A through I are given in the second column of Table 5.4. Also, for $\lambda_{vf}=1\times10^{-7}$, $\lambda_{mv}=1\times10^{-10}$, $\omega_f=1\times10^{-4}$, $\omega_v=1\times10^{-5}$ and $1\times10^{-7}<\lambda_{mf}<1\times10^{-4}$, the resulting expression is:

$$\ln\left(\lambda_{mf}\right)=A+B\times\left(\ln\left(t_{Dus}\right)\right)^2+\frac{C}{t_{Dusi}^{0.5}}+\frac{D\times\ln\left(t_{Dus}\right)}{t_{Dus}}+E\times\left(t_D*P_D'\right)_{us}^{0.5}\times\ln\left(\left(t_D*P_D'\right)_{us}\right)+$$

$$F\times\left(\ln\left(\left(t_D*P_D'\right)_{us}\right)\right)^2+G\times\left(t_D*P_D'\right)_{us}^{0.5}+H\times\ln\left(\left(t_D*P_D'\right)_{us}\right)+I\times e^{-(t_D*P_D')_{us}} \qquad (5.19)$$

Constants A through I are given in the third column of Table 5.4.

4) Another way to use the Equations (5.18) and (5.19) is to read the intercept point between the unit-slope line formed at the end of the transition period with the line of the second radial flow regime, taking t_{Dus} as t_{Dusi} and $(t_D*P_D')_{us}$ as $(t_D*P_D')_{usi}$. Since any point on the radial flow has a dimensionless pressure derivative of 0.5 at this intersection point, $(t_D*P_D')_{usi}=0.5$, replacing this value and the respective constant yields:

$$\ln\left(\lambda_{mf}\right)=A+B\times t_{Dusi}^{0.5}+\frac{C}{t_{Dusi}^{0.5}}+\frac{D\times\ln\left(t_{Dusi}\right)}{t_{Dusi}} \qquad (5.20)$$

for $\lambda_{vf}=1\times10^{-6}$, $\lambda_{mv}=1\times10^{-9}$, $\omega_f=1\times10^{-3}$, $\omega_v=1\times10^{-4}$ and $1\times10^{-6}<\lambda_{mf}<1\times10^{-4}$, with constants A, B, C and D provided in the second column of Table 5.5.

For $\lambda_{vf}=1\times10^{-7}$, $\lambda_{mv}=1\times10^{-10}$, $\omega_f=1\times10^{-4}$, $\omega_v=1\times10^{-5}$ and $1\times10^{-7}<\lambda_{mf}<1\times10^{-4}$, use:

$$\ln(\lambda_{mf})=A+B\times[\ln(t_{Dusi})]^2+\frac{C}{t_{Dusi}^{0.5}}+\frac{D\times\ln(t_{Dusi})}{t_{Dusi}} \qquad (5.21)$$

The constants for Equation (5.21) are given in the third column of Table 5.5.

Matrix-Vugs Interporosity Flow Parameter, λ_{mv}

Figure 5.4 and Figure 5.5 show the effect of the matrix-fracture interporosity flow parameter, λ_{mv}, on the transient pressure behavior for constant values of $\lambda_{mf}=1\times10^{-7}$, $\lambda_{vf}=1\times10^{-10}$, $\omega_f=1\times10^{-7}$ and $\omega_v=1\times10^{-8}$. Unlike the observation of λ_{mf}, the starting time of the second radial flow converges at the same point for different λ_{mv} values, and the unit-slope line at the end of the pressure derivative depression varies in length, though it is the same in terms of location (no parallel displacement along the time axis). These observations led to developing the expressions given below:

5) Using the second minimum point, $P_{Dmin2}/(t_D*P_D')_{min2}$ and t_{Dmin2}:

Table 5.5. Constants for Equations (5.20) and (5.21)

Constant	Value	Value
A	1082.188281	380.1675508
B	-0.0031036453	-0.043074373
C	348.57011	39.852572
D	-1018.2157	0.48824847

$$\lambda_{mv} = A + B \times \frac{P_{D\min 2}}{(t_D * P_D')_{\min 2}} + \frac{C}{t_{D\min 2}} + \frac{D}{t_{D\min 2}^2} + \frac{E}{t_{D\min 2}^3} + \frac{F}{t_{D\min 2}^4} \quad (5.22)$$

The constants depending on λ_{mf}, λ_{vf}, ω_f and ω_v are provided in Table 5.6.

6) Using the dimensionless delta time, Δt_D, defined as the difference between t_{Dusi} and t_{Dmin2}:

$$\lambda_{mv} = \frac{A}{\lambda_{mf}} + \frac{B \times \Delta t_D}{\lambda_{mf}} + \frac{C \times \Delta t_D^2}{\lambda_{mf}} + \frac{D \times \Delta t_D^3}{\lambda_{mf}} \quad (5.23)$$

The constants depending upon λ_{mf}, λ_{vf}, ω_f and ω_v values are given in Table 5.7.

Vugs-Fractures Interporosity Flow Parameter, λ_{vf}

Figure 5.6 and Figure 5.7 show the effect of the vugs-fractures interporosity flow parameter λ_{vf} on the behavior of the dimensionless pressure and the dimensionless pressure derivative, having constant values of $\lambda_{mf}=1\times10^{-2}$, $\lambda_{mv}=1\times10^{-7}$, $\omega_f=1\times10^{-2}$ and $\omega_v=1\times10^{-3}$.

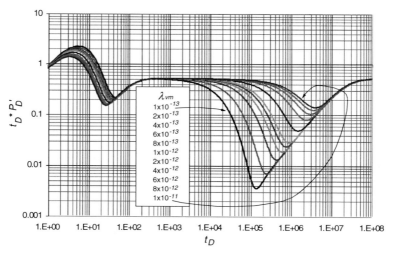

Figure 5.4. Effect of λ_{mv} on the dimensionless pressure derivative for $\lambda_{mf}=1\times10^{-7}$, $\lambda_{vf}=1\times10^{-10}$, $\omega_f=1\times10^{-7}$ and $\omega_v=1\times10^{-8}$; adapted from Escobar et al. (2014).

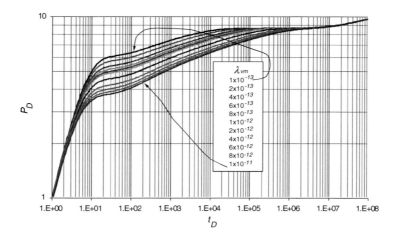

Figure 5.5. Effect of λ_{mv} on the dimensionless pressure for $\lambda_{mf}=1\times10^{-7}$, $\lambda_{vf}=1\times10^{-10}$, $\omega_f=1\times10^{-7}$ and $\omega_v=1\times10^{-8}$; adapted from Escobar et al. (2014).

Table 5.6. Constants for Equation (5.22)

Rank λ_{mv}	1×10^{-11} to 1×10^{-8}	1×10^{-12} to 1×10^{-9}	1×10^{-12} to 1×10^{-10}	1×10^{-13} to 1×10^{-11}
λ_{mf}	1×10^{-4}	1×10^{-5}	1×10^{-6}	1×10^{-7}
λ_{vf}	1×10^{-7}	1×10^{-8}	1×10^{-9}	1×10^{-10}
ω_f	1×10^{-4}	1×10^{-5}	1×10^{-6}	1×10^{-7}
ω_v	1×10^{-5}	1×10^{-6}	1×10^{-7}	1×10^{-8}
A	-1.6415236×10^{-9}	-1.4333864×10^{-11}	-1.7301336×10^{-13}	-5.1835462×10^{-16}
B	5.9010132×10^{-14}	8.3945406×10^{-17}	1.2107007×10^{-17}	8.4049346×10^{-21}
C	1.5850872×10^{-7}	1.6214372×10^{-8}	1.7214656×10^{-9}	2.3157263×10^{-10}
D	6.008467×10^{-6}	1.5473226×10^{-6}	1.9670959×10^{-7}	2.610502×10^{-8}
E	9.4481581×10^{-6}	-3.3702849×10^{-5}	-3.5827173×10^{-6}	-5.6665697×10^{-7}
F	0.0019066336	0.0011408245	0.00017086007	3.1146649×10^{-5}

Table 5.7. Constants for Equation (5.23)

Rank λ_{mv}	1×10^{-11} to 1×10^{-8}	1×10^{-12} to 1×10^{-9}	1×10^{-12} to 1×10^{-10}	1×10^{-13} to 1×10^{-11}
λ_{mf}	1×10^{-4}	1×10^{-5}	1×10^{-6}	1×10^{-7}
λ_{vf}	1×10^{-7}	1×10^{-8}	1×10^{-9}	1×10^{-10}
ω_f	1×10^{-4}	1×10^{-5}	1×10^{-6}	1×10^{-7}
ω_v	1×10^{-5}	1×10^{-6}	1×10^{-7}	1×10^{-8}
A	1.0013273×10^{-9}	-1.1127195×10^{-12}	8.3458729×10^{-17}	-2.1318805×10^{-17}
B	-1.510529×10^{-13}	1.6910126×10^{-17}	9.8924541×10^{-22}	4.2037988×10^{-24}
C	7.5990051×10^{-18}	-8.4508641×10^{-23}	-9.1889747×10^{-28}	-2.471287×10^{-31}
D	-1.2747192×10^{-22}	1.3923415×10^{-28}	2.0543111×10^{-34}	4.5497101×10^{-39}

As for the case of λ_{mf}, the variation of the interporosity flow parameter between a vugs fracture, λ_{vf}, has the same effect on the behavior of the dimensionless pressure and the dimensionless pressure derivative. It affects the second minimum point, the onset of the unit-slope line occurring prior to the second radial flow regime. The last one occurs later (location of the second depression in the pressure derivative). However, λ_{vf} does not alter to the same extent as λ_{mf} does. For example, comparing Figure 5.6 with Figure 5.2, the second minimum time point has a smaller variation, but their corresponding pressure derivative values vary largely with changes of λ_{vf} (as compared to λ_{mf}). Based on these considerations, the generated expressions for calculating the interporosity flow parameter between vugs fractures are as follows:

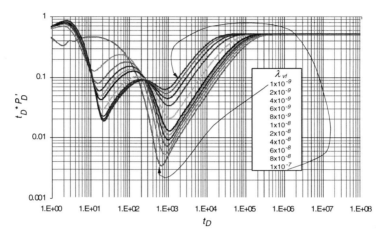

Figure 5.6. Effect of λ_{vf} on the dimensionless pressure derivative for $\lambda_{mf}=1\times10^{-2}$, $\lambda_{mv}=1\times10^{-7}$, $\omega_f=1\times10^{-2}$ and $\omega_v=1\times10^{-3}$; adapted from Escobar et al. (2014).

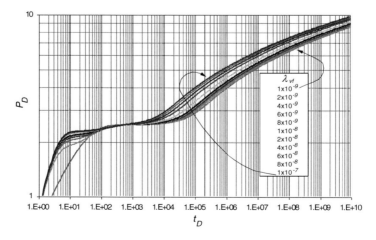

Figure 5.7. Effect of λ_{vf} on the dimensionless pressure for $\lambda_{mf}=1\times10^{-2}$, $\lambda_{mv}=1\times10^{-7}$, $\omega_f=1\times10^{-2}$ and $\omega_v=1\times10^{-3}$; adapted from Escobar et al. (2014).

7) Using the second minimum point, $P_{Dmin2}/(t_D*P_D')_{min2}$ and t_{Dmin2}:

$$\lambda_{vf} = A + \frac{B}{t_{Dmin2}} + C \times \ln\left(\frac{P_{Dmin2}}{(t_D*P_D')_{min2}}\right) + \frac{D}{(t_{Dmin2})^2} + E \times \left(\ln\left(\frac{P_{Dmin2}}{(t_D*P_D')_{min2}}\right)\right)^2 + \frac{F \times \ln\left(\frac{P_{Dmin2}}{(t_D*P_D')_{min2}}\right)}{t_{Dmin2}} + \quad (5.24)$$

$$\frac{G}{(t_{Dmin2})^3} + H \times \left(\ln\left(\frac{P_{Dmin2}}{(t_D*P_D')_{min2}}\right)\right)^3 + \frac{I \times \left(\ln\left(\frac{P_{Dmin2}}{(t_D*P_D')_{min2}}\right)\right)^2}{t_{Dmin2}} + \frac{J \times \ln\left(\frac{P_{Dmin2}}{(t_D*P_D')_{min2}}\right)}{(t_{Dmin2})^2}$$

The constants for Equation (5.24) are presented in Table 5.8. For another range:

$$\lambda_{vf} = \frac{A + (B \times t_{Dmin2}) + \left(C \times \frac{P_{Dmin2}}{(t_D*P_D')_{min2}}\right) + \left(D \times \left(\frac{P_{Dmin2}}{(t_D*P_D')_{min2}}\right)^2\right)}{1 + (E \times t_{Dmin2}) + \left(F \times \frac{P_{Dmin2}}{(t_D*P_D')_{min2}}\right) + \left(G \times \left(\frac{P_{Dmin2}}{(t_D*P_D')_{min2}}\right)^2\right)} \quad (5.25)$$

The constants depending on the values of λ_{mf}, λ_{mv}, ω_f and ω_v are given in Table 5.9.

8) Using the unit-slope line developed prior to the second radial flow regime; this point is read with the same conditions as for the estimation of λ_{mf}. Thus, the equation is:

$$\lambda_{vf} = A + B \times \ln(t_{Dus}) + C \times (\ln(t_{Dus}))^2 + D \times (\ln(t_{Dus}))^3 + E \times (t_D*P_D')_{us} + $$

$$F \times ((t_D*P_D')_{us})^2 + G \times ((t_D*P_D')_{us})^3 + H \times ((t_D*P_D')_{us})^4 \quad (5.26)$$

Table 5.8. Constants for Equation (5.24)

Rank λ_{vf}	5×10^{-8} to 9×10^{-7}	3×10^{-9} to 3×10^{-8}
λ_{mf}	1×10^{-4}	1×10^{-5}
λ_{mv}	1×10^{-10}	1×10^{-11}
ω_f	1×10^{-4}	1×10^{-5}
ω_v	1×10^{-5}	1×10^{-6}
A	$3.7301428 \times 10^{-06}$	$2.8589362 \times 10^{-06}$
B	0.00014429885	0.0064676422
C	2.66054×10^{-06}	$-1.5822898 \times 10^{-06}$
D	0.21684084	3.6608159
E	$6.4429743 \times 10^{-07}$	$2.9058185 \times 10^{-07}$
F	-0.00023134716	-0.0023232953
G	2.9644256	703.42261
H	$5.2651235 \times 10^{-08}$	$-1.7712087 \times 10^{-08}$
I	$5.4043502 \times 10\text{-}05$	0.00020931574
J	0.010538647	-0.66612539

Triple-Porosity. Single-Permeability Reservoirs

Table 5.9. Constants for Equation (5.25)

Rank λ_{vf}	1×10^{-9} to 1×10^{-7}	5×10^{-7} to 1×10^{-5}
λ_{mf}	1×10^{-2}	$1\times10\text{-}1$
λ_{mv}	1×10^{-8}	$1\times10\text{-}7$
ω_f	1×10^{-2}	$1\times10\text{-}1$
ω_v	1×10^{-3}	$1\times10\text{-}2$
A	-4.6443423×10^{-07}	-3.4145192×10^{-05}
B	8.0324488×10^{-11}	-4.9380799×10^{-08}
C	9.936457×10^{-10}	-5.6541379×10^{-7}
D	-6.2787366×10^{-13}	-9.0007149×10^{-12}
E	0.0010076483	-0.002093633
F	-0.13825481	-0.092186966
G	0.00015750301	-0.0038416544

Table 5.10. Constants for Equation (5.26)

Rank λ_{vf}	5×10^{-8} to 9×10^{-7}	3×10^{-9} to 3×10^{-8}
λ_{mf}	1×10^{-4}	1×10^{-5}
λ_{mv}	1×10^{-10}	1×10^{-11}
ω_f	1×10^{-4}	1×10^{-5}
ω_v	1×10^{-5}	1×10^{-6}
A	9.1288254×10^{-5}	4.6546782×10^{-6}
B	-2.7841968×10^{-5}	-1.2091635×10^{-6}
C	2.8397741×10^{-6}	1.0455704×10^{-7}
D	-9.6854292×10^{-8}	-3.0237203×10^{-9}
E	9.0567788×10^{-7}	7.0123633×10^{-7}
F	-3.574932×10^{-6}	-7.8123529×10^{-6}
G	5.1720753×10^{-6}	3.8047196×10^{-5}
H	-2.3694035×10^{-6}	-6.7541516×10^{-5}

with the respective constants given in Table 5.10. For another range:

$$\lambda_{vf} = \frac{A + B\times\ln\left(t_{Dus}\right) + C\times\left(\ln\left(t_{Dus}\right)\right)^2 + D\times\ln\left(\left(t_D * P_D{}'\right)_{us}\right)}{1 + E\times\ln\left(t_{Dus}\right) + F\times\left(\ln\left(t_{Dus}\right)\right)^2 + G\times\ln\left(\left(t_D * P_D{}'\right)_{us}\right)} \tag{5.27}$$

with the constants depending on λ_{mf}, λ_{mv}, ω_f and ω_v values given in Table 5.11.

9) Another way of using Equation (5.27) and taking into account the development of Equations (5.20) and (5.21), Equation (5.27) is rewritten below with its constants provided in Table 5.12:

$$\lambda_{vf} = \frac{A + B \times \ln(t_{Dusi}) + C \times (\ln(t_{Dusi}))^2}{D + E \times \ln(t_{Dusi}) + F \times (\ln(t_{Dusi}))^2} \tag{5.28}$$

Dimensionless Fracture Storativity Coefficient, ω_f

The same analysis and aspects considered in the study of interporosity flow parameters are used for the dimensionless storativity coefficients. Basically, the effect on the dimensionless pressure derivative due to variation in the dimensionless storativity coefficients depends upon the size of the transition periods (size of the depression).

In the same fashion as λ_{mv}, the starting time of the second radial flow regime converges at the same point for different ω_f values, and the unit-slope line during the transition period varies in length but in terms of location is the same (no parallel displacement occurs along the time axis). The developed correlations for the calculation of the dimensionless fracture storativity coefficient are presented below:

10) Using the second minimum point, $P_{Dmin2}/(t_D*P_D')_{min2}$ and t_{Dmin2}:

$$\omega_f = \frac{A + (B \times \ln(t_{Dmin2})) + \left(C \times \left(\frac{P_{Dmin2}}{(t_D * P_D')_{min2}}\right)\right)}{1 + (D \times \ln(t_{Dmin2})) + \left(E \times \left(\frac{P_{Dmin}}{(t_D * P_D')_{min2}}\right)\right)} \tag{5.29}$$

For range of $1\times10^{-3} < \omega_f < 1\times10^{-2}$, the constants are provided in Table 5.13, and for $1\times10^{-2} < \omega_f < 1\times10^{-1}$, the constants are given in Table 5.14.

Table 5.11. Constants for Equation (5.27)

Rank λ_{vf}	1×10^{-9} to 1×10^{-7}	5×10^{-7} to 1×10^{-5}
λ_{mf}	1×10^{-2}	1×10^{-1}
λ_{mv}	1×10^{-8}	1×10^{-7}
ω_f	1×10^{-2}	1×10^{-1}
ω_v	1×10^{-3}	1×10^{-2}
A	9.473719×10^{-9}	-2.3554034×10^{-6}
B	-1.9452657×10^{-9}	6.2511182×10^{-7}
C	1.0186085×10^{-10}	-4.1006933×10^{-8}
D	6.5972841×10^{-11}	6.9129679×10^{-9}
E	-0.21933576	-0.29487292
F	0.012033019	0.021752823
G	-0.0010661593	0.0011396252

Table 5.12. Constant for Equation (5.28)

Rank λ_{vf}	1×10^{-9} to 1×10^{-7}	5×10^{-7} to 1×10^{-5}
λ_{mf}	1×10^{-2}	1×10^{-1}
λ_{mv}	1×10^{-8}	1×10^{-7}
ω_f	1×10^{-2}	1×10^{-1}
ω_v	1×10^{-3}	1×10^{-2}
A	9.42799011×10^{-9}	-2.3601951×10^{-6}
B	-1.9452657×10^{-9}	6.2511182×10^{-7}
C	1.0186085×10^{-10}	-4.1006933×10^{-8}
D	1.000739005	0.999210072
E	-0.21933576	-0.29487292
F	0.012033019	0.021752823

11) Using the dimensionless vugs storativity coefficient and the second minimum point, $t_{Dmin2}*(t_D*P_D')_{min2}$:

$$\omega_f = \frac{A + B\times Z + C\times Z^2 + D\times Z^3 + (E\times\omega_v)}{1 + F\times Z + (G\times\omega_v) + (H\times\omega_v^2) + (I\times\omega_v^3)} \tag{5.30}$$

Table 5.13. Constants for Equation (5.29)

λ_{vf}	1×10^{-10}	1×10^{-9}	1×10^{-8}
λ_{mv}	1×10^{-13}	1×10^{-12}	1×10^{-11}
λ_{mf}	1×10^{-7}	1×10^{-6}	1×10^{-5}
ω_v	1×10^{-7}	1×10^{-6}	1×10^{-5}
A	-0.001704	-0.00972031	0.006821921
B	-0.00037989	0.000383259	-0.00026108
C	2.83459×10^{-06}	6.25012×10^{-06}	-0.000027956
D	-0.06035366	-0.07376984	-0.09143191
E	-0.00275305	-0.00230932	4.51481×10^{-05}

Table 5.14. Constants for Equation (5.29)

λ_{vf}	1×10^{-6}	1×10^{-7}	1×10^{-8}	1×10^{-9}	1×10^{-10}
λ_{mv}	1×10^{-9}	1×10^{-10}	1×10^{-11}	1×10^{-12}	1×10^{-13}
λ_{mf}	1×10^{-3}	1×10^{-4}	1×10^{-5}	1×10^{-6}	1×10^{-7}
ω_v	1×10^{-4}	1×10^{-4}	1×10^{-5}	1×10^{-6}	1×10^{-7}
A	-0.1954122	1.08574413	0.563443198	0.025335109	0.031881466
B	-0.10102445	-0.0694529	-0.05643964	-0.00389316	-0.00352166
C	0.2754074	-0.0173495	-0.0002592	1.21575×10^{-05}	2.55978×10^{-06}
D	0.0363188	-0.1278309	-0.07562235	-0.0641791	-0.0559242
E	-0.7619847	0.05136569	-0.01871387	-0.00854222	-0.00647706

Table 5.15. Constants for Equation (5.30)

λ_{vf}	1×10^{-6}	1×10^{-7}	1×10^{-8}
λ_{mv}	1×10^{-9}	1×10^{-10}	1×10^{-11}
λ_{mf}	1×10^{-3}	1×10^{-4}	1×10^{-5}
ω_v	1×10^{-4} a 1×10^{-5}	1×10^{-4} a 1×10^{-5}	1×10^{-4} a 1×10^{-5}
A	0.008088891	0.009042676	0.009173307
B	0.001645593	0.000166906	1.67034×10^{-05}
C	-5.2357×10^{-06}	-5.7774×10^{-08}	-5.8567×10^{-10}
D	2.62323×10^{-08}	2.91282×10^{-11}	2.97594×10^{-14}
E	-1678.67637	-1639.2486	-1637.05428
F	0.003283867	0.000319506	3.29846×10^{-05}
G	-10161.9333	-12050.8585	-13037.4575
H	34220600	92257700	106184000
I	2.27831×10^{11}	-1.4116×10^{11}	-2.3217×10^{11}

Table 5.16. Constants for Equation (5.31)

λ_{vf}	1×10^{-5}	1×10^{-6}	1×10^{-7}	1×10^{-8}	1×10^{-9}
λ_{mv}	1×10^{-8}	1×10^{-9}	1×10^{-10}	1×10^{-11}	1×10^{-12}
λ_{mf}	1×10^{-2}	1×10^{-3}	1×10^{-4}	1×10^{-5}	1×10^{-6}
ω_f	1×10^{-2}	1×10^{-3}	1×10^{-4}	1×10^{-5}	1×10^{-6}
A	0.007676832	6.14386×10^{-05}	-0.00028323	-0.00047454	-0.00041341
B	-1.9204×10^{-06}	1.79456×10^{-07}	1.91542×10^{-08}	2.1017×10^{-09}	1.97731×10^{-10}
C	2.5944×10^{-08}	-8.8781×10^{-12}	-2.2493×10^{-13}	-7.9535×10^{-14}	-8.155×10^{-16}
D	-1.8222×10^{-10}	6.99042×10^{-14}	3.34262×10^{-17}	3.17546×10^{-17}	9.03711×10^{-20}
E	1.15661×10^{-11}	-4.2625×10^{-15}	-6.4638×10^{-19}	-1.4331×10^{-18}	-3.6323×10^{-21}

Z being the product of $t_{Dmin2}*(t_D*P_D')_{min2}$. For the range of $1\times10^{-2}<\omega_f<1\times10^{-1}$, the constants for Equation (5.30) are shown in Table 5.15.

12) Using the beginning of the second radial flow, t_{Db2}:

$$\omega_v = A + Bt_{Db2} + Ct_{Db2}^{1.5} + Dt_{Db2}^2 + Et_{Db2}^2 \ln(t_{Db2}) \qquad (5.31)$$

The above expression applies for $1\times10^{-2}<\omega_f<1\times10^{-1}$, and its constants are given in Table 5.16.

12) Using the second minimum point, $(t_D*P_D')_{min2}$ and t_{Dmin2}:

$$\omega_v = \frac{A+(B\times t_{Dmin2})+(C\times t_{t_{Dmin2}}^2)+D\times(t_D*P_D')_{min2}}{1+(B\times t_{Dmin2})+(F\times t_{t_{Dmin2}}^2)+G\times(t_D*P_D')_{min2}} \qquad (5.32)$$

which applies for $1\times10^{-6}<\omega_f<1\times10^{-4}$, with the constants given in Table 5.17.

Table 5.17. Constants for Equation (5.32)

λ_{vf}	1×10^{-6}	1×10^{-7}	1×10^{-8}	1×10^{-9}
λ_{mv}	1×10^{-9}	1×10^{-10}	1×10^{-11}	1×10^{-12}
λ_{mf}	1×10^{-3}	1×10^{-4}	1×10^{-5}	1×10^{-6}
ω_f	1×10^{-3}	1×10^{-4}	1×10^{-5}	1×10^{-6}
A	-1.2604×10^{-06}	-6.4072×10^{-07}	-5.507×10^{-07}	-6.0548×10^{-07}
B	-2.8684×10^{-08}	-3.092×10^{-09}	-4.7228×10^{-10}	-3.327×10^{-11}
C	2.53119×10^{-10}	2.81413×10^{-12}	2.62234×10^{-14}	2.60903×10^{-16}
D	0.000281329	0.000283662	0.000342072	0.000295697
E	0.004368705	0.000474133	4.24526×10^{-05}	4.43181×10^{-06}
F	-1.1735×10^{-06}	-1.3626×10^{-08}	-1.2531×10^{-10}	-1.222×10^{-12}
G	-10.5345235	-10.8263553	-9.74235409	-10.449659

Example 5.1

This is a synthetic example given by Escobar et al. (2014). Pressure and pressure derivative for this test are provided in Figure 5.8. Other relevant data are given in the second column of Table 5.18 and below:

q = 210 STB $s = 0$ $C = 0$ bbl/psi
h = 160 ft $c_t = 1.4\times10^{-6}$ psi^{-1} $r_w = 0.21$ ft
ϕ = 37% μ = 1.2 cp k = 231 md
B = 1.3 bbl/STB

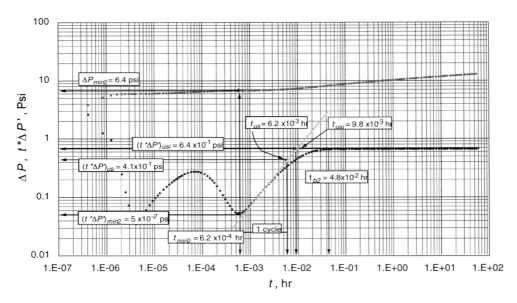

Figure 5.8. Pressure and pressure derivative versus time for example 5.1; adapted from Escobar et al. (2014).

Freddy Humberto Escobar

Table 5.18. Results for example 5.1

Parameters	Actual	Escobar et al. (2014)	Equation
λ_{mf}	1×10^{-4}	9.8537×10^{-5}	5.15
λ_{mv}	1×10^{-10}	1.18034×10^{-12}	5.20
λ_{vf}	1×10^{-7}	1.03007×10^{-7}	5.24
ω_f	1×10^{-4}	1.839317×10^{-3}	5.25
ω_v	1×10^{-5}	9.85352×10^{-6}	5.26

Solution

The following information was obtained from Figure 5.8.

$t_{min2} = 6.2\times10^{-4}$ hr $\quad$ $(t*\Delta P')_{us}= 0.41$ psi $\quad$ $\Delta P_{min2} = 6.4$ psi

$t_{usi} = 9.8\times10^{-3}$ hr $\quad$ $(t*\Delta P')_{min2}=0.05$ psi $\quad$ $t_{b2}=0.048$ hr

$t_{us} = 6.2\times10^{-3}$ hr $\quad$ $\Delta t_D = 0.00918$ hr

Although, the naturally fractured parameters were estimated several times from different sources, for space-saving purposes, only one estimation is provided here. Using the second minimum point, $P_{Dmin2}/(t_D*P_D')_{min2}$ and t_{Dmin2}, λ_{mf} is calculated with Equation (5.15) and λ_{vf} is calculated with Equation (5.24). λ_{mv} was estimated with Equation (5.20) using the point between the intersection of the unit-slope line with the second radial flow regime extrapolated line, t_{Dusi}. In order to calculate ω_v, a point during the unit-slope line is read and substituted into Equation (5.26). Finally, ω_f is calculated with Equation (5.25), which uses the second lowest or minimum point on the pressure derivative curves. Needless to say, the readings from the plot are in oil-field units, and they need to be translated into their dimensionless form. The results and their comparison with the actual parameters are given in Table 5.18.

Nomenclature

B	Volumetric factor, rb/STB
C	Storage coefficient, bbl/psi
c_t	Total compressibility, 1/psi
h	Formation thickness, ft
k	Permeability, md
q	Flow rate, STB/D
t	Time, hr
r	Radius, ft
S	Skin
$t*\Delta P'$	Pressure derivative, psi
t_D*P_D'	Derivative of the dimensionless pressure psi/BPD
P	Pressure, psi
P_D	Dimensionless pressure

Greek

Δ	Change, drop
ϕ	Porosity, fraction
μ	Viscosity, cp
λ	Storage coefficient
ω	Interporosity flow parameter

Suffixes

D	Dimensionless
min	First minimum
$min2$	Second minimum
us	Unitary slope
usi	Intercept between unitary slope and the second radial flow
$r2$	Second radial
$b2$	Beginning second radial
f	Fractures
v	Vugs
mf	Matrix fractures
mv	Matrix vugs
vf	Vugs fractures

REFERENCES

Abdassah, D. & Ershaghi, I. (1986). "Triple-Porosity Systems for Representing Naturally Fractured Reservoirs." SPEFE (April 1986) *113, Trans. AIME,* 281.

Camacho-Velazquez, R., Vazquez-Cruz, M., Castejón-Alvar, R. & Arana-Ortiz, V. 2005. "Pressure-Transient and Decline-Curve Behavior in Naturally Fractured Vuggy Carbonate Reservoirs." *SPE Formation Evaluation & Engineering.* 95-111. April.

Escobar, F. H., Saavedra, N. F., Escorcia, G. D. & Polania, J. H. (2004, January 1). "Pressure and Pressure Derivative Analysis without Type-Curve Matching for Triple Porosity Reservoirs." *Society of Petroleum Engineers.* doi:10.2118/88556-MS.

Escobar, F. H., Rojas, J. D. & Rojas, R. F. (2014). "Pressure and Pressure Derivative Analysis for Triple-Porosity and Single-Permeability Systems in Naturally Fractured Vuggy Reservoirs." *Journal of Engineering and Applied Sciences.* Vol. *9.* Nro. 8. P. 1323-1335. August.

Mirshekari, B., Modarress, H., Hamzehnataj, Z. & Momennasab, E. (2007, January 1). "Application of Pressure Derivative Function for Well Test Analysis of Triple Porosity Naturally Fractured Reservoirs." *Society of Petroleum Engineers.* doi:10.2118/110943-MS

Tiab, D. (1993). "Analysis of Pressure and Pressure Derivative without Type-Curve Matching: 1- Skin and Wellbore Storage." *Journal of Petroleum Science and Engineering. 12,* 171-181.

Warren, J. E. & Root, P. J. (1963). "The Behavior of Naturally Fractured Reservoirs." *SPEJ.* 245-255.

Wu, Y. S., Liu, H. H. & Bodvarson, G. S. (2004). "A triple-continuum approach for modeling flow and transport processes in fractured rock." *Journal of Contaminant Hydrology,* Volume *73,* Issues 1–4, September 2004, Pages 145-179, ISSN 0169-7722, http://dx.doi.org/10.1016/j.jconhyd.2004.01.002. (http://www.sciencedirect.com/science/article/pii/S016977220400004X).

Chapter 6

TRIPLE-POROSITY, DOUBLE-PERMEABILITY RESERVOIRS

BACKGROUND

Novel and sophisticated mathematical models for describing transient pressure behavior in heterogeneous hydrocarbon-bearing formations are introduced almost daily in the literature; however, there is a gap between a model's release and its use in commercial software, so engineers have to write computer codes and use nonlinear regression to interpret pressure test data. In some cases, this is not a practical task since either the analyst lacks computer-coding abilities and/or time to do the job. Thus, the need surges for a practical interpretation technique to facilitate the interpreter's milestone. In chapter 5, the attention was devoted to triple-porosity systems with single permeability, meaning no primary flow occurs though vugs. The characterization was given by Escobar, Rojas and Rojas (2014a). In this chapter, the *TDS* methodology (Tiab 1993) for characterization of heterogeneous naturally fractured vuggy reservoirs with triple porosity and double permeability (meaning a dissolution process has created an interconnected system of vugs) is extended. This interpretation methodology is based upon finding the fingerprints presented on the pressure and pressure derivative versus time plot to obtain expressions for the estimation of such parameters as dimensionless storativity coefficients, ω_v and ω_f, for the systems of fractures and vugs and the interporosity flow parameters: matrix fracture, matrix vugs, fractures vugs, λ_{mf}, λ_{vf} and λ_{mv}.

The presence of cavities or vugs in naturally fractured reservoirs, according to studies by different authors, influences the well pressure behavior because in these systems, especially in carbonate reservoirs, the dissolution of the pore throats has created a system of interconnected cavities and/or vugs, creating two systems that provide the fluid (fractures and cavities or vugs interconnected) to the well. This generates a complex pore system known as triple-porosity and dual-permeability reservoirs, which are common in naturally fractured vuggy reservoirs. As a consequence, strange anomalies are observed in the slope of the semilog plot during the transition period. The behavior of the dimensionless pressure versus dimensionless time alters the normal slope, which is reflected as an additional depression in the pressure curve. These abnormalities in the slope changes are caused by the presence of an additional pore system with different petrophysical properties in the reservoir due to the presence of fractures, vugs and matrices or large fractures, small fractures and matrices, which are especially seen in naturally fractured carbonate reservoirs. Consequently, the flux is affected by the presence of two systems that provide the primary flow toward the well (permeability ratio parameter).

This chapter is devoted to studying the recent work of Escobar, Camacho and Rojas (2014b) using the triple-porosity, double-permeability model also presented by Camacho et al. (2005) to model secondary porosities, mainly in naturally fractured vuggy carbonate reservoirs. The *TDS* technique is extended for such cases.

MATHEMATICAL MODELING

Camacho et al. (2005) developed a complex model for the well pressure behavior of a triple-porosity and double-permeability reservoir:

$$P_D(t_D) = \sqrt{\frac{t_D \kappa}{\omega_f \pi}} \left[\frac{2}{(1-\kappa)s_f/s_v+\kappa} \right] + s_f \left[\frac{1}{(1-\kappa)s_f/s_v+\kappa} \right] \tag{6.1}$$

Camacho et al. (2005) defined the following dimensionless quantities as:

$$P_{Dj} = \frac{2\pi k_f h \left(P_i - P_j \right)}{q\mu B} \tag{5.8}$$

with j = fractures or vugs and

$$t_D = \frac{k_f t}{\left[\left(\phi_f c_f + \phi_m c_m + \phi_v c_v \right) \mu r_w^2 \right]} \tag{5.9}$$

They also defined the interporosity flow parameters as:

$$\lambda_{mf} = \frac{\sigma_{mf} k_m r_w^2}{k_f} \tag{5.10}$$

$$\lambda_{mv} = \frac{\sigma_{mv} k_m r_w^2}{k_f} \tag{5.11}$$

$$\lambda_{vf} = \frac{\sigma_{vf} k_{vf} r_w^2}{k_f + k_v} \tag{5.12}$$

where $k_{vf} = k_v$ if $P_v > P_f$ and $k_{vf} = k_f$ in the contrary case; σ is the shape factor between the media i and j. The dimensionless storativity coefficients are given as:

$$\omega_f = \frac{\phi_f c_f}{\phi_f c_f + \phi_m c_m + \phi_v c_v} \qquad (5.13)$$

$$\omega_v = \frac{\phi_v c_v}{\phi_f c_f + \phi_m c_m + \phi_v c_v} \qquad (5.14)$$

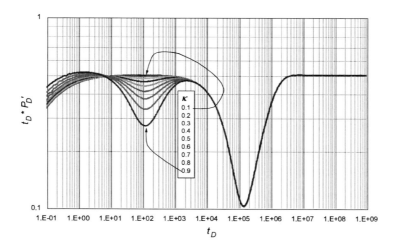

Figure 6.1. Effect of κ on the dimensionless pressure derivative, $\lambda_{mf}=1\times10^{-6}$, $\lambda_{vf}=1\times10^{-4}$, $\lambda_{mv}=1\times10^{-7}$, $\omega_f=5\times10^{-3}$ and $\omega_v=5\times10^{-2}$, $\omega_f < \omega_v$; adapted from Escobar et al. (2014b).

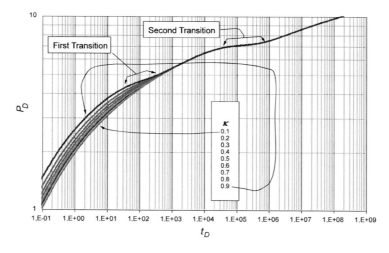

Figure 6.2. Effect of κ on the dimensionless pressure, $\lambda_{mf}=1\times10^{-6}$, $\lambda_{vf}=1\times10^{-4}$, $\lambda_{mv}=1\times10^{-7}$, $\omega_f=5\times10^{-3}$ and $\omega_v=5\times10^{-2}$, $\omega_f < \omega_v$; adapted from Escobar et al. (2014b).

With the model presented by Equation (6.1), Escobar et al. (2014b) studied the dimensionless pressure and pressure derivative behavior to find appropriate fingerprints for developing the expressions for characterization following the philosophy of the *TDS* technique (Tiab 1993). The main features of the dimensionless pressure and pressure derivative versus time log-log plot are given in Figure 6.1.

Figure 6.1 shows the effect of the permeability ratio parameter, κ, on the behavior of the dimensionless pressure derivative versus dimensionless time log-log plot for reservoirs with triple porosity and dual permeability (naturally fractured vuggy reservoirs), keeping constant the values of $\lambda_{mf}=1\times10^{-6}$, $\lambda_{vf}=1\times10^{-4}$, $\lambda_{mv}=1\times10^{-7}$, $\omega_f=5\times10^{-3}$ and $\omega_v=5\times10^{-2}$. The pressure behavior for the same conditions is reported in Figure 6.2.

The permeability ratio parameter, κ, is defined as the ratio between the fracture network permeability and the permeability of the systems contributing to fluid flow toward the well; these are calculated by fracture network permeability plus interconnected vugs system permeability, which is mathematically expressed by:

$$\kappa = \frac{k_f}{k_f+k_v} \tag{6.2}$$

Also, expressing dimensionless pressure and dimensionless time in more convenient oil-field units yields:

$$t_D = \frac{0.0002637\,kt}{(\phi c_t)_{m+f+v}\,\mu r_w^2} \tag{6.3}$$

$$P_D = \frac{kh\Delta P}{141.2q\mu B} \tag{6.4}$$

The pressure derivative is given by Equation (3.2). A triple-porosity system is expected to have two transition periods, which may reduce to a single one in certain cases where the interporosity flow parameters have the same value; thus, a double-porosity system is seen. As observed in Figure 6.1, κ only affects the first transition period.

Two cases are analyzed: when $\omega_f < \omega_v$ (Figure 6.1) and when $\omega_f > \omega_v$ (Figure 6.3). In the first case ($\omega_f < \omega_v$), when κ approaches zero, the first transition does not occur and the behavior of the dimensionless pressure derivative curve resembles a dual-porosity reservoir. A different situation occurs when the κ value increases and approaches the unity; one can clearly see two typical transitions of a triple-porosity reservoir. In the second case, when $\omega_f > \omega_v$, both transitions appear for values of κ close to 0, and when κ approaches 1, the dimensionless pressure derivative curve resembles a dual-porosity reservoir. The same situation applies to the behavior of the dimensionless pressure curve; the two characteristic slope changes representing transitions in triple-porosity reservoirs are presented in the same conditions mentioned for the pressure derivative curve. However, the features are less pronounced in this case.

Even though the values of ω_f or ω_v are small in the above mentioned cases ($\omega_f < \omega_v$ and $\omega_f > \omega_v$), the influence of κ on the pressure and pressure derivative curves is only seen if these systems have a high permeability; otherwise, having low permeability values and a small dimensionless storativity coefficient reduces the system's ability to generate the first transition period.

Taking into account the above considerations and making use of certain characteristic points and/or features found in the pressure and pressure derivative versus time log-log plot leads to several expressions for obtaining the naturally fractured vuggy reservoir parameters.

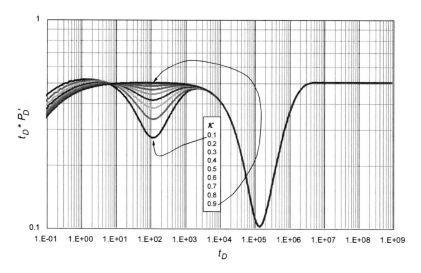

Figure 6.3. Effect of κ on the dimensionless pressure derivative, $\lambda_{mf}=1\times10^{-6}$, $\lambda_{vf}=1\times10^{-4}$, $\lambda_{mv}=1\times10^{-7}$, $\omega_f=5\times10^{-2}$ and $\omega_v=5\times10^{-3}$, $\omega_f > \omega_v$; adapted from Escobar et al. (2014b).

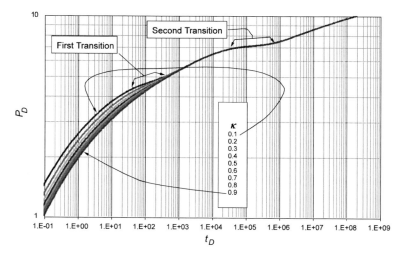

Figure 6.4. Effect of κ on the dimensionless pressure, $\lambda_{mf}=1\times10^{-6}$, $\lambda_{vf}=1\times10^{-4}$, $\lambda_{mv}=1\times10^{-7}$, $\omega_f=5\times10^{-2}$ and $\omega_v=5\times10^{-3}$, $\omega_f > \omega_v$; adapted from Escobar et al. (2014b).

TDS Technique

1) When $\omega_f < \omega_v$, using the first minimum point, $(t_D*P_D')_{min}$ and t_{Dmin} allows obtaining an expression for estimating κ:

$$\kappa = \frac{A+B\ln(t_{Dmin})+C(\ln(t_{Dmin}))^2+D(\ln(t_{Dmin}))^3+E(\ln(t_D*P_D')_{min})}{1+F\ln(t_{Dmin})+G(\ln(t_{Dmin}))^2+H(\ln(t_D*P_D')_{min})} \quad (6.5)$$

The constants in Equation (6.5) are given in Table 6.1.

Equation (6.5) applies to the range $1\times10^{-5}>\lambda_{mf.}>1\times10^{-9}$, $1\times10^{-3} > \lambda_{vf.} >1\times10^{-7}$, $1\times10^{-6} >\lambda_{mv}>1\times10^{-10}$, $1\times10^{-2}>\omega_{f.} >1\times10^{-4}$, $1\times10^{-1}>\omega_v>1\times10^{-3}$, $0.3< \kappa< 0.9$ and $\omega_f< \omega_v$.

2) When $\omega_f > \omega_v$, κ can be calculated with following expression, developed using the first minimum point, $(t_D*P_D')_{min}$ and t_{Dmin}:

$$\kappa = \frac{A+B\ln(t_{Dmin})+C(\ln(t_{Dmin}))^2+D(\ln(t_{Dmin}))^3+E(\ln(t_D*P_D')_{min})}{1+F\ln(t_{Dmin})+G(\ln(t_{Dmin}))^2+H(\ln(t_D*P_D')_{min})} \quad (6.6)$$

The constants A through H are given in Table 6.2.

Equation (6.6) applies to $1\times10^{-5}>\lambda_{mf.}>1\times10^{-9}$, $1\times10^{-3} >\lambda_{vf.}>1\times10^{-7}$, $1\times10^{-6}>\lambda_{mv}>1\times10^{-10}$, $1\times10^{-1}>\omega_{f.}>1\times10^{-3}$, $1\times10^{-2}>\omega_v>1\times10^{-4}$, $0.1< \kappa < 0.7$ and $\omega_f > \omega_v$. Equations (6.5) and (6.6) show similar application ranges; the actual value is the one given by equation (6.5) if $\omega_f <\omega_v$ or the one given by equation (6.6) if $\omega_f >\omega_v$. Usually, $\omega_f <<\omega_v$ for most naturally fractured vuggy systems.

Table 6.1. Constants for Equation (6.5)

Constant	Value	Constant	Value
A	1.510916247	E	2.011693342
B	-0.07286147	F	-0.1173159
C	0.007625919	G	0.007772186
D	-0.00020029	H	1.592874436

Table 6.2. Constants for Equation (6.6)

Constant	Value	Constant	Value
A	-0.50589379	E	-0.41219133
B	-0.03966166	F	-0.11857324
C	-0.00074603	G	0.007906332
D	0.00025665	H	1.574599604

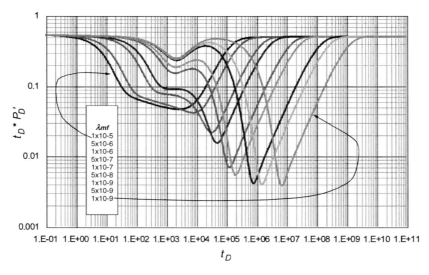

Figure 6.5. Effect of λ_{mf} on the dimensionless pressure derivative, $\lambda_{vf}=1\times10^{-7}$, $\lambda_{mv}=1\times10^{-10}$, $\omega_f=1\times10^{-4}$, $\omega_v=1\times10^{-3}$ and $\kappa=0.9$; adapted from Escobar et al. (2014b).

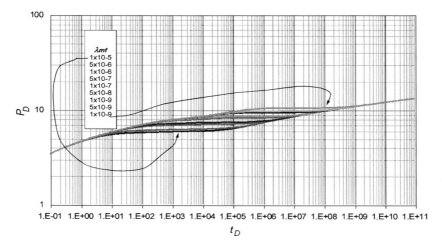

Figure 6.6. Effect of λ_{mf} on the dimensionless pressure, $\lambda_{vf}=1\times10^{-7}$, $\lambda_{mv}=1\times10^{-10}$, $\omega_f=1\times10^{-4}$, $\omega_v=1\times10^{-3}$ and $\kappa=0.9$; adapted from Escobar et al. (2014b).

Matrix-Fracture Interporosity Flow Parameter, λ_{mf}

The effect of the matrix-fracture interporosity flow parameter, λ_{mf}, on the behavior of the dimensionless pressure derivative with constant values of $\lambda_{vf}=1\times10^{-7}$, $\lambda_{mv}=1\times10^{-10}$, $\omega_f=1\times10^{-4}$, $\omega_v=1\times10^{-3}$ and $\kappa=0.9$ is shown in Figure 6.5. The pressure behavior for these conditions is given in Figure 6.6. As seen in Figure 6.5, basically, λ_{mf} affects the occurrence of the transition period. When the value of λ_{mf} decreases, the presence of the second minimum point, the unit-slope behavior just before the second radial flow regime (the radial flow formed in the homogenous system once the transition period is no longer felt) and the same second radial flow occur later. The first transition zone is also affected by this interporosity flow parameter; however, for high values of λ_{mf}, the first minimum point is not formed, and for small values of this parameter, the time for the first minimum point tends to remain constant. The two slope changes in the behavior of the dimensionless pressure (Figure 6.6) are better appreciated at small values of λ_{mf}, as a consequence of the typical transitions generated by the influence of different porous media affecting fluid flow in naturally fractured vuggy reservoirs. Given these considerations, the expressions generated to calculate λ_{mf} are as follows:

1) Using the point of interception between the unit-slope line of the second transition period and the radial flow regime, t_{Dusi}:

$$\lambda_{mf} = \lambda_{mf}{}^{x} \times (t_{e}{}^{x})\qquad(6.7)$$

where $t_{e}{}^{x}$ is equal to ten to the same power of t_{Dusi}. An example for $t_{Dusi} = 3.586\times10^{-6}$, 3.586×10^{-6} is applied in Equation (6.8). In this case, $t_{e}{}^{x}$ only contains the order of magnitude, which, for this case, is 1×10^{-6}. $\lambda_{mf}{}^{x}$ is calculated as follows:

$$\lambda_{mf}{}^{x} = A \times \left(\frac{t_{Dusi}}{t_{e}{}^{x}}\right)^{B}\qquad(6.8)$$

Constants A and B are given in Table 6.3.
Equation (6.7) is applicable to $1\times10^{-5}>\lambda_{mf}>1\times10^{-9}$, $1\times10^{-5}>\lambda_{vf}>1\times10^{-7}$, $1\times10^{-8}>\lambda_{mv}>1\times10^{-10}$, $1\times10^{-3}>\omega_{f}>1\times10^{-4}$, $1\times10^{-2}>\omega_{v}>1\times10^{-3}$ and $0.1< \kappa < 0.9$.

2) Using the second minimum point, t_{Dmin2} and $(t_{D}*P_{D}')_{min2}$:

Table 6.3. Constants for Equation (6.8)

Constant	Value
A	4.19526×10^{-08}
B	-0.00487301

Table 6.4. Constants for Equation (6.9)

Constant	Value	Constant	Value
A	4.19526×10^{-08}	F	1.917864631
B	-0.00487301	G	927175.7136
C	-4.2135×10^{-06}	H	-0.00082525
D	63.17929161	I	-1.34413549
E	0.000106335	J	-3272.68436

Table 6.5. Ranges of application of Equation (6.9)

	λ_{vf}	λ_{mv}	ω_f	ω_v	κ
1×10^{-5} to 1×10^{-6}	1×10^{-3}	1×10^{-6}	1×10^{-2}	1×10^{-1}	0.1
1×10^{-5} to 1×10^{-7}	1×10^{-4}	1×10^{-7}	5×10^{-3}	5×10^{-2}	0.3
1×10^{-5} to 1×10^{-8}	1×10^{-5}	1×10^{-8}	1×10^{-3}	1×10^{-2}	0.5
1×10^{-5} to 1×10^{-9}	1×10^{-6}	1×10^{-9}	5×10^{-4}	5×10^{-3}	0.7
1×10^{-5} to 1×10^{-9}	1×10^{-7}	1×10^{-10}	1×10^{-4}	1×10^{-3}	0.9

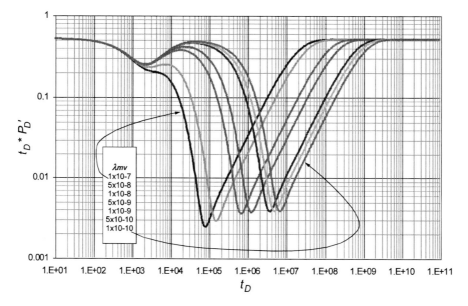

Figure 6.8. Effect of λ_{mv} on the dimensionless pressure derivative, $\lambda_{mf}= 1\times10^{-9}$, $\lambda_{vf}= 1\times10^{-7}$, $\omega_f=1\times10^{-4}$, $\omega_v=1\times10^{-3}$ and $\kappa=0.9$; adapted from Escobar et al. (2014b).

$$\lambda_{mf} = A + \frac{B}{t_{D\min2}} + C(t_D*P_D')_{\min2} + \frac{D}{(t_{D\min2})^2} + \left[(t_D*P_D')_{\min2}\right]^2 + F\frac{(t_D*P_D')_{\min2}}{t_{D\min2}} + E\frac{G}{(t_{D\min2})^3} + H((t_D*P_D')_{\min2})^3 + I\frac{((t_D*P_D')_{\min2})^2}{t_{D\min2}} + J\frac{(t_D*P_D')_{\min2}}{(t_{D\min2})^2} \quad (6.9)$$

Constants A through J are given in Table 6.4. Equation (6.9) applies according to the ranges given in Table 6.5. For the application of Equation (6.9), the interpreter does not need to know the values of the parameters given in Table 6.5. Once all the parameters are estimated, one should use the table to verify the equation's applicability for a given case.

A similar situation will be observed in Equations (6.10) through (6.17), which are also subjected to specific ranges of application, given in their respective tables. There are cases in which the same parameters can be obtained from different equations; in these cases, one should use them all and choose the best result according to the ranges. An average value may even be used if all results fit in their respective application ranges.

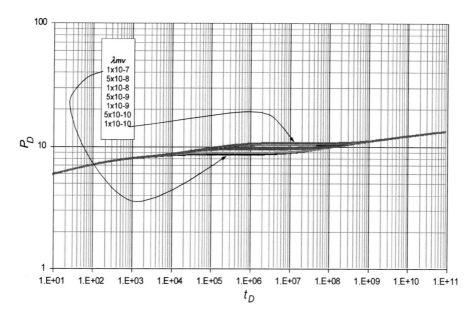

Figure 6.9. Effect of λ_{mv} on the pressure derivative, $\lambda_{mf}=1\times10^{-9}$, $\lambda_{vf}=1\times10^{-7}$, $\omega_f=1\times10^{-4}$, $\omega_v=1\times10^{-3}$ and $\kappa=0.9$; adapted from Escobar et al. (2014b).

Table 6.6. Constants for Equation (6.10)

Constant	Value	Constant	Value
A	-1.245×10^{-07}	F	-0.00001885
B	5.85434×10^{-11}	G	259931.4604
C	-7.9543×10^{-15}	H	$-1.1964\times10^{+12}$
D	3.32652×10^{-19}	I	$1.22363\times10^{+18}$
E	1.671007544		

Table 6.7. Ranges of application of Equation (6.10)

λ_{mf}	λ_{vf}	λ_{mv}	ω_f	ω_v	κ
1×10^{-7}	1×10^{-5}	1×10^{-6} to 1×10^{-8}	1×10^{-3}	1×10^{-2}	0.5
1×10^{-8}	1×10^{-6}	1×10^{-7} to 1×10^{-9}	5×10^{-4}	5×10^{-3}	0.7
1×10^{-9}	1×10^{-7}	5×10^{-8} to 1×10^{-10}	1×10^{-4}	1×10^{-3}	0.9

Matrix-Vugs Interporosity Flow Parameter, λ_{mv}

Figures 6.8 and 6.9 show the effect of the matrix-vugs interporosity flow parameter, λ_{mv}, on the transient behavior for constant values of $\lambda_{mf}=1\times10^{-9}$, $\lambda_{vf}=1\times10^{-7}$, $\omega_f=1\times10^{-4}$, $\omega_v=1\times10^{-3}$ and $\kappa=0.9$. As for the case of λ_{mf}, the variation of the interporosity flow parameter between matrix-vugs (λ_{mv}) has the same effect on the behavior of the dimensionless pressure derivative. This parameter affects the occurrence of the transition period; when the value of

Triple-Porosity, Double-Permeability Reservoirs

λ_{mv} decreases, the presence of the second minimum point, the unit-slope behavior just before the second radial flow and the second radial flow occur later. In the case of dimensionless pressure behavior, as in λ_{mf}, the changes of the two slopes (Figure 6.8) are better appreciated for the small values of this parameter. These observations lead to the development of the expressions given below.

1) Using the first and second minimum points, $(t_{Dmin}, (t_D*P_D')_{min})$ and $((t_D*P_D')_{min2}, t_{Dmin2})$:

$$\lambda_{mv} = \frac{A + B\left(\dfrac{t_{D\min}}{(t_D*P_D')_{\min}}\right) + C\left(\dfrac{t_{D\min}}{(t_D*P_D')_{\min}}\right)^2 + D\left(\dfrac{t_{D\min}}{(t_D*P_D')_{\min}}\right)^3 + E\left(\dfrac{(t_D*P_D')_{\min2}}{t_{D\min2}}\right)}{1 + F\left(\dfrac{t_{D\min}}{(t_D*P_D')_{\min}}\right) + G\left(\dfrac{(t_D*P_D')_{\min2}}{t_{D\min2}}\right) + H\left(\dfrac{(t_D*P_D')_{\min2}}{t_{D\min2}}\right)^2 + I\left(\dfrac{(t_D*P_D')_{\min2}}{t_{D\min2}}\right)^3} \tag{6.10}$$

Constants A through I are given in Table 6.6. Equation (10) is applicable to the ranges indicated in Table 6.7.

2) Another expression developed for the determination of λ_{mv} using the first and second minimum points, $t_{Dmin}/(t_D*P_D')_{min}$ and $(t_D*P_D')_{min2}/t_{Dmin2}$ is given below:

$$\lambda_{mv} = A + \frac{B}{\dfrac{t_{D\min}}{(t_D*P_D')_{\min}}} + \frac{C}{\left(\dfrac{t_{D\min}}{(t_D*P_D')_{\min}}\right)^2} + \frac{D}{\left(\dfrac{t_{D\min}}{(t_D*P_D')_{\min}}\right)^3} + E\frac{(t_D*P_D')_{\min2}}{t_{D\min2}} +$$

$$F\left(\dfrac{(t_D*P_D')_{\min2}}{t_{D\min2}}\right)^2 + G\left(\dfrac{(t_D*P_D')_{\min2}}{t_{D\min2}}\right)^3 + H\left(\dfrac{(t_D*P_D')_{\min2}}{t_{D\min2}}\right)^4 + I\left(\dfrac{(t_D*P_D')_{\min2}}{t_{D\min2}}\right)^5 \tag{6.11}$$

Constants A through I are given in Table 6.8. Equation (6.11) is applicable in the ranges described in Table 6.9.

Table 6.8. Constants for Equation (6.11)

Constant	Value	Constant	Value
A	1.40042×10^{-08}	F	-2770000
B	-0.00014802	G	$6.16641\times10^{+12}$
C	0.230679819	H	$-5.6158\times10^{+18}$
D	-95.2152707	I	$1.74285\times10^{+24}$
E	2.018418377		

Table 6.9. Ranges of application of Equation (6.11)

λ_{mf}	λ_{vf}	λ_{mv}	ω_f	ω_v	κ
1×10^{-6}	1×10^{-4}	1×10^{-6} to 1×10^{-7}	5×10^{-3}	5×10^{-2}	0.3
1×10^{-7}	1×10^{-5}	1×10^{-6} to 1×10^{-8}	1×10^{-3}	1×10^{-2}	0.5
1×10^{-8}	1×10^{-6}	1×10^{-7} to 1×10^{-9}	5×10^{-4}	5×10^{-3}	0.7

Vugs-Fractures Interporosity Flow Parameter, λ_{vf}

Figure 6.10 shows the effect of the vugs-fracture interporosity flow parameter, λ_{vf}, on the behavior of the dimensionless pressure derivative with constant values of $\lambda_{mf}=1\times10^{-6}$, $\lambda_{mv}=1\times10^{-7}$, $\omega_f=5\times10^{-3}$, $\omega_v=5\times10^{-2}$ and $\kappa=0.7$. The pressure behavior for these conditions is presented in Figure 6.11. Unlike λ_{mf} and λ_{mv}, the vugs-fracture interporosity flow parameter, λ_{vf}, mainly affects the first transition. When the value of λ_{vf} decreases, the presence of the first transition occurs later.

The second transition is almost unaffected by this parameter. The starting time of the second radial flow converges at the same point for different λ_{vf} values, and a unit-slope line at the end of the pressure derivative depression varies little in length, but its location is the same (no displacement along the time axis). These observations lead to the development of the expressions given below.

1) Using the first and second minimum points, $(t_{Dmin}, (t_D*P_D')_{min})$ and $((t_D*P_D')_{min2}, t_{Dmin2})$:

$$\ln\left(\lambda_{mv}\right) = A + B\ln\left(\frac{t_{Dmin}}{(t_D*P_D')_{min}}\right) + \frac{C}{\dfrac{t_{Dmin}}{(t_D*P_D')_{min}}} + \frac{D}{\left(\dfrac{t_{Dmin}}{(t_D*P_D')_{min}}\right)^{1.5}} + E\left(\frac{(t_D*P_D')_{min2}}{t_{Dmin2}}\right)^2 \tag{6.12}$$

Constants A through E are given in Table 6.10. Equation (6.12) applies in the ranges given in Table 6.11.

2) Another good expression for estimating λ_{mv} is developed using the first and second minimum point, $(t_{Dmin}, (t_D*P_D')_{min})$ and $((t_D*P_D')_{min2}, t_{Dmin2})$:

Table 6.10. Constants for Equation (6.12)

Constant	Value
A	-3.74946872
B	-0.94844128
C	15.88614414
D	-51.8756665
E	$1.96913\times10^{+10}$

Table 6.11. Ranges of application of Equation (6.12)

λ_{mf}	λ_{vf}	λ_{mv}	ω_f	ω_v	κ
1×10^{-5}	1×10^{-3} to 3×10^{-5}	1×10^{-6}	1×10^{-2}	1×10^{-1}	0.9
1×10^{-6}	1×10^{-3} to 3×10^{-6}	1×10^{-7}	5×10^{-3}	5×10^{-2}	0.7

Triple-Porosity, Double-Permeability Reservoirs

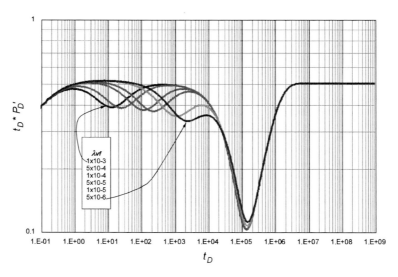

Figure 6.10. Effect of λ_{vf} on the dimensionless pressure derivative, $\lambda_{mf}=1\times10^{-6}$, $\lambda_{mv}=1\times10^{-7}$, $\omega_f=5\times10^{-3}$, $\omega_v=5\times10^{-2}$ and $\kappa=0.7$; adapted from Escobar et al. (2014b).

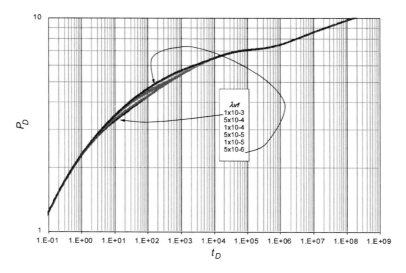

Figure 6.11. Effect of λ_{vf} on the dimensionless pressure, $\lambda_{mf}=1\times10^{-6}$, $\lambda_{mv}=1\times10^{-7}$, $\omega_f=5\times10^{-3}$, $\omega_v=5\times10^{-2}$ and $\kappa=0.7$; adapted from Escobar et al. (2014b).

$$\ln(\lambda_{mv}) = A + B \ln\left(\frac{t_{D\min}}{(t_D{}^*P_D{}')_{\min}}\right) + \frac{C}{\dfrac{t_{D\min}}{(t_D{}^*P_D{}')_{\min}}} + \frac{D}{\left(\dfrac{t_{D\min}}{(t_D{}^*P_D{}')_{\min}}\right)^{1.5}} + E\left(\ln\left(\frac{(t_D{}^*P_D{}')_{\min 2}}{t_{D\min 2}}\right)\right)^2 \quad (6.13)$$

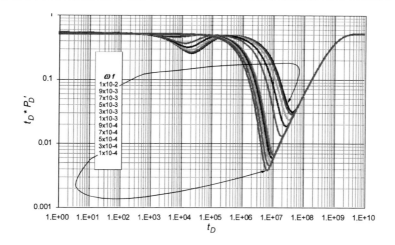

Figure 6.12. Effect of ω_f on the dimensionless pressure derivative, $\lambda_{mf}=1\times10^{-9}$, $\lambda_{vf}=1\times10^{-7}$, $\lambda_{mv}=1\times10^{-10}$, $\omega_v=1\times10^{-3}$ and $\kappa=0.1$; adapted from Escobar et al. (2014b).

Constants *A* through *E* are given in Table 6.12. The ranges of application of Equation (6.13) are provided in Table 6.13.

Dimensionless Fracture Storativity Coefficient, ω_f

The effect of the dimensionless fracture storativity coefficient, ω_f, on the behavior of the dimensionless pressure derivative with constant values of $\lambda_{mf}=1\times10^{-9}$, $\lambda_{vf}=1\times10^{-7}$, $\lambda_{mv}=1\times10^{-10}$, $\omega_v=1\times10^{-3}$ and $\kappa=0.1$ is given in Figure 6.12. The pressure behavior for these conditions is presented in Figure 6.13.

The same analysis and aspects considered in the study of interporosity flow parameters are used for the dimensionless storativity coefficients. Unlike the interporosity flow parameters, which affect the times at which transitions occur, the dimensionless storativity coefficient basically affects the size of the transition periods.

Table 6.12. Constants for Equation (6.13)

Constant	Value
A	-0.93976037
B	-1.01662627
C	0.320447
D	-0.27735296
E	-0.01328292

Table 6.13. Ranges of application of Equation (6.13)

λ_{mf}	λ_{vf}	λ_{mv}	ω_f	ω_v	κ
1×10^{-8}	1×10^{-3} to 1×10^{-7}	1×10^{-9}	5×10^{-4}	5×10^{-3}	0.7
1×10^{-9}	1×10^{-3} to 1×10^{-7}	1×10^{-10}	1×10^{-5}	1×10^{-4}	0.9

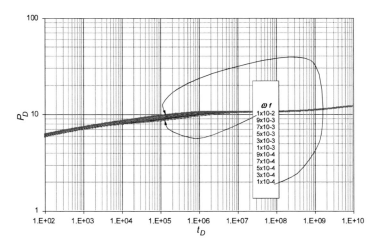

Figure 6.13. Effect of ω_f on the dimensionless pressure, $\lambda_{mf}=1\times10^{-9}$, $\lambda_{vf}=1\times10^{-7}$, $\lambda_{mv}=1\times10^{-10}$, $\omega_v=1\times10^{-3}$ and $\kappa=0.1$; adapted from Escobar et al. (2014b).

In the same fashion as for λ_{vf}, the starting time of the second radial flow regime converges at the same point for different ω_f values, and the unit-slope line varies in length during the transition period, but its location remains the same (no displacement occurs along the time axis). The developed correlations for the calculation of the dimensionless fracture storativity coefficient are presented below.

1) The first minimum point, t_{Dmin} and $(t_D*P_D')_{min}$ are used to find ω_f.

$$\omega_f = \frac{A+B(t_{Dmin})+C(t_{Dmin})^2+D(t_{Dmin})^3+E(t_D*P_D')_{min}+F((t_D*P_D')_{min})^2}{1+G(t_{Dmin})+H(t_D*P_D')_{min}} \qquad (6.14)$$

Constants A through H in Equation (6.14) are given in Table 6.14, with its ranges of application given in Table 6.15.

Table 6.14. Constants for Equation (6.14)

Constant	Value	Constant	Value
A	-0.02325878	E	0.19159594
B	1.37488×10^{-05}	F	-0.38546643
C	3.31535×10^{-07}	G	-0.00025739
D	-1.2728×10^{-09}	H	-3.26386202

Table 6.15. Ranges of application of Equation (6.14)

λ_{mf}	λ_{vf}	λ_{mv}	ω_f	ω_v	κ
1×10^{-5}	1×10^{-3}	1×10^{-6}	1×10^{-2} to 1×10^{-3}	1×10^{-1}	0.9
1×10^{-6}	1×10^{-4}	1×10^{-7}	1×10^{-2} to 1×10^{-3}	5×10^{-2}	0.7

Table 6.16. Constants for Equation (6.15)

Constant	Value	Constant	Value
A	0.010526247	F	0.009081477
B	0.000403017	G	1.02498082
C	0.007882236	H	0.329843138
D	0.001160169	I	0.032986927
E	5.22774×10^{-05}		

Table 6.17. Constants for Equation (6.15) for another range of application

Constant	Value	Constant	Value
A	0.011397275	F	0.008783797
B	0.000397396	G	1.015745712
C	0.008486408	H	0.325783293
D	0.001309282	I	0.032497064
E	6.41607×10^{-05}		

2) The second minimum point, t_{Dmin2} and $(t_D*P_D')_{min2}$ are used to find ω_f.

$$\omega_f = \frac{A + B\ln(t_{Dmin2}) + C\ln\left((t_D*P_D')_{min2}\right) + D\left(\ln\left((t_D*P_D')_{min2}\right)\right)^2 + E\left(\ln\left((t_D*P_D')_{min2}\right)\right)^3}{1 + F\ln(t_{Dmin2}) + G\ln\left((t_D*P_D')_{min2}\right) + H\left(\ln\left((t_D*P_D')_{min2}\right)\right)^2 + I\left(\ln\left((t_D*P_D')_{min2}\right)\right)^3} \tag{6.15}$$

Constants A through I are given in Table 6.16. Equation (6.15) is applicable to $1\times10^{-7} > \lambda_{mf} > 1\times10^{-9}$, $1\times10^{-5} > \lambda_{vf} > 1\times10^{-7}$, $1\times10^{-8} > \lambda_{mv} > 1\times10^{-10}$, $1\times10^{-2} > \omega_f > 1\times10^{-3}$, $1\times10^{-2} > \omega_v > 1\times10^{-3}$ and $0.5 > \kappa > 0.1$. For a range of $1\times10^{-7} > \lambda_{mf} > 1\times10^{-9}$, $1\times10^{-5} > \lambda_{vf} > 1\times10^{-7}$, $1\times10^{-8} > \lambda_{mv} > 1\times10^{-10}$, $1\times10^{-2} > \omega_f > 1\times10^{-3}$, $1\times10^{-2} > \omega_v > 1\times10^{-3}$ and $0.5 < \kappa < 0.9$. Equation (6.15) is also applied with the constant values given in Table 6.17.

Dimensionless Vugs Storativity Coefficient, ω_v

By the same token as for ω_f, the starting time of the second radial flow regime converges at the same point for different ω_v values, and the unit-slope line varies in length during the transition period, but its location remains the same (no displacement occurs along the time axis).

1) Using the dimensionless delta time Δt_D, defined as $(t_{Dusi} - t_{Dmin2}) / t_{Dusi}$:

$$\omega_v = A + B(\Delta t_D) + \frac{C}{\Delta t_D} + \frac{D}{\Delta t_D^{1.5}} + \frac{E}{\Delta t_D^2} \tag{6.16}$$

Constants A through E are given in Table 6.18. Equation (6.16) applies in the range of $1\times10^{-5} > \lambda_{mf} > 1\times10^{-8}$, $1\times10^{-3} > \lambda_{vf} > 1\times10^{-6}$, $1\times10^{-6} > \lambda_{mv} > 1\times10^{-9}$, $1\times10^{-2} > \omega_f > 5\times10^{-4}$, $1\times10^{-1} > \omega_v > 1\times10^{-3}$ and $0.9 > \kappa > 0.3$.

2) The second minimum point, t_{Dmin2} and $(t_D*P_D')_{min2}$ are used to find ω_v.

Table 6.18. Constants for Equation (6.16)

Constant	Value
A	-104.00979
B	23.87052799
C	272.2569597
D	-272.007418
E	79.88613675

Table 6.19. Constants for Equation (6.17)

Constant	Value	Constant	Value
A	-0.00360451	F	161.779671
B	7.41077×10^{-07}	G	-2.0549×10^{-06}
C	-1.6835×10^{-12}	H	163.3901537
D	-1.97786996	I	-243.859592
E	68.1607616	J	203.1526803

$$\omega_f = \frac{A+B(t_{D\min 2})+C(t_{D\min 2})^2+D(t_D{}^*P_D{}')_{\min 2}+E((t_D{}^*P_D{}')_{\min 2})^2+F((t_D{}^*P_D{}')_{\min 2})^3}{1+G(t_{D\min 2})+H(t_D{}^*P_D{}')_{\min 2}+I((t_D{}^*P_D{}')_{\min 2})^2+J((t_D{}^*P_D{}')_{\min 2})^3} \quad (6.17)$$

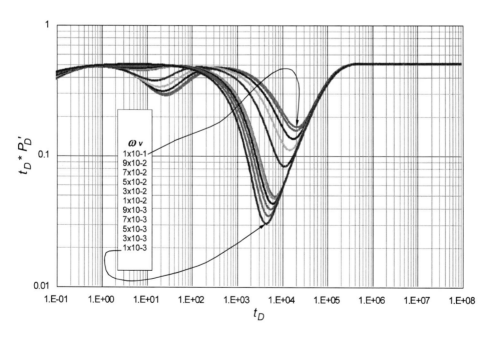

Figure 6.14. Effect of ω_v on the dimensionless pressure derivative, $\lambda_{mf}=1\times 10^{-5}$, $\lambda_{vf}=1\times 10^{-3}$, $\lambda_{mv}=1\times 10^{-6}$, $\omega_f=1\times 10^{-2}$ and $\kappa=0.9$; adapted from Escobar et al. (2014b).

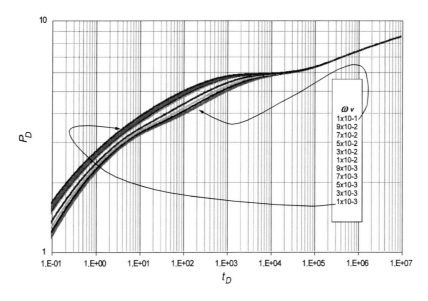

Figure 6.15. Effect of ω_v on the dimensionless pressure, $\lambda_{mf}=1\times10^{-5}$, $\lambda_{vf}=1\times10^{-3}$, $\lambda_{mv}=1\times10^{-6}$, $\omega_f=1\times10^{-2}$ and $\kappa=0.9$; adapted from Escobar et al. (2014b).

Constants *A* through *J* are given in Table 6.19. The ranges of application of Equation (6.17) are $1\times10^{-5}>\lambda_{mf}>1\times10^{-6}$, $1\times10^{-3}>\lambda_{vf}>1\times10^{-4}$, $1\times10^{-6}>\lambda_{mv}>1\times10^{-7}$, $1\times10^{-2}>\omega_f>5\times10^{-3}$, $1\times10^{-1}>\omega_v>1\times10^{-3}$ and $0.9>\kappa>0.7$. With the constants given in Table 6.20, Equation (6.17) applies in the ranges of $1\times10^{-6}>\lambda_{mf}>1\times10^{-7}$, $1\times10^{-4}>\lambda_{vf}>1\times10^{-5}$, $1\times10^{-7}>\lambda_{mv}>1\times10^{-8}$, $5\times10^{-3}>\omega_f>1\times10^{-4}$, $1\times10^{-1}>\omega_v>1\times10^{-3}$ and $0.7>\kappa>0.5$. Additionally, with the constants given in Table 6.21, Equation (6.17) applies in the ranges of $1\times10^{-7}>\lambda_{mf}>1\times10^{-8}$, $1\times10^{-5}>\lambda_{vf}>1\times10^{-6}$, $1\times10^{-8}>\lambda_{mv}>1\times10^{-9}$, $1\times10^{-3}>\omega_f>5\times10^{-4}$, $1\times10^{-1}>\omega_v>1\times10^{-3}$ and $0.5>\kappa>0.3$.

It is important to clarify that for each estimated parameter, the application ranges may coincide. This is due to the fact that, for verification purposes, a given parameter should be estimated from different sources. Although, in some cases, the correlation is the same, the coefficients may be different for a specific range of some parameters, which provides a more accurate determination of the parameter for which the correlation is provided.

Finally, it is necessary to take into account the ranges of application of a given correlation in order to establish the appropriate set of values to be chosen.

Table 6.20. Constants for Equation (6.17) for other ranges of application

Constant	Value	Constant	Value
A	-0.00094874	F	377.6641219
B	7.6135×10^{-08}	G	-1.9253×10^{-07}
C	-1.0856×10^{-14}	H	222.3130093
D	-1.37093556	I	-83.4129396
E	82.05481856	J	-146.488449

Table 6.21. Constants for Equation (6.17) for other ranges of application

Constant	Value	Constant	Value
A	-0.00056126	F	172.1592979
B	4.39813×10^{-10}	G	-4.1688×10^{-09}
C	-1.7051×10^{-17}	H	68.0892155
D	0.143171944	I	75.21139296
E	24.78533848	J	-231.048025

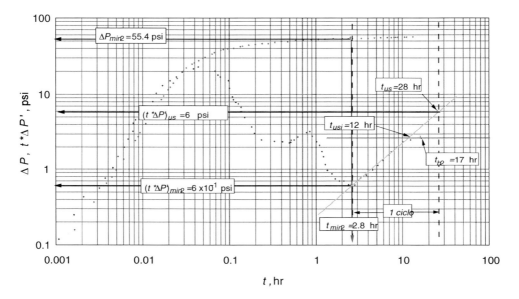

Figure 6.16. Pressure and pressure derivative versus time for example 6.1; adapted from Camacho et al. (2005).

Example 6.1

Camacho et al. (2005) presented a well test from an offshore, naturally fractured, vuggy reservoir located in southeastern Mexico. The test is digitized from Camacho et al. (2005) and reported in Figure 6.16. The field structure is an anticline affected by a normal fault, parallel to the major axis of the structure. This fact facilitated a saline intrusion at the middle of the field, dividing it into two blocks. The well was completed in a Cretaceous formation with a thickness between 558 and 912 ft. Other information is given below:

q = 75 STB $\quad\quad s = 2$ $\quad\quad\quad\quad\quad\quad\quad\quad C_D = 4500$
h = 912.07 ft $\quad c_t = 1.4 \times 10^{-6}$ psi^{-1} (assumed)
r_w = 0.21 ft $\quad\quad \phi = 40\%$ $\quad\quad\quad\quad\quad\quad\quad \mu = 0.6$ cp
k = 524 md

Table 6.22. Results for example 6.1

Parameters	Equation	Escobar et al. (2014b)	Camacho et al. (2005)
λ_{mf}	6.7	8.24×10^{-9}	1×10^{-7}
λ_{mv}	6.11	2.5×10^{-6}	1×10^{-8}
λ_{vf}	6.12	1.4×10^{-6}	1×10^{-5}
ω_f	6.15	5.8×10^{-3}	1×10^{-3}
ω_v	6.16	0.12	0.2

Solution.

The following information is read from Figure 6.16.

$t_{min2} = 2.8$ hr $(t^*\Delta P')_{us} = 6$ psi $\Delta P_{min2} = 55.4$ psi

$t_{usi} = 12$ hr $(t^*\Delta P')_{min2} = 0.6$ psi $t_{b2} = 17$ hr

$t_{us} = 28$ hr $\Delta t_D = 9$ hr

Once the data read from Figure 6.16 are converted into dimensionless form, the calculations are performed and reported in Table 6.22. The data read from Figure 6.16 is converted into dimensionless form, with which the results reported in Table 6.22 are obtained.

Equations (6.11), (6.12) and (6.15) are used since the values are known (input parameters), as are the ranges of applications. Normally, all of the equations should be used, and then the best option should be chosen at the end of the calculations, once the application ranges are verified. Although in some cases the results differ in order of magnitude with the actual values, the authors believe that the results are acceptable, especially if these values are used as starting values for a fitting algorithm. For the given example, it is necessary to assume the system compressibility since it was not provided. Additionally, the accuracy of reading the characteristic points influences the results.

Nomenclature

B	Volumetric factor, rb/STB
C	Storage coefficient, bbl/psi
c_t	Total compressibility, 1/psi
h	Formation thickness, ft
k	Permeability, md
q	Flow rate , STB/D
t	Time, hr
r	Radius, ft
S	Skin
$t^*\Delta P'$	Pressure derivative, psi
$t_D^*P_D'$	Derivative of the dimensionless pressure psi/BPD
P	Pressure, psi
P_D	Dimensionless pressure

Greeks

Δ Change, drop

ϕ Porosity, fraction

μ Viscosity, cp

λ Storage coefficient

ω Interporosity flow parameter

Suffixes

D Dimensionless

min First minimum

$min2$ Second minimum

us Unitary slope

usi Intercept between unitary slope and the second radial flow

$r2$ Second radial

$b2$ Beginning second radial

f Fractures

v Vugs

mf Matrix-fractures

mv Matrix-vugs

vf Vugs-fractures

REFERENCES

Camacho-Velazquez, R., Vazquez-Cruz, M., Castejón-Alvar, R. & Arana-Ortiz, V. 2005. "Pressure-Transient and Decline-Curve Behavior in Naturally Fractured Vuggy Carbonate Reservoirs." *SPE Formation Evaluation & Engineering*. 95-111. April.

Escobar, F. H., Rojas, J. D. & Rojas, R. F. (2014a). "Pressure and Pressure Derivative Analysis for Triple-Porosity and Single-Permeability Systems in Naturally Fractured Vuggy Reservoirs." *Journal of Engineering and Applied Sciences*. Vol. 9. Nro. 8. P. 1323-1335. August.

Escobar, F. H., Camacho, R. G. & Rojas, J. D. (2014b). "Pressure and Pressure Derivative Analysis for Triple-Porosity and Dual-Permeability Systems in Naturally Fractured Vuggy Reservoirs." *Journal of Engineering and Applied Sciences*. Vol. 9. Nro. 12. P. 2500-2512.

Tiab, D. (1993, January 1). "*Analysis of Pressure and Pressure Derivatives Without Type-Curve Matching: I-Skin and Wellbore Storage.*" Society of Petroleum Engineers. This paper was prepared for presentation at the *Production Operations Symposium* held in Oklahoma City, OK, U.S.A., March 21-23. doi:10.2118/25426-MS.

PART TWO: NON-NEWTONIAN FLUIDS

Chapter 7

INFINITE RESERVOIRS

BACKGROUND

It is customary for the radial flow regime to display a zero slope in the natural logarithm pressure derivative. However, that is not the case when dealing with either non-Newtonian fluids or fractal reservoirs, in which the slope is different than zero. Fractal reservoirs will always have an increasing slope, depending on the nature of the porous media. During the radial flow regime, non-Newtonian fluids the pressure derivative will have either an increasing slope (if the fluid is pseudoplastic) or a negative slope (if the fluid is dilatant). Therefore, the traditional method for interpreting pressure tests in such systems completely fails. Cases in which non-Newtonian fluids are dealt with include such completion and stimulation treatment fluids as polymer solutions, foams, drilling muds (which should not be considered as a reservoir fluid, since before testing one should clean the well to remove all drilling invasion fluids; however, it obeys the power law), etc., and some paraffinic oils and *heavy crude oils*. Non-Newtonian fluids are generally classified as time independent, time dependent and viscoelastic. Examples of the first classification are Bingham, pseudoplastic and dilatant fluids, which are commonly dealt by petroleum engineers. More information on the classification and behavior of non-Newtonian fluids can be found in the classic book by Bird, Stewart and Lightfoot (2007).

As seen in Figure 7.1, there is a special kind of non-Newtonian fluid, Bingham fluids (or plastics), which exhibit a finite yield stress at zero shear rates. There is no gross movement of fluids until the yield stress, τ_y, is exceeded. Once this is accomplished, it is also necessary to cut efforts to increase the shear rate; that is, they then behave as Newtonian fluids. These fluids behave as a straight line crossing the y axis in $\tau = \tau_y$, with the shear stress, τ, plotted against the shear rate, γ, in Cartesian coordinates. The characteristics of these fluids are defined by two constants: the yield, τ_y, which is the stress that must be exceeded for flow to begin, and the Bingham plastic coefficient, μ_B. The rheological equation for a Bingham plastic is:

$$\tau = \tau_y + \mu_B \gamma \tag{7.1}$$

The Bingham plastic concept has been found to approximate closely many real fluids existing in porous media, such as paraffinic oils, heavy oils, drilling muds and fracturing fluids, which are suspensions of finely divided solids in liquids. Laboratory investigations have indicated that the flow of heavy oil in some fields has non-Newtonian behavior and approaches the Bingham type.

Pseudoplastic and dilatant fluids have no yield point. The slope of shear stress versus shear rate decreases progressively and tends to become constant for high values of shear stress for pseudoplastic fluids. The simplest model is power law,

$$\tau = k\gamma^n; \quad n < 1 \qquad (7.2)$$

k and n are constants that differ for each particular fluid. k measures the flow consistency, and n measures the deviation from the Newtonian behavior (flow behavior index or power-law parameter), in which $k = \mu$ and $n = 1$. Dilatants fluids are similar to pseudoplastic fluids, except that the apparent viscosity increases as the shear stress increases. The power-law model also describes the behavior of dilatant fluids, but with $n > 1$.

Currently, unconventional reservoirs are the most impactful subject in the oil industry. Shale reservoirs, coalbed gas, tight gas, gas hydrates, gas storage, geothermal energy, coal conversion to gas, coal-to-gas, in-situ gasification and heavy oil are considered unconventional reservoirs. In the well-testing literature, several analytical and numerical models that take Bingham, pseudoplastic and dilatant non-Newtonian behavior into account have been introduced, so that their transient nature in porous media can be studied for better reservoir characterization. Most of them deal with fracture wells, homogeneous and double-porosity formations, and well-test interpretation is conducted via the straight-line conventional analysis or type-curve matching. Recently, some studies involving the pressure derivative have also been introduced. Escobar (2012) has compiled in a book chapter the most recent research in well-test interpretation of non-Newtonian fluids.

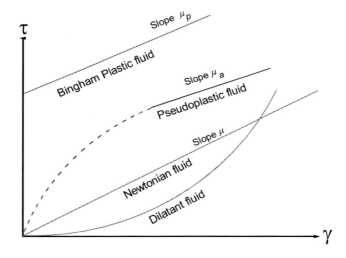

Figure 7.1. Schematic representation of time-independent fluid; adapted from Bird, Stewart and Lightfoot (2007).

CONVENTIONAL ANALYSIS

Dr. Chinyere Ukeagumo Ikoku has been the most outstanding researcher on the study of the behavior of non-Newtonian fluid in porous media. Their remarkable contributions include Ikoku (1978), Ikoku (1979), Ikoku and Ramey (1978), Ikoku and Ramey (1979) and Ikoku and Ramey (1980). Ikoku and his coauthors arrived at the following partial differential equation for the fluid flow of a non-Newtonian power-law fluid through porous material under the common assumptions always considered in well-test analysis:

$$\frac{\partial^2 P}{\partial r^2} + \frac{n}{r}\frac{\partial P}{\partial r} = G r^{1-n}\frac{\partial P}{\partial t} \tag{7.3}$$

G in oil-field units being:

$$G = \frac{3792.188 n \phi \mu_{eff}}{k}\left(96681.605\frac{h}{qB}\right)^{1-n} \tag{7.4}$$

The effective viscosity given in Equation (7.4) is defined in oil-field units by:

$$\mu_{eff} = \frac{H}{12}\left(9+\frac{3}{n}\right)^n \left(1.59344\times10^{-12}k\phi\right)^{(1-n)/2} \tag{7.5}$$

The dimensionless parameters are also given in oil-field units:

$$P_{DNN} = \frac{P-P_i}{141.2(96681.605)^{1-n}\left(\dfrac{qB}{h}\right)^n \dfrac{\mu_{eff}r_w^{1-n}}{k}} \tag{7.6}$$

$$r_D = \frac{r}{r_w} \tag{7.7}$$

$$t_{DNN} = \frac{t}{G r_w^{3-n}} \tag{7.8}$$

The dimensionless form of Equation (7.3) is, therefore:

$$\frac{\partial^2 P_{DNN}}{\partial r_D^{\,2}} + \frac{n}{r_D}\frac{\partial P_{DNN}}{\partial r_D} = r_D^{\,1-n}\frac{\partial P_{DNN}}{\partial t_{DNN}} \tag{7.9}$$

The well-pressure solution for the constant-rate case in an infinite reservoir was also provided by Ikoku (1978, 1979) as:

$$P_{DNN} = \left[\frac{(3-n)^{\frac{2(1-n)}{3-n}} t_{DNN}^{\frac{1-n}{3-n}}}{(1-n)\Gamma\left(\frac{2}{3-n}\right)} - \left(\frac{1}{1-n}\right) \right] \quad (7.10)$$

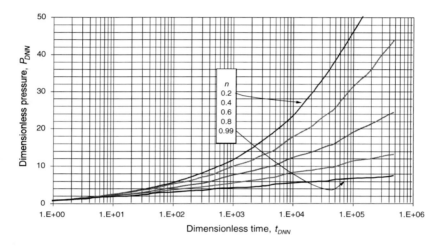

Figure 7.2. Dimensionless pressure versus dimensionless time semilog plot for a non-Newtonian fluid with different n values.

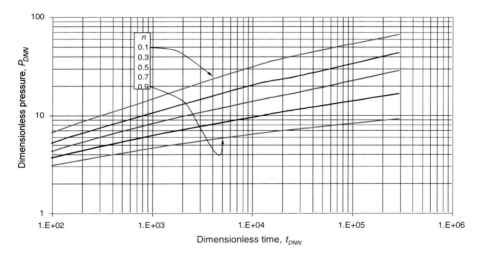

Figure 7.3. Dimensionless pressure versus dimensionless time log-log plot for a non-Newtonian fluid with different n values.

$$t_{DNN} \simeq 98200 n^{2.7} \qquad (7.11)$$

The solution provided by Equation (7.10) allowed Ikoku to generate the dimensionless semilog plot given in Figure 7.2. As observed in that plot, the only straight line behavior is given for $n \approx 1$, which corresponds to the Newtonian behavior, as expected. It is also observed for all values of n that ascending behavior is exhibited at early times, and that a linear tendency is exhibited at very late times. According to Ikoku and Ramey (1979), the starting of those linear behaviors is governed by the following:

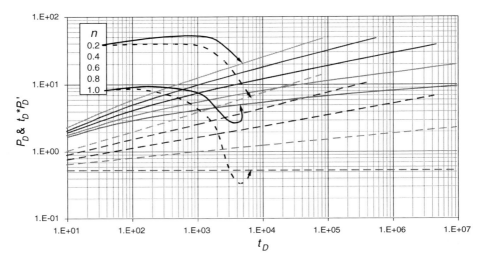

Figure 7.4. Dimensionless pressure and pressure derivative versus dimensionless time log-log plot for a non-Newtonian fluid with different n values.

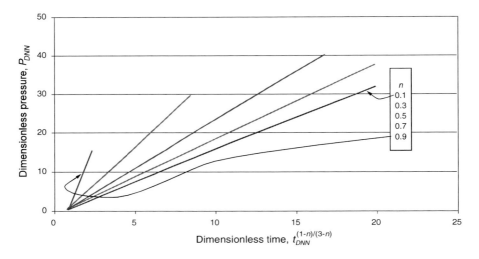

Figure 7.5. Dimensionless pressure versus dimensionless time to the power $(1-n)/(3-n)$ Cartesian plot for a non-Newtonian fluid with different n values.

A plot like the one given in Figure 7.3 may be used directly for a type-curve matching procedure. However, the slopes of the lines for different n values do not differ much. Therefore, the possibility of a skin factor suggests the use of alternative interpretation methods. Figure 7.4 presents the log-log behavior of both pressure and pressure derivative versus time without considering wellbore storage effects. As noted previously, a zero slope line is given for the Newtonian case, $n = 1$. For the non-Newtonian case ($n < 1$), both pressure and pressure derivative show an increasing slope, and the flow behavior index decreases.

For the application of conventional analysis, Ikoku and Ramey (1979) built a Cartesian plot of dimensionless pressure versus time to the power $(1-n)/(3-n)$. They show that a Cartesian plot of either ΔP or P_{wf} vs. $t^{(1-n)/(3-n)}$ gives a straight line which slope, m_{NN}, can be used to find mobility:

$$m_{NN} = \frac{\left(\dfrac{q}{2\pi h}\right)^{\frac{1+n}{3-n}} \left(\dfrac{\mu_{eff}}{k}\right)^{\frac{2}{3-n}}}{(1-n)\Gamma\left(\dfrac{2}{3-n}\right)\left[\dfrac{n\phi c_t}{(3-n)^2}\right]^{\frac{1-n}{3-n}}} \qquad (7.12)$$

From the intercept at $t = 0$ s, $\Delta P = \Delta P_o$, the skin factor can be found from:

$$s = \left(\frac{\Delta P_o}{r_w^{1-n}}\right)\left(\frac{2\pi h}{q}\right)^n \left(\frac{k_r}{\mu_{eff}}\right) + \left(\frac{1}{1-n}\right) \qquad (7.13)$$

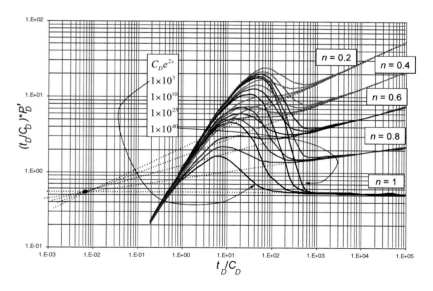

Figure 7.6. Pressure derivative for a pseudoplastic, non-Newtonian fluid in an infinite reservoir with wellbore storage and skin effects; adapted from Katime-Meindel and Tiab (2001).

Notice that Equations (7.12) and (7.13) are given in international units. The radius of investigation can be found from:

$$r_{inv} = \left[\Gamma\left(\frac{2}{3-n}\right)\right]^{\frac{1}{n-1}} \left[\frac{(3-n)^2 t}{G}\right]^{\frac{1}{3-n}} \qquad (7.14)$$

t is the time at which the non-Newtonian straight line ends. If the flow consistency, H, is known, the following expression can be used to find reservoir permeability:

$$k = \left(\frac{q}{2\pi h}\right)\left\{\frac{\left[(1-n)\Gamma\left(\frac{2}{3-n}\right)\right]^{n-3}\left[\frac{H}{12}\left(9+\frac{3}{n}\right)^n (150)^{\frac{1-n}{2}}\right]^2 (3-n)^{2(1-n)}}{(nc_t)^{1-n}(m_{NN})^{3-n}}\right\}^{\frac{1}{1+n}} \qquad (7.15)$$

TDS TECHNIQUE

Katime-Meindl and Tiab (2001) applied the pressure derivative to non-Newtonian fluids for the second time. They used the well-pressure behavior mathematical solution presented by Ikoku and Ramey (1979), which considers wellbore storage and skin effects:

$$\bar{P}_D(z) = \frac{K_v\left(\beta\sqrt{z}r_D^{1/\beta}\right) + s\sqrt{z}K_\beta\left(\beta\sqrt{z}\right)}{z\left(\sqrt{z}K_\beta\left(\beta\sqrt{z}\right) + zC_D\left[K_v\left(\beta\sqrt{z}\right) + s\sqrt{z}K_\beta\left(\beta\sqrt{z}\right)\right]\right)} \qquad (7.16)$$

where $\beta = 2/(3-n)$ and $v = (1-n)/(3-n)$.

Katime-Mendel and Tiab (2001) developed several expressions for pressure-test interpretation in formations with non-Newtonian fluids, based on observation of the pressure-derivative plot, Figure 7.6. As for the Newtonian case, the wellbore storage coefficient is calculated with a point during the early unit-slope line:

$$C = \frac{qt_N}{\Delta P_N} \qquad (7.17)$$

The coordinates of the maximum point can also be used to find the wellbore storage coefficient:

$$C = \frac{0.36qt_x}{(t\Delta P')_x - \beta \dfrac{\mu_{eff}}{k}\left(\dfrac{q}{2\pi h}\right)nr_w^{1-n}}$$ (7.18)

$$P_D' = \frac{2\pi n\phi c_t hr_w^2}{q}\Delta P'$$ (7.19)

Mobility can be found from the value of the pressure derivative, read at any arbitrary time, t_r, during the radial flow regime:

$$\frac{k}{\mu_{eff}} = \left[0.5 \frac{t_r^\alpha}{C^\alpha (t*\Delta P')_r} \frac{(2\pi h)^{n(\alpha-1)}}{q^{n(\alpha-1)-\alpha}} r_w^{(1-n)(1-\alpha)} \right]^{\frac{1}{1-\alpha}}$$ (7.20)

α is defined as follows:

$$\alpha = -0.1486n^2 - 0.178n + 0.3269$$ (7.21)

The coordinates of the intersection point between radial flow and the early unit-slope line can be used to find the mobility or the wellbore storage coefficient:

$$t_i = \frac{(3.13e^{-1.85n})C}{(2\pi h)^n}\frac{\mu_e}{k}\left(\frac{q}{r_w}\right)^{n-1}$$ (7.22)

$$(t*\Delta P')_i = \frac{\left(4.36e^{-2.14n}\right)}{r_w^{n-1}}\frac{\mu_e}{k}\left(\frac{q}{2\pi h}\right)^n$$ (7.23)

The coordinates of the maximum point during the transition from wellbore storage to the radial flow regime can be used to find other parameters, such as mobility, wellbore storage coefficient and skin factor:

$$\frac{k}{\mu_{eff}} = \frac{\beta\left(\dfrac{q}{2\pi h}\right)nr_w^{1-n}}{0.36\dfrac{qt_x}{C} - (t*\Delta P')_x}$$ (7.24)

$$s = \frac{1}{2}\left[5.27\Psi\frac{(t*\Delta P')_x}{(t*\Delta P')_i} - 2.303\delta - \ln(C_D) \right]$$ (7.25)

$$s = \frac{1}{2}\left[1.91\Gamma\frac{t_x}{t_i} - 2.303\delta - \ln(C_D)\right] \tag{7.26}$$

$$\Psi = 4.36e^{-2.14n} \tag{7.27}$$

$$\Gamma = 3.13e^{-1.85n} \tag{7.28}$$

If the unit-slope line is not defined, the skin factor can be calculated using the following equations:

$$s = \frac{1}{2}\left[5.27\left(\frac{2\pi h}{q}\right)^n \frac{k}{\mu_e} r_w^{n-1}(t*\Delta P')_x - 2.303\delta - \ln(C_D)\right] \tag{7.29}$$

$$s = \frac{1}{2}\left[1.91(2\pi h)^n \frac{k}{\mu_e}\left(\frac{r_w}{q}\right)^{n-1}\frac{t_x}{C} - 2.303\gamma - \ln(C_D)\right] \tag{7.30}$$

δ and γ, introduced in Equations (7.25) and (7.30), respectively, are defined by:

$$\delta = 25.341n^2 - 48.336n + 23.447 \tag{7.31}$$

$$\gamma = 42.268n^2 - 80.879n + 40.097 \tag{7.32}$$

Katime-Meindl and Tiab (2001) also found that the starting time of the radial flow regime is governed by:

$$t_{SR} = 9.5689\times10^4 n^{2.932}\frac{\mu_{eff}}{k}\frac{C}{(2\pi h)^n}\left(\frac{q}{r_w}\right)^{n-1} \tag{7.33}$$

They also found an expression to find the distance from the well to a linear boundary, respectively:

$$b_x = \frac{1}{2}\left[\frac{1}{3\times10^6 e^{6.7962n}}\frac{k}{\mu_{eff}}\frac{(2\pi h)^n q^{1-n}}{C}r_w^2 t_{erf}\right]^{\frac{1}{3-n}} \tag{7.34}$$

"*erf*" stands for "end of radial flow line." As a reminder, all of the above expressions are given in international units.

Escobar, Martinez and Montealegre (2010) found that the dimensionless pressure derivative during radial flow regime is governed by:

$$\log[t_D * P_D']_r = \alpha \log(t_{DNN})_r + \log 0.5 \tag{7.35}$$

or:

$$(t_D * P_D')_{rNN} = 0.5 t_{DNN}^{\alpha} \tag{7.36}$$

Taking the derivative of Equation (7.6) yields:

$$(t_D * P_D')_r = 0.00708(96681.605)^{n-1}\left(\frac{h}{qB}\right)^n \frac{k}{\mu_{eff}} r_w^{n-1}(t * \Delta P')_r \tag{7.37}$$

A combination of Equations (7.35) and (7.37) provides:

$$k = \left[70.6(96681.605)^{(1-\alpha)(1-n)}\left(\frac{0.0002637 t_r}{n\phi c_t}\right)^{\alpha}\left(\frac{qB}{h}\right)^{n-\alpha(n-1)}\left(\frac{1}{(t*\Delta P')_r}\right)\right]^{\frac{1}{1-\alpha}} \tag{7.38}$$

To find formation permeability directly, Escobar et al. (2010) substituted Equation (7.5) into Equation (7.38):

$$k = \left\{\left[\frac{\left[70.6(96681.605)^{(1-\alpha)(1-n)}\left(\frac{0.0002637 t_r}{n\phi c_t}\right)^{\alpha}\left(\frac{qB}{h}\right)^{n-\alpha(n-1)}\right]^{\frac{1}{1-\alpha}}}{\left[\frac{r_w^{\alpha(n-3)+(1-n)}}{(t*\Delta P')_r}\right]\left[\left(\frac{H}{12}\right)\left(9+\frac{3}{n}\right)^n\left(1.59344\times10^{-12}\phi\right)^{\left(\frac{1-n}{2}\right)}\right]}\right]\right\}^{\frac{2}{1+n}} \tag{7.39}$$

α can be found using Equation (7.21). However, is the slope of the pressure-derivative curve, and it is also defined by:

$$\alpha = \frac{1-n}{3-n} \tag{7.40}$$

Escobar et al. (2010) also found that during the non-Newtonian radial flow regime, the governing well-pressure expression is given by:

$$P_{DNNr} = 0.5\left[\ln t_{DNNr} + 0.80907 + 2s\right] \tag{7.41}$$

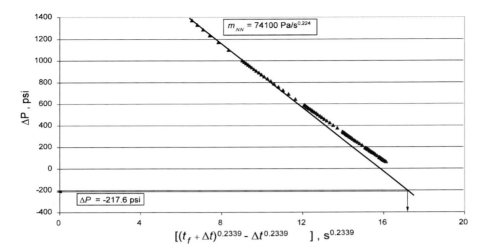

Figure 7.7. Cartesian plot of pressure ΔP versus $(t_f+\Delta t)^{0.2239}-\Delta t^{0.2239}$ for example 7.1.

Following the philosophy of the *TDS* technique, Tiab (1993); Escobar et al. (2010); and Martinez, Escobar and Cantillo (2011) found an expression to compute skin factor from the division of Equation (7.41) by Equation (7.35):

$$s = \frac{1}{2}\left[\frac{\left[\left[\dfrac{0.0002637 k t_r}{n\phi\mu_{eff}c_t r_w^{3-n}}\left(96681.605\dfrac{h}{qB}\right)^{n-1}\right]^{\alpha}\right]\left(\dfrac{\Delta P}{(t*\Delta P)}\right)_r -}{\ln\left(\dfrac{kt_r}{n\phi\mu_{eff}c_t r_w^{3-n}}\left(96681.605\dfrac{h}{qB}\right)^{n-1}\right)+7.43}\right] \quad (7.42)$$

Escobar, Martinez and Bonilla (2012a) found a much more practical expression to estimate the skin factor by dividing Equation (7.41) by Equation (7.36):

$$s = \frac{1}{2}\left(\frac{(\Delta P)_{rNN}}{(t*\Delta P')_{rNN}} - \frac{1}{\alpha}\right)\left(\frac{t_{rNN}}{G\, r_w^{3-n}}\right)^{\alpha} \quad (7.43)$$

Example 7.1

Escobar, Ascencio and Real (2013) used the example provided by Ikoku (1979) of a field case taken from a polymer demonstration enhanced oil recovery project. Actual pressure-versus-time data and their dimensionless quantities are reported in Table 7.1. Unit conversions are given in Table 7.2, and relevant information is given below:

$q = 138.1$ cm^3/s $k = 34.4$ md $\phi = 0.228$
$r_w = 2.41$ cm $\mu = 5$ cp $h = 518.2$ cm

Freddy Humberto Escobar

$n = 0.423$
$c_t = 7.567 \times 10^{-6}$ psi^{-1}

$H = 65$ cp*seg^{n-1}
$t_f = 82$ hr

$B = 1.0$ rb/STB

Escobar, Ascencio and Real found the reservoir permeability and skin factor for this injection test using both conventional and *TDS* techniques.

Solution by Conventional Analysis

Since n is equal 0.423, a Cartesian plot of pressure drop versus $(t_f+\Delta t)^{0.2239}-\Delta t^{0.2239}$ was built and reported in Figure 7.7.

After entering the value of $t_f = 82$ hr (295200 s) and 0 s into the time function, the following value is obtained:

$$(295200+0)^{0.2239}-0^{0.2239} = 16.78 \text{ s}^{0.2239}$$

Tabla 7.1. Pressure data for example 7.1; adapted from Ikoku (1979)

t, s	ΔP, psi	$(t^*\Delta P')$, psi	t, s	ΔP, psi	$(t^*\Delta P')$, psi
0.15	62.03	17.16	95	337.6	72.84
0.2	67.69	18.36	100	341.76	73.8
0.25	72.34	19.32	150	376.43	80.64
0.3	76.32	20.04	200	403.01	86.16
0.35	79.81	20.76	300	443.5	94.32
0.4	82.93	21.36	400	474.54	100.56
0.45	85.76	21.96	450	487.83	103.32
0.5	88.36	22.56	500	500.03	105.72
0.55	90.76	23.04	550	511.31	108
0.6	93	23.4	600	521.82	110.16
0.65	95.09	23.88	650	531.68	112.08
0.7	97.07	24.24	700	540.96	114
0.75	98.94	24.6	750	549.74	115.8
0.8	100.72	24.96	800	558.08	117.48
0.85	102.41	25.32	850	566.02	119.04
0.9	104.02	25.68	900	573.6	120.6
1	107.06	26.28	950	580.87	122.04
1.5	119.42	28.68	1000	587.85	123.6
2	128.9	30.72	1500	645.9	135
2.5	136.68	32.28	2000	690.41	144.24
3	143.34	33.6	2500	726.96	151.56
3.5	149.18	34.8	3000	758.21	157.92
4	154.4	35.88	3500	785.64	163.44
4.5	159.15	36.84	4000	810.19	168.48
5	163.5	37.68	4500	832.45	172.92
5.5	167.52	38.52	5000	852.87	177.12
6	171.27	39.24	5500	871.77	180.84
6.5	174.78	39.96	6000	889.37	184.44

Infinite Reservoirs

t, s	ΔP, psi	$(t*\Delta P')$, psi	t, s	ΔP, psi	$(t*\Delta P')$, psi
7	178.09	40.68	6500	905.87	187.8
7.5	181.22	41.28	7000	921.41	190.92
8	184.2	41.88	7500	936.11	193.92
8.5	187.03	42.48	8000	950.08	196.68
9	189.73	42.96	8500	963.37	199.44
9.5	192.32	43.56	9000	976.08	201.96
10	194.81	44.04	9500	988.25	204.36
20	231.39	51.48	10000	999.93	206.88
25	244.42	54	15000	1097.15	225.96
30	255.56	56.28	20000	1171.67	241.56
40	274.1	60.12	25000	1232.88	253.8
45	282.04	61.68	30000	1285.21	264.48
50	289.32	63.12	35000	1331.15	273.72
55	296.06	64.56	40000	1372.25	282
60	302.34	65.76	45000	1409.54	289.56
65	308.22	66.96	50000	1443.74	296.52
70	313.76	68.04	55000	1475.37	302.88
75	319.01	69.12	60000	1504.85	308.88
80	323.99	70.2	65000	1532.48	314.4
85	328.73	71.16	70000	1558.51	319.68
90	333.26	72	75000	1583.13	324.6

After entering the above value into Figure 7.1, an intercept ΔP_0 equals to -217.6 psi (-1.5×10^6) Pa is obtained. A slope m_{NN} of 74100 Pa/s$^{0.224}$ is also obtained. Since the effective viscosity is needed for the calculations, it is found with the MKS version of Equation (7.5):

$$\mu_{eff} = \frac{H}{12}(9+3/n)^n(150k\phi)^{(1-n)/2} \qquad (7.44)$$

$$u_{eff} = \frac{6.5 \times 10^{-02} \text{ Pa.seg}}{12}\left(9+\frac{3}{0.423}\right)^{0.423}\left(150\left(3.4 \times 10^{-14} \text{ m}^2\right) 0.228\right)^{\frac{(1-0.423)}{2}}$$

$$u_{eff} = 6.322 \times 10^{-06} \text{ Pa.s}^{0.423}.\text{m}^{0.577}$$

Permeability is then found with Equation (7.15) and skin factor with Equation (7.13):

$$k_r = \left(\frac{1.381 \times 10^{-04} \frac{\text{m}^3}{\text{s}}}{2\pi 5.182 \text{ m}}\right)\left\{\frac{\left[\left[(1\text{-}0.423)\Gamma\left(\frac{2}{3\text{-}0.423}\right)\right]^{0.423\text{-}3}\right]\left[\frac{6.5 \times 10^{-02} \text{ Pa.s}^{0.423}}{12}\left(9+\frac{3}{0.423}\right)^{0.423}(150)^{\frac{1\text{-}0.423}{2}}\right]^2 (3\text{-}0.423)^{2(1\text{-}0.423)}}{\left(0.423*1.0975 \times 10^{-09} \text{ Pa}^{-1}\right)^{1\text{-}0.423}\left(74100\frac{\text{Pa}}{\text{s}^{0.423}}\right)^{3\text{-}0.423}}\right\}^{\frac{1}{1+0.423}}$$

Freddy Humberto Escobar

Table 7.2. Unit conversion for example 7.1

Parameter	Lab units	International units
q	138.1 cm^3/seg	1.381x10^{-04} m^3/seg
h	518.2 cm	5.182 m
k	0.0344 D	3.4x10^{-14} m^2
μ	5 cp	0.005 Pa.seg
r_w	2.41 cm	0.0241 m
c_t	7.567x10^{-06} psi^{-1}	1.0975x10^{-09} Pa^{-1}
ΔP_0	-217.6 psi	-1.5x10^{+06} Pa

$$k = 3.4 \times 10^{-14} \text{ m}^2 \ (34.4 \text{ md})$$

$$s = \left(\frac{-1.5 \times 10^{06} \text{ Pa}}{0.0241 \text{ m}^{1-0.423}} \right) \left(\frac{2 \pi \ 5.182 \text{ m}}{1.381 \times 10^{-0.4} \text{m}^3 / \text{s}} \right)^{0.423} \left(\frac{3.4 \times 10^{-14} \text{ m}^2}{6.322 \times 10^{-06} \text{ Pa.s}^{0.423}.\text{m}^{0.577}} \right) + \left(\frac{1}{1-0.423} \right) = -9.8$$

Solution by TDS Technique

The following information is read at an arbitrary time of 3600 s (1 hr) from the pressure and pressure-derivative plot given in Figure 7.8:

$$t_{rNN} = 1 \text{ hr} \qquad \Delta P_{rNN} = 730 \text{ psi} \qquad (t^*\Delta P')_{rNN} = 185 \text{ psi}$$

Equation (7.40) is used to find the slope of the pressure-derivative curve:

$$\alpha = \frac{1-n}{3-n} = \frac{1-0.423}{3-0.423} = 0.2239$$

Equation (7.39) is used to find the reservoir permeability:

$$k = \left\{ \left[\frac{\left[70.6(96681.605)^{(1-0.423)(1-0.423)} \left(\frac{0.0002637*1 \text{ hr}}{0.423*0.228*7.567 \times 10^{-6} \text{ psi}^{-1}} \right)^{0.2239} \right]^{\frac{1}{1-0.2239}}}{\left(\frac{75 \frac{\text{bbl}}{\text{D}} *1.0 \frac{\text{rb}}{\text{STB}}}{17 \text{ ft}} \right)^{0.423-0.2239(0.423-1)} \left(\frac{0.0791 \text{ ft}^{0.2239(0.423-3)+(1-0.423)}}{185 \text{ psi}} \right)}{\left[\left(\frac{65 \text{ cp.s}^{0.423}}{12} \right) \left(9 + \frac{3}{0.423} \right)^{0.423} \left(1.59344 \times 10^{-12} *0.228 \right)^{\left(\frac{1-0.423}{2} \right)} \right]} \right\}^{\frac{2}{1+0.423}} = 39.01 \text{ md}$$

Equation (7.5) is used to find the effective viscosity:

$$\mu_{eff} = \left(\frac{65 \text{ cp.s}^{0.423}}{12} \right) \left(9 + \frac{3}{0.423} \right)^{0.423} \left[1.59344 \times 10^{-12} (39.01) \text{ md}(0.228) \right]^{(1-0.423)/2}$$

Infinite Reservoirs

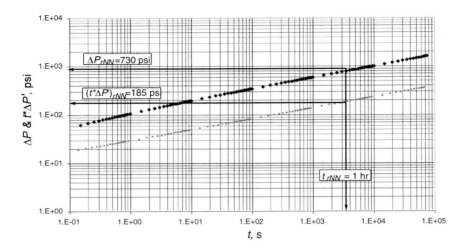

Figure 7.8. Pressure and pressure derivative versus time log-log plot for example 7.1.

$$\mu_{eff} = 0.012548 \text{ cp}\left(\frac{\text{s}}{\text{ft}}\right)^{0.423}$$

Equation (7.4) is used to find G:

$$G = \frac{3792.188(0.423)(0.228)(7.567\times10^{-6} \text{ psi}^{-1})(0.012548 \text{ cp.s}^{0.423}/\text{ft})}{39.01 \text{ md}}$$

$$\left(96681.605 \frac{17 \text{ ft}}{75\frac{\text{STB}}{\text{D}}*1.0\frac{\text{rb}}{\text{STB}}}\right)^{1-0.423} = 3.22\times10^{-04} \frac{\text{hr}}{\text{ft}^{3-n}}$$

Finally, Equation (7.43) is used to estimate skin factor:

$$s = \frac{1}{2}\left(\frac{730 \text{ psi}}{185 \text{ psi}} - \frac{1}{0.2239}\right)\left(\frac{1 \text{ hr}}{3.22\times10^{-04}\frac{\text{hr}}{\text{ft}^{3-0.423}} 0.0791 \text{ ft}^{3-0.423}}\right)^{0.2239} = -11.84$$

The results from both methods are reported and compared in Table 7.3.

Table 7.3. Comparison of results for example 7.1

Parameter	Conventional Analysis	*TDS* Technique
k	34.4 mD	39.01 mD
s	-9.8	-11.84

SPHERICAL FLOW

Heavy oil reservoirs are now considered some of the most important unconventional reservoirs in the oil industry. Some of them display non-Newtonian, pseudoplastic behavior, for which mathematical modeling differs from the conventional case, and, therefore, the flow regimes display some particular behaviors. Fracturing fluids, foams, some fluids for enhanced oil recovery (EOR) and drilling muds can also fall into this category. The spherical/hemispherical flow mainly caused by partial completion/penetration deserves particular attention for pseudoplastic flow. The only study of this case was presented by Ci-qun (1988) to introduce a mathematical model. No interpretation technique was provided.

Escobar et al. (2012a) used the model presented by Ci-qun (1988) to analyze the pressure behavior of transient spherical pseudoplastic. It is observed that the pressure derivative changes its slope from -½ (Newtonian case) to about 0.22 as the flow index varies from 1 to 0.1. For $n = 0.5$, the pressure derivative becomes flat, similarly to the radial Newtonian flow regime. Escobar et al. (2012a) extended both the straight-line conventional method and the *TDS* technique for well-test interpretation. Their new equations were applied to simulated cases, providing very low deviation errors with respect to the input simulation values. Since only two works are presented on this issue, no field examples have been reported yet. However, partial completion/penetration is everywhere, and the applicability of this work is valuable. In fact, some wells with sand deposition easily develop hemispherical flow. If a well under these conditions is used for the injection of foams, fracturing fluids or tertiary recovery fluids and is then tested, the tools for interpretation were reported by Escobar et al. (2012a) and are given here. This section is also very useful for understanding the pressure and pressure-derivative behavior of hemispherical/spherical non-Newtonian flow, which has a unique behavior. As observed in Figure 7.9, the pressure derivative displays a slope of -½ for the Newtonian case ($n = 1$), as expected. As the value of coefficient n decreases gradually to 0.1, the slope of the pressure derivative also increases gradually to a value of 0.22. Interestingly, at an n value of 0.5, the pressure derivative becomes flat, resembling the radial flow regime of a Newtonian fluid.

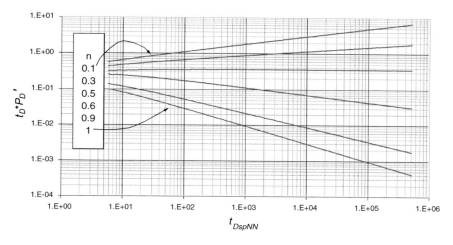

Figure 7.9. Dimensionless pressure derivative versus time log-log behavior for a non-Newtonian fluid under spherical flow conditions with different values of n; adapted from Escobar et al. (2012a).

Infinite Reservoirs

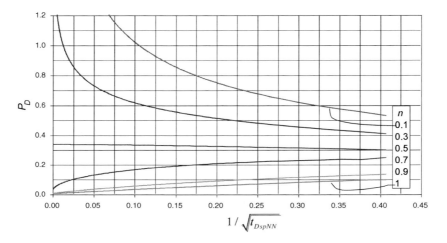

Figure 7.10. Dimensionless Cartesian plot of pressure against the inverse-square-root of time; adapted from Escobar et al. (2012a).

The mathematical solution to the diffusivity equation for a spherical wellbore in an infinite reservoir was presented by Ci-qun (1988), for $n < 0.5$, $n = 0.5$ and $0.5 < n \leq 1$, each of which is given below:

$$\overline{P}_D(z) = K_{\frac{1-2n}{4-2n}}\left(\frac{\sqrt{z}}{2-n}\right) \bigg/ z^{\frac{3}{2}} K_{\frac{3}{4-2n}}\left(\frac{\sqrt{z}}{2-n}\right) \quad (7.45)$$

$$\overline{P}_D(z) = K_0\left(\frac{2\sqrt{z}}{3}\right) \bigg/ z^{\frac{3}{2}} K_1\left(\frac{2\sqrt{z}}{3}\right) \quad (7.46)$$

$$\overline{P}_D(z) = \frac{K_{n-0.5}(\sqrt{z})}{z^{\frac{3}{2}} K_{n+0.5}(\sqrt{z})} \quad (7.47)$$

Redefining Equations (7.4) to (7.8) for spherical flow conditions:

$$P_{DspNN} = \frac{\Delta P}{70.6(96681.605)^{1-n}(qB)^n \dfrac{\mu_{eff} r_{sw}^{1-2n}}{k_{sp}}} \quad (7.48)$$

$$(t_D * P_D')_{spNN} = \frac{(t*\Delta P')_{spNN}}{70.6(96681.605)^{1-n}(qB)^n \dfrac{\mu_{eff} r_{sw}^{1-2n}}{k_{sp}}} \quad (7.49)$$

$$(t_D * P_D')_{spNN} = \frac{(t * \Delta P')_{spNN}}{70.6(96681.605)^{1-n}(qB)^n \dfrac{\mu_{eff} r_{sw}^{1-2n}}{k_{sp}}} \tag{7.49}$$

$$t_{DspNN} = \frac{t}{G_1 r_{sw}^{3-n}} \tag{7.50}$$

where

$$G_1 = \frac{3792.188 n\phi\mu_{eff}c_t}{k_{sp}}\left(96681.605\frac{r_{sw}}{qB}\right)^{1-n} \tag{7.51}$$

CONVENTIONAL ANALYSIS FOR NON-NEWTONIAN SPHERICAL FLOW

In gas or crude oils displaying Newtonian behavior, the pressure derivative is recognized by a slope of -½ during hemispherical, spherical or parabolic behavior, according to Escobar et al. (2005). Therefore, this flow regime is represented by a Cartesian behavior of pressure versus either the inverse of the square root of time for drawdown or $1/(t_p+\Delta t)^{0.5} - \Delta t^{0.5}$ for buildup cases. Such a plot exhibits straight-line behavior, as observed in Figure 7.10 for the case of $n = 1$, but as can be inferred from Figure 7.10, that fails to reproduce a straight line for $n < 1$. To overcome this situation, Escobar et al. (2012a) provided a solution by plotting pressure versus t^β, β being the slope of the Cartesian plot, which depends upon the value of n and is given by:

Table 7.4. Constant values for Equation (7.51)

Constant	Value	Constant	Value
a	0.42935158	f	7053.57725
b	1116.18599	g	9437.07854
c	225.272219	h	25686.2003
d	-6572.71512	i	-8348.94782
e	-2517.29978	j	4322.10228
		k	-14599.4614

Table 7.5. Constant values for Equation (7.52)

Constant	$n > 0.5$	$n < 0.5$
a	0.30696567	0.35144873
b	1.93824052	-0.36832669
c	-3.69373734	-0.14199263
d	1.44854403	0.13283018

$$\beta = \frac{a + cn^2 + en^4 + gn^6 + in^8 + kn^{10}}{1 + bn^2 + dn^4 + fn^6 + hn^8 + jn^{10}} \tag{7.52}$$

This expression has a correlation coefficient of $R^2 = 0.9999998$. The values of the coefficients are given in Table 7.4. The governing pressure equation needs a correction factor, FC, given by:

$$FC = e^{a + bn + cn^2 + dn^3} \tag{7.52}$$

for which the constants are given in Table 7.5. The dimensionless pressure for spherical flow in a non-Newtonian fluid is expressed according to the value of n:

For $n \neq 0.5$ and $n \leq 1$:

$$P_{DspNN} = 1 + \frac{FC}{2\beta\sqrt{\pi}} t_{DspNN}^{\beta} + s_{spNN} \tag{7.53}$$

Entering the dimensionless quantities given by Equations (7.48) and (7.50) into Equation (7.53) yields:

$$P_{wf} = P_i - 70.6 \left[1 + \frac{FC}{2\beta\sqrt{\pi}} \left(\frac{t}{G_t r_{sw}^{3-n}} \right)^{\beta} + s_{spNN} \right] (96681.605)^{1-n} (qB)^n \frac{\mu_{eff} r_{sw}^{1-2n}}{k_{sp}} \tag{7.54}$$

As indicated before, notice that Equation (7.54) suggests that a Cartesian plot of either P_{wf} versus t^{β} for drawdown or P_{ws} versus $(t_p + \Delta t)^{\beta} - \Delta t^{\beta}$ for buildup gives a linear trend, which slope and intercept allow for the estimation of the spherical permeability and spherical skin factor, such as:

$$m_{sp} = -19.9158 \frac{FC}{\beta k_{sp}^{1-\beta}} \left(\frac{[96681.605 / (qB)]^{n-1}}{3792.188 n\phi\mu_{eff} c_t r_{sw}^{4-2n}} \right)^{\beta} (96681.605)^{1-n} (qB)^n \mu_{eff} r_{sw}^{1-2n} \tag{7.55}$$

where

$$k_{sp} = \left[-19.9158 \frac{FC}{m_{sp}\beta} \left(\frac{[96681.605 / (qB)]^{n-1}}{3792.188 n\phi\mu_{eff} c_t r_{sw}^{4-2n}} \right)^{\beta} (96681.605)^{1-n} (qB)^n \mu_{eff} r_{sw}^{1-2n} \right]^{\frac{1}{1-\beta}} \tag{7.56}$$

It must be taken into account that if $\beta > 0.5$, the slope is taken as positive, and if $\beta < 0.5$, the slope is taken as negative. This behavior is clearly seen in Figure 7.10, in which the curves have either a negative or a positive tendency, according to the n value.

$$S_{spNN} = (96681.605)^{n-1} \left(\frac{1}{qB}\right)^n \frac{k_{sp}\left(P_i - P_{wf(0hr)}\right)}{70.6\mu_{eff}r_{sw}^{1-2n}} - 1 \qquad (7.57)$$

For $n = 0.5$, the governing equation is given by:

$$P_{DspNN} = 0.3333\left[\ln\left(t_{DspNN}\right) + (1 + s_{spNN})\right] \qquad (7.58)$$

Again, by entering the dimensionless quantities given by Equations (7.48) and (7.50) into Equation (7.58), we obtain:

$$P_i - P_{wf} = 54.2(96681.605)^{0.5}(qB)^{0.5}\frac{\mu_{eff}}{k_{sp}}\left[\log\left(\frac{t}{G_1 r_{sw}^{2.5}}\right) + (1 + s_{spNN})\right] \qquad (7.59)$$

Equation (7.59) suggests that a semilog plot of P_{wf} vs. t^β or P_{wf} vs. $(t_p + \Delta t)^\beta - \Delta t^\beta$ gives a linear trend, which slope and intercept allow for the estimation of the spherical permeability and spherical skin factor, such as:

$$m_{sp} = 54.2(96681.605)^{1-n}(qB)^n\frac{\mu_{eff}}{k_{sp}} \qquad (7.60)$$

where

$$k_{sp} = 54.2(96681.605)^{0.5}(qB)^{0.5}\frac{\mu_{eff}}{m_{sp}} \qquad (7.61)$$

and

$$s_{spNN} = \left[\frac{P_i - P_{wf(1hr)}}{m_{sp}} - \log\left(\frac{(96681.605/qB)^{-0.5}}{3792.188n\phi\mu_{eff}c_t r_{sw}^3}\right)\right] - 1 \qquad (7.62)$$

TDS TECHNIQUE FOR NON-NEWTONIAN SPHERICAL FLOW

Escobar et al. (2012a) found several characteristic behaviors from the pressure derivative log-log plot, Figure 7.9, which are outlined below.

The dimensionless pressure derivative for spherical flow in a non-Newtonian fluid is given according to the value of n. For $n \neq 0.5$ and $n \leq 1$, the derivative of Equation (7.53) is:

$$(t_D * P_D')_{spNN} = \frac{FC}{2\sqrt{\pi}} t_{DspNN}^{\beta} \tag{7.63}$$

Entering the dimensionless quantities given by Equations (7.48) and (7.50) into Equation (7.63) yields an expression to find the spherical (3D) permeability by reading the pressure-derivative value, $(t*\Delta P')_{sp}$ at any arbitrary point, t_{sp}, during spherical non-Newtonian flow behavior.

$$k_{sp} = \left[\frac{19.9158 FC}{(t*\Delta P')_{spNN}} (96681.605)^{1-n} (qB)^n \mu_{eff} r_{sw}^{1-2n} \left(\frac{t_{sp} \left(96681.605 \frac{1}{qB} \right)^{n-1}}{3792.188 n \phi \mu_{eff} c_t r_{sw}^{4-2n}} \right)^{\beta} \right]^{\frac{1}{1-\beta}} \tag{7.64}$$

For $n = 0.5$, the natural log derivative of Equation (7.58) is a constant, as observed in Figure 7.9:

$$\left(t_D * P_D' \right)_{spNN} = 0.3333 \tag{7.65}$$

Once the dimensionless quantities are replaced in the above expression, the spherical permeability can be found from:

$$k_{sp} = 23.53 (96681.605)^{0.5} (qB)^{0.5} \frac{\mu_{eff}}{(t*\Delta P')_{spNN}} \tag{7.66}$$

The spherical skin factor is obtained by dividing the dimensionless pressure equation by its dimensionless pressure derivative; therefore, for $n \neq 0.5$ and $n \leq 1$:

$$s_{spNN} = \frac{FC}{2\sqrt{\pi}} \left(\frac{0.0002637 k_{sp} t_{sp}}{n \phi \mu_{eff} c_t r_{sw}^{4-2n}} \left(96681.605 \frac{1}{qB} \right)^{n-1} \right)^{\beta} \left[\frac{(\Delta P)_{spNN}}{(t*\Delta P')_{spNN}} - \frac{1}{\beta} \right] - 1 \tag{7.67}$$

For $n = 0.5$, the skin factor equation is:

$$s_{spNN} = \left(\frac{\Delta P}{t*\Delta P'} \right)_{spNN} - \left[\ln \left(\frac{k_{sp} t_{sp}}{n \phi \mu_{eff} c_t r_{sw}^2} \right) + 7.24 \right] \tag{7.68}$$

Since the equations for the pressure and the pressure derivative—Equations (7.53), (7.59), (7.63) and (7.65)—are given in spherical symmetry, it is necessary to transform them to a radial symmetry. For this procedure, consider the dimensionless variables of radial symmetry—Equations (7.6), (7.7) and (7.8)—and spherical symmetry—Equations (7.48), (7.49) and (7.50). Combining Equations (7.48) and (7.6) yields:

$$P_{DspNN} = 2\left(\frac{r_{sw}}{h}\right)^{n} \frac{k_{sp}}{k} \left(\frac{r_{w}}{r_{sw}}\right)^{1-n} P_{DrNN} \qquad (7.69)$$

Combining Equations (7.49) and (7.7) yields:

$$(t_D * P_D')_{spNN} = 2\left(\frac{r_{sw}}{h}\right)^{n} \frac{k_{sp}}{k} \left(\frac{r_{w}}{r_{sw}}\right)^{1-n} (t_D * P_D')_{rNN} \qquad (7.70)$$

Combining Equations (7.50) and (7.8) yields:

$$t_{DspNN} = \left(\frac{r_{w}}{r_{sw}}\right)^{3-n} \frac{k_{sp}}{k} \left(\frac{h}{r_{sw}}\right)^{1-n} t_{rDNN} \qquad (7.71)$$

The equation for the dimensionless pressure of the spherical flow non-Newtonian in radial symmetry is given by the insertion of Equations (7.69) and (7.71) into Equation (7.53) and of Equation (7.70) into Equation (7.65).

For $n \neq 0.5$ and $n \leq 1$:

$$P_{DrNN} = \frac{1}{2}\left(\frac{h}{r_{sw}}\right)^{n} \frac{k}{k_{sp}} \left(\frac{r_{w}}{r_{sw}}\right)^{n-1} (1+s_{spNN}) + \frac{FC}{4\beta\sqrt{\pi}} t_{DrNN}^{\beta} \left(\frac{h}{r_{sw}}\right)^{\beta(1-n)+n} \left(\frac{k_{sp}}{k}\right)^{\beta-1} \left(\frac{r_{w}}{r_{sw}}\right)^{\beta(3-n)+n-1} \qquad (7.72)$$

For $n = 0.5$:

$$P_{DrNN} = 0.16665 \frac{k}{k_{sp}}\left(\frac{h}{r_{w}}\right)^{0.5} \left[\ln\left(\left(\frac{r_{w}}{r_{sw}}\right)^{2.5} \frac{k_{sp}}{k} \left(\frac{h}{r_{sw}}\right)^{0.5} t_{rDNN}\right) + (1+s_{spNN}) \right] \qquad (7.73)$$

The equation for the pressure derivative of the spherical flow non-Newtonian in radial symmetry is given by substitution of Equations (7.70) and (7.71) in Equation (7.63) and of Equation (7.70) in Equation (7.65).

For $n \neq 0.5$ and $n \leq 1$:

$$(t_D * P_D')_{rNN} = \frac{1}{4} t_{DrNN}^{\beta} \frac{FC}{\sqrt{\pi}}\left(\frac{h}{r_{sw}}\right)^{\beta(1-n)+n} \left(\frac{r_{w}}{r_{sw}}\right)^{\beta(3-n)+n-1} \left(\frac{k_{sp}}{k}\right)^{\beta-1} \qquad (7.74)$$

For $n = 0.5$:

$$(t_D * P_D')_{rNN} = 0.16665 \frac{k}{k_{sp}} \left(\frac{h}{r_w} \right)^{0.5} \tag{7.75}$$

Escobar et al. (2012a) presented a new dimensionless pressure and pressure derivative for radial flow in a non-Newtonian fluid, not valid for $n = 1$:

$$P_{DrNN} = \frac{1}{2\alpha} t_{DrNN}^{\alpha} + s_{rNN} \tag{7.76}$$

$$(t_D * P_D')_{irNN} = 0.5 t_{DrNN}^{\alpha} \tag{7.36}$$

By replacing the dimensionless quantities given by Equations (7.7) and (7.8) in Equation (7.36) the radial permeability can be obtained using the pressure derivative, read at any arbitrary time during the radial flow regime:

$$k = \left\{ \left[\begin{array}{c} 70.6 (96681.605)^{(1-\alpha)(1-n)} \left(\dfrac{0.0002637 t_r}{n\phi c_t} \right)^{\alpha} \left(\dfrac{qB}{h} \right)^{n-\alpha(n-1)} \\ \left(\dfrac{r_w^{\alpha(n-3)+(1-n)}}{(t*\Delta P')_r} \right) \left[\left(\dfrac{H}{12} \right) \left(9 + \dfrac{3}{n} \right)^n \left(1.59344 \times 10^{-12} \phi \right)^{\left(\frac{1-n}{2} \right)} \right] \end{array} \right]^{\frac{1}{1-\alpha}} \right\}^{\frac{2}{1+n}} \tag{7.39}$$

Dividing Equation (7.76) by Equation (7.36) and solving for skin factor yields:

$$s = \frac{1}{2} \left(\frac{(\Delta P)_{rNN}}{(t*\Delta P')_{rNN}} - \frac{1}{\alpha} \right) \left(\frac{t_{rNN}}{G \, r_w^{3-n}} \right)^{\alpha} \tag{7.43}$$

which is not valid for $n = 1$.

The line corresponding to the spherical flow derivative, Equation (7.74), and the late radial flow line of the dimensionless pressure derivative in radial symmetry, Equation (7.36), intersect each other at:

$$\frac{1}{4} t_{DrNN}^{\beta} \frac{FC}{\sqrt{\pi}} \left(\frac{h}{r_{sw}} \right)^{\beta(1-n)+n} \left(\frac{r_w}{r_{sw}} \right)^{\beta(3-n)+n-1} \left(\frac{k_{sp}}{k_r} \right)^{\beta-1} = 0.5 t_{DrNNi}^{\alpha} \tag{7.77}$$

which is valid for $n \neq 0.5$ and $n \leq 1$. By replacing the dimensionless parameters and solving for the spherical permeability, we obtain:

$$k_{sp} = k \left[\cfrac{t_i}{G \, r_w^{3-n} \left[\cfrac{1}{2} \cfrac{FC}{\sqrt{\pi}} \left(\cfrac{h}{r_{sw}} \right)^{\beta(1-n)+n} \left(\cfrac{r_w}{r_{sw}} \right)^{\beta(3-n)+n-1} \right]^{\frac{1}{\alpha-\beta}}} \right]^{\frac{\alpha-\beta}{\beta-1}} \qquad (7.78)$$

When $n = 0.5$, the intercept of Equation (7.75) will be equal to Equation (7.36):

$$0.16665 \frac{k}{k_{sp}} \left(\frac{h}{r_w} \right)^{0.5} = 0.5 t_{DrNNi}^{0.2} \qquad (7.79)$$

Again, replacing the dimensionless time and solving for time of intersection and also solving for spherical permeability yields:

$$k_{sp} = 0.3333k \left[\frac{G h}{t_i} \right]^{0.2} \qquad (7.80)$$

Finally, the equivalent wellbore radius is found using an expression introduced by Chatas (1966):

$$r_{sw} = \frac{h_p}{2 \ln \left(\cfrac{h_p}{r_w} \right)} \qquad (7.81)$$

Both spherical permeability and radial permeability are used to find the vertical permeability:

$$k_z = \frac{k_{sp}^3}{k^2} \qquad (7.82)$$

Moncada et al. (2005) presented a complete study for characterization pressure tests in Newtonian oil and gas reservoirs using the *TDS* technique.

Infinite Reservoirs

NON-NEWTONIAN HEMISPHERICAL FLOW

For this case, the dimensionless quantities defined by Escobar et al. (2012a) are given below:

$$P_{DhsNN} = \frac{\Delta P}{141.2(96681.605)^{1-n}(qB)^n \dfrac{\mu_{eff} r_{sw}^{1-2n}}{k_{hs}}} \tag{7.83}$$

$$(t_D * P_D')_{hsNN} = \frac{(t*\Delta P')_{hsNN}}{141.2(96681.605)^{1-n}(qB)^n \dfrac{\mu_{eff} r_{sw}^{1-2n}}{k_{hs}}} \tag{7.84}$$

$$t_{DhsNN} = \frac{t}{G_1 r_{sw}^{3-n}} \tag{7.85}$$

For $n \neq 0.5$ and $n \leq 1$, entering the dimensionless quantities given by Equations (7.83) and (7.85) into Equation (7.53) and solving for k_{hs} and skin factor yields:

$$k_{hs} = \left[-39.8316 \frac{FC}{m_{hs}\beta} \left(\frac{\left(96681.605 \dfrac{1}{qB}\right)^{n-1}}{3792.188 n\phi\mu_{eff}c_t r_{sw}^{4-2n}} \right)^{\beta} (96681.605)^{1-n}(qB)^n \mu_{eff} r_{sw}^{1-2n} \right]^{\frac{1}{1-\beta}} \tag{7.86}$$

$$s_{hsNN} = (96681.605)^{n-1} \left(\frac{1}{qB} \right)^n \frac{k_{hs}\left(P_i - P_{wf(0hr)}\right)}{141.2\mu_{eff} r_{sw}^{1-2n}} - 1 \tag{7.87}$$

Again, if $\beta > 0.5$, the slope is taken as positive. If $\beta < 0.5$, the slope is taken as negative. For $n = 0.5$, entering the dimensionless quantity given by equations (7.83) and (7.85) into Equation (7.58) and solving for k_{hs} yields:

$$k_{hs} = 108.4(96681.605)^{0.5}(qB)^{0.5}\frac{\mu_{eff}}{m_{hs}} \tag{7.88}$$

For $n \neq 0.5$ and $n \leq 1$, entering the dimensionless quantities given by Equations (7.84) and (7.85) into Equation (7.63) and solving for k_{hs} yields:

$$k_{hs} = \left[\frac{39.8316FC}{(t * \Delta P')_{hsNN}} (96681.605)^{1-n} (qB)^n \mu_{eff} r_{sw}^{1-2n} \left(\frac{t_{sp} \left(96681.605 \frac{1}{qB} \right)^{n-1}}{3792.188 n\phi \mu_{eff} c_t r_{sw}^{4-2n}} \right)^{\beta} \right]^{\frac{1}{1-\beta}}$$ (7.89)

For $n = 0.5$, entering the dimensionless quantity given by Equation (7.84) into Equation (7.65) and solving for k_{hs} yields:

$$k_{hs} = 47.06(96681.605)^{0.5} (qB)^{0.5} \frac{\mu_{eff}}{(t * \Delta P')_{hsNN}}$$ (7.90)

Again, the equations for the pressure and the pressure derivative—Equations (7.53), (7.58), (7.63) and (7.65)—are given in hemispherical symmetry, so it is necessary to transform them to a radial symmetry. For this purpose, the equations in hemispherical geometry, Equations (7.83), (7.84) and (7.85), are used with the dimensionless variables from Equations (7.6), (7.7) and (7.8) in radial symmetry. Combining Equations (7.6) and (7.83) yields:

$$P_{DhsNN} = \left(\frac{r_{sw}}{h} \right)^n \frac{k_{hs}}{k} \left(\frac{r_w}{r_{sw}} \right)^{1-n} P_{DrNN}$$ (7.91)

Combining Equation (7.7) with Equation (7.84) and Equation (7.8) with Equation (7.85) yields:

$$(t_D * P_D')_{hsNN} = \left(\frac{r_{sw}}{h} \right)^n \frac{k_{hs}}{k} \left(\frac{r_w}{r_{sw}} \right)^{1-n} (t_D * P_D')_{rNN}$$ (7.92)

$$t_{DhsNN} = \left(\frac{r_w}{r_{sw}} \right)^{3-n} \frac{k_{hs}}{k} \left(\frac{h}{r_{sw}} \right)^{1-n} t_{rDNN}$$ (7.93)

Thus, the equations of the dimensionless pressure and pressure derivative for hemispherical non-Newtonian flow in radial symmetry for $n \neq 0.5$ and $n \leq 1$ are:

$$P_{DrNN} = \left(\frac{h}{r_{sw}} \right)^n \frac{k}{k_{hs}} \left(\frac{r_w}{r_{sw}} \right)^{n-1} (1 + s_{hsNN}) + \frac{FC}{2\beta\sqrt{\pi}} t_{DrNN}^{\beta} \left(\frac{h}{r_{sw}} \right)^{\beta(1-n)+n} \left(\frac{k_{hs}}{k} \right)^{\beta-1} \left(\frac{r_w}{r_{sw}} \right)^{\beta(3-n)+n-1}$$ (7.94)

$$(t_D * P_D')_{rNN} = \frac{1}{2} t_{DrNN}^{\beta} \frac{FC}{\sqrt{\pi}} \left(\frac{h}{r_{sw}} \right)^{\beta(1-n)+n} \left(\frac{r_w}{r_{sw}} \right)^{\beta(3-n)+n-1} \left(\frac{k_{hs}}{k_r} \right)^{\beta-1}$$ (7.95)

Infinite Reservoirs

For $n = 0.5$:

$$P_{DrNN} = 0.3333 \frac{k}{k_{hs}} \left(\frac{h}{r_w} \right)^{0.5} \left[\ln \left(\left(\frac{r_w}{r_{sw}} \right)^{2.5} \frac{k_{hs}}{k} \left(\frac{h}{r_{sw}} \right)^{0.5} t_{rDNN} \right) + (1 + s_{hsNN}) \right] \quad (7.96)$$

$$(t_D * P_D')_{rNN} = 0.3333 \frac{k}{k_{hs}} \left(\frac{h}{r_w} \right)^{05} \quad (7.97)$$

Solving for the hemispherical permeability with the time of intersection between hemispherical flow and radial flow, for $n \neq 0.5$ y $n \leq 1$, yields:

$$k_{hs} = k \left[\frac{t_i}{G_2 r_w^{3-n} \left[\frac{FC}{\sqrt{\pi}} \left(\frac{h}{r_{sw}} \right)^{\beta(1-n)+n} \left(\frac{r_w}{r_{sw}} \right)^{\beta(3-n)+n-1} \right]^{\frac{1}{\alpha-\beta}}} \right]^{\frac{\alpha-\beta}{\beta-1}} \quad (7.98)$$

For $n = 0.5$:

$$k_{hs} = 0.6666 k \left[\frac{G_2 h}{t_i} \right]^{0.2} \quad (7.99)$$

Example 7.2

It is necessary to use both the *TDS* technique and conventional analysis to estimate spherical permeability, areal permeability and skin factor for a synthetic pressure injection test, the data for which are provided in Figures 7.11 and 7.12. The input data for the simulation are given below:

$q = 300$ BPD	$k = 25$ md	$\phi = 0.2$
$r_w = 0.3$ ft	$h = 100$ cm	$n = 0.6$
$H = 2$ cp*s^{n-1}	$B = 1.2$ rb/STB	$c_t = 1\text{x}10^{-5}$ psi^{-1}
$k_{sp} = 15$ md	$h_p = 13$ ft	

Solution by TDS Technique
The following information is read from Figure 7.11.

$t_{spNN} = 0.02$ hr $\Delta P_{spNN} = 169$ psi $(t*\Delta P')_{spNN} = 17.8$ psi

$t_{rNN} = 10$ hr $\Delta P_{rNN} = 271.2$ psi $(t*\Delta P')_{rNN} = 22.03$ psi

$t_i = 0.44$ hr

First, α is evaluated and the non-Newtonian effective fluid permeability is estimated with Equation (7.40).

$$\alpha = \frac{1-n}{3-n} = \frac{1-0.6}{3-0.6} = 0.16667$$

Next, radial permeability is found with Equation (7.39);

$$k = \left(\frac{\left[\left[70.6(96681.605)^{(1-0.1667)(1-0.6)} \left(\frac{0.0002637(10)}{0.6(0.2)(1\times10^{-5})} \right)^{0.1667} \right]^{\frac{1}{1-0.1667}} \right]}{\left(\frac{300*1.2}{100} \right)^{0.6-0.1667(0.6-1)} \left(\frac{1}{22.03} \right)} \right)^{\frac{2}{1+0.6}} = 25.48 \text{ md}$$

The effective viscosity and the G parameter are found with Equations (7.5) and (7.4), respectively:

$$\mu_{eff} = \left(\frac{2}{12} \right) \left(9 + \frac{3}{0.6} \right)^{0.6} \left(1.59344\times10^{-12}(25.48)(0.2) \right)^{(1-0.6)/2} = 0.00491 \text{ cp*(s/ft)}^{n-1}$$

$$G = \frac{3792.188(0.6)(0.2)(1\times10^{-5})(0.00491)}{25.48} \left(96681.605\frac{100}{300(1.2)} \right)^{1-0.6} = 5.183\times0^{-5} \text{hr}/\text{ft}^{2.4}$$

The skin factor is estimated with Equation (7.43):

$$S_{rNN} = \frac{1}{2} \left(\frac{271.12}{22.03} - \frac{1}{0.1667} \right) \left(\frac{10}{(5.183\times10^{-5})(0.3)^{3-0.6}} \right)^{0.1667} = 38.83$$

Equation (7.52) is used to find the slope of spherical flow, and Equation (7.53) is used to find the correction factor:

$$\beta = \frac{a+c(0.6)^2 + e(0.6)^4 + g(0.6)^6 + i(0.6)^8 + k(0.6)^{10}}{1+b(0.6)^2 + d(0.6)^4 + f(0.6)^6 + h(0.6)^8 + j(0.6)^{10}} = -0.0975$$

Infinite Reservoirs

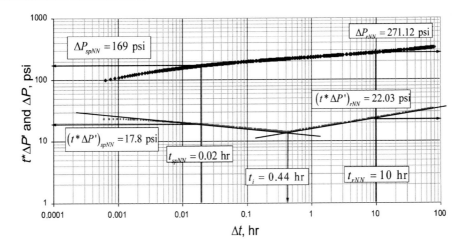

Figure 7.11. Pressure and pressure derivative versus time log-log plot for example 7.2.

$$FC = e^{a+b0.6+c0.6^2+d0.6^3} = 1.573$$

The spherical radius is found with Equation (7.81) and spherical permeability with Equation (7.64):

$$r_{sw} = \frac{h_p}{2\ln\left(\frac{h_p}{r_w}\right)} = \frac{13}{2\ln\left(\frac{13}{0.3}\right)} = 1.725 \text{ ft}$$

$$k_{sp} = \left[\frac{\frac{19.9158(1.573)}{17.8}(96681.605)^{1-0.6}(300*1.2)^{0.6}(0.00491)}{(1.725)^{1-2*0.6}\left(\frac{0.02\left(96681.605\frac{1}{300*1.2}\right)^{0.6-1}}{3792.188(0.6)(0.2)(0.0491)(1\times10^{-5})(1,725)^{4-2*0.6}}\right)^{-0.0975}}\right]^{\frac{1}{1+0.0975}} = 14.93 \text{ md}$$

The spherical skin factor is determined with Equation (7.67):

$$S_{spNN} = \frac{1.573}{2\sqrt{\pi}}\left(\frac{0.0002637(14.93)(0.02)}{(0.6)(0.2)(0.00491)(1\times10^{-5})(1.725)^{4-2*0.8}}\right)^{-0.0975}\left[\frac{169}{17.8}+\frac{1}{0.0975}\right]-1$$

$$S_{spNN} = 4$$

The vertical permeability is estimated using Equation (7.82):

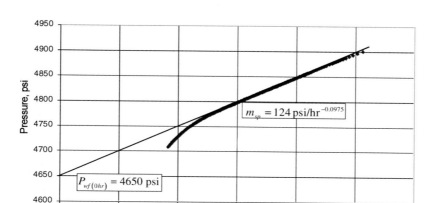

Figure 7.12. Cartesian plot for example 7.2.

$$k_z = \frac{k_{sp}^3}{k^2} = \frac{14.93^3}{25.48^2} = 5.13 \text{ md}$$

The point of intersection between radial and spherical flow regimes in Equation (7.78) is used to verify the spherical permeability:

$$k_{sp} = 25.48 \left[\frac{0.44}{5.183 \times 10^{-5} (0.3)^{3-0.6} \left[\frac{1.573}{2\sqrt{\pi}} \left(\frac{100}{1.725} \right)^{-0.0975(1-0.6)+0.6} \left(\frac{0.3}{1.725} \right)^{-0.0975(3-0.6)+0.6-1} \right]^{\frac{1}{0.1667+0.0975}}} \right]^{\frac{0.1667+0.0975}{-0.0975-1}}$$

k_{sp} = 15.03 md

Solution by TDS Conventional Analysis

Previously, Equation (7.52) provided a β value of -0.0975. A Cartesian plot of injected pressure against $t^{-0.0975}$ is reported in Figure 7.12. The slope and intersect from this plot are 124 psi/hr$^{-0.0975}$ and 4650 psi, respectively. The spherical permeability is estimated with Equation (7.56) and skin factor with Equation (7.57), respectively:

$$k_{sp} = \left[-19.9158 \frac{1.573}{(124)(-0.0975)} \left(\frac{\left(96681.605 \frac{1}{(300)(1.2)}\right)^{0.6-1}}{3792.188(0.6)(0.2)(0.00491)(1\times10^{-5})1.725^{4-2*0.6}} \right)^{-0.0975} \right]^{\frac{1}{1+0.0975}} = 15 \text{ md}$$

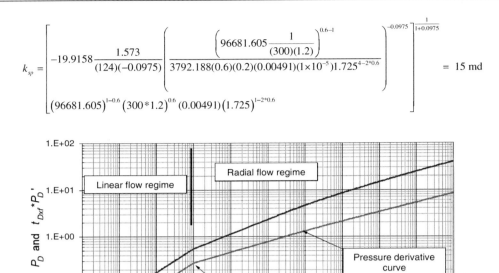

Figure 7.13. Dimensionless pressure and pressure-derivative behavior for a vertical infinite-conductivity fractured well with a non-Newtonian pseudoplastic fluid with $n = 0.5$.

$$s_{spNN} = (96681.605)^{0.6-1} \left(\frac{1}{300*1.2}\right)^{0.6} \frac{15(5000-4650)}{70.6(0.00491)(1.725)^{1-2*0.6}} - 1 = 4.008$$

HYDRAULICALLY FRACTURED WELLS

As noted before, petroleum engineers often deal with non-Newtonian fluids in many oil industry activities. Some of these fluids are used as fracturing, EOR and drilling mud fluids. If one of these fluids is used to fracture a well and a post-fracture test is run afterwards, the interpretation cannot be conducted with the conventional models. A pseudoplastic model has to be used. The oil literature presents only the work of Odeh and Yang (1979) on well-test analysis for wells fractured with non-Newtonian fluids. The application was specific for fall-off testing, and the interpretation for determination of the half-fractured length was conducted using both the conventional straight-line method and type-curve matching. However, the application of the pressure derivative for these types of systems was performed later by Escobar (2012) and Escobar, Bonilla and Ciceri (2012).

Odeh and Yang (1979) linearized the partial-differential equation for the problem of a well intercepted by a vertical fracture. Their dimensionless pressure solution is given below:

$$P_D\left(t_D\right)=\frac{(3-n)^{2v}t_D^v}{\left(1-n\right)\Gamma(1-v)}-\frac{1}{1-n} \tag{7.100}$$

where $v=(1-n)/(3-n)$.

Vongvuthipornchai and Raghavan (1987) presented two interpretation methodologies: type-curve matching and conventional straight-line for characterization of fall-off tests in vertically hydraulic wells with pseudoplastic fluid. Conventional analysis will not be covered here, since their formulation requires some correction factors depending on the n value, which are not very practical. Therefore, only the *TDS* technique will be provided. Vongvuthipornchai and Raghavan (1987) indicated that at early times, a well-defined straight line with slope equal to 0.5 on log-log coordinates will be evident; thus,

$$P_D=\left(\frac{\pi}{2}\right)^{\frac{n-1}{2}}\sqrt{\pi t_{Dxf}} \tag{7.101}$$

where the dimensionless time as a function of half-fracture length is given by:

$$t_{Dxf}=\frac{0.0002637kt}{\phi c_t\mu_{eff}x_f^2} \tag{7.102}$$

Equation (7.100) was used by Escobar et al. (2012b) to develop pressure and pressure-derivative curves for vertically infinite-conductivity fractured wells. An example for $n = 0.5$ is given in Figure 7.13. Notice that the slope of the pressure derivative during radial flow regime is greater than zero. The zero value of the pressure derivative corresponds to the case for $n = 1$ (Newtonian). It is important to observe that during the formation linear flow to the fracture, the slope of the pressure and pressure derivative is still 0.5, as in the Newtonian case. The governing equation for the pressure derivative is:

$$t_{Dxf}*P_D{}'=0.5\left(\frac{\pi}{2}\right)^{\frac{n-1}{2}}\sqrt{\pi t_{Dxf}} \tag{7.103}$$

Entering Equations (7.37) and (7.102) into Equation (7.103) and solving for the half-fracture length yields:

$$x_f=\frac{(96681.605)^{n-1}r_w^{n-1}}{(t*\Delta P')_{LNN}}\left(\frac{h}{qB}\right)^n\sqrt{\frac{\mu_{eff}t_{LNN}}{kc_t}} \tag{7.104}$$

Equation (7.104) performs better if the pressure derivative is read at the time of 1 hr. This ensures that the procedure will have an average value of pressure derivative for the whole linear flow regime instead of a single point. Using the intersect point of the pressure derivative during linear flow regime, Equation (7.103), with the radial flow regime's

Infinite Reservoirs 181

governing pressure-derivative equation (Equation (7.36), called t_{RLi}) an expression to obtain the half-fracture length is presented:

$$x_f = \left[0.028783(\pi/2)^{\frac{n-1}{2}} \left(\frac{\phi c_t \mu_{eff}}{0.0002637 kt_{LRi}} \right)^{\alpha} \sqrt{\frac{t_{LRi} k}{\phi c_t \mu_{eff}}} \right]^{(3-n)/(1+n)} \tag{7.104}$$

Escobar, Vega and Bonilla (2012c) showed that the late-time pressure-derivative plot during pseudosteady state is the same for Newtonian and non-Newtonian fluids. The expression governing the late-time, pseudosteady-state flow regime, derivative of Equation (1.12), is:

$$t_D * P_D' = 2\pi t_{DA} \tag{7.105}$$

Using the point of intersection of the pressure derivatives during linear flow and pseudosteady state (not shown here), called t_{LPi}, the well-drainage area can be obtained. In other words, equating Equations (7.103) and (7.105) and solving for the well-drainage area leads to the following expression:

$$A = \pi \left[\frac{t_{LPi}}{0.0625 \left(\frac{\pi}{2} \right)^{n-1} G} \right]^{2/(3-n)} \tag{7.106}$$

Example 7.3

Fan (1998) presented a pressure test conducted in a hydraulic fractured well with the information given below. Pressure and pressure-derivative data for this test are reported in Figure 7.14.

$n = 0.4$	$h = 70$ ft	$k = 0.65$ md
$q = 507.5$ BPD	$\phi = 10\%$	$B = 1$ rb/STB
$\mu_{eff} = 0.00065$ cp	$c_t = 0.00001$ psi^{-1}	$r_w = 0.26$ ft
$H = 20$ cp*s^{n-1}		

Solution
The following information is read from the pressure and pressure-derivative plot, Figure 7.14,

$t_{LRi} = 0.4495$ hr	$t_r = 0.7217$ hr
$\Delta P_r = 762$ psi	$(t*\Delta P')_r = 522.06$ psi

Using Equation (7.40), a value of 0.23 is found for α:

$$\alpha = \frac{1-n}{3-n} = \frac{1-0.4}{3-0.4} = 0.2307692$$

Reservoir permeability, parameter G, skin factor and half-fracture length are estimated using Equations (7.39), (9.4), (7.43) and (7.104), respectively,

$$k = \left[\frac{70.6(96681.605)^{(1-0.2307692)(1-0.4)} \left(\frac{0.0002637 * 0.7214}{0.1 * 0.00001 * 0.4} \right)^{0.2307692}}{\left(\frac{211*1}{70} \right)^{0.4-0.2307692(0.4-1)} \left(\frac{(0.26)^{0.2307692(0.4-3)+(1-0.4)}}{522.06} \right)} \right]^{\left(\frac{1}{1-0.2307692}\right)} * 0.00065 = 0.65 \text{ md}$$

$$G = \frac{3792.188(0.4)(0.1)(0.00001)(0.00065)}{0.6504} \left(96681.605 \frac{70}{211*1} \right)^{1-0.4} = 7.66 \times 10^{-4} \text{ hr/ft}^{2.6}$$

$$s = \frac{1}{2} \left(\frac{762}{522.06} - \frac{1}{0.2307692} \right) \left(\frac{0.7217}{0.0007662638489(0.26)^{2.6}} \right)^{0.23} = 16.2$$

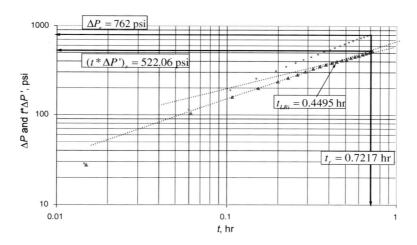

Figure 7.14. Pressure and pressure derivative versus time log-log plot for example 7.3

$$x_f = \left[0.028783 \frac{(1.570796)^{\frac{0.4-1}{2}}}{\left(\frac{0.0002637(0.65)(0.445)}{0.1(0.00001)0.00065} \right)^{0.23}} \sqrt{\frac{0.445 * 0.65}{0.1(0.00001)0.00065}} \right]^{-\left(\frac{3-0.4}{1+0.4}\right)} = 771.12 \text{ ft}$$

Infinite Reservoirs

Table 7.6. Pressure and pressure-derivative data for example 7.3

t, hrs	P, psi	ΔP, psi	$t*\Delta P'$, psi	t, hrs	P, psi	ΔP, psi	$t*\Delta P'$, psi
0	5990	0	0	0.3833	5504	486	366.863
0.015	5963	27	27.512	0.415	5474	516	384.202
0.0617	5882	108	102.99	0.445	5447	543	396.822
0.1083	5811	179	157.105	0.4767	5418	572	409.402
0.1533	5748	242	199.084	0.5067	5392	598	423.181
0.2	5694	296	235.248	0.5383	5364	626	437.878
0.23	5659	331	259.235	0.5683	5338	652	451.007
0.2617	5626	364	282.553	0.6	5314	676	464.868
0.2917	5594	396	306.371	0.63	5291	699	477.998
0.3233	5564	426	329.859	0.66	5269	721	492.023
0.3533	5534	456	350.081	0.6917	5247	743	507.11
				0.7217	5228	762	522.06

Nomenclature

B	Volumetric factor, RB/STB
b_x	Distance from well to a linear boundary, ft
c_t	System total compressibility, 1/psi
C	Wellbore storage, bbl/psi
FC	Correction factor
h	Formation thickness, ft
h_p	Length of perforations, ft
H	Consistency (power-law parameter), $cp*s^{n-1}$
G	Group defined by Equation (7.4)
G_1	Group defined by Equation (7.51)
G	Minimum pressure gradient, psi/ft
G_D	Dimensionless pressure gradient
k	Permeability, md
k	Flow consistency parameter
m	Slope
n	Flow behavior index (power-law parameter)
P	Pressure, psi
q	Flow/injection rate, STB/D
t	Time, hr
t_f	Injection time, hr
r	Radius, ft
$t*\Delta P'$	Pressure derivative, psi
t_D*P_D'	Dimensionless pressure derivative
xf	Half-fracture length, ft
z	Laplace parameter

Greeks

α Slope of the pressure-derivative curve during radial flow regime

β Exponent of Cartesian plot

Δ Change, drop

ϕ Porosity, Fraction

Γ Gamma function

γ Shear rate, s^{-1}

σ Shear stress, N/m

μ Viscosity, cp

μ_{eff} Effective viscosity for power-law fluids, $cp*(s/ft)^{n-1}$

μ_B Bingham plastic coefficient, cp

μ^* Characteristic viscosity, cp/ft^{1-n}

τ Shear stress, N/m

Suffixes

app	Apparent
D	Dimensionless
DA	Dimensionless as function of reservoir area
Dxf	Dimensionless as function of half-fracture length
DhsNN	Dimensionless, non-Newtonian, hemispherical
DspNN	Dimensionless, non-Newtonian, spherical
eff	Effective
erf	End of radial flow
hs	Hemispherical
hsNN	Hemispherical, non-Newtonian
i	Initial
i	Intersection between early unit slope and radial lines
inv	Investigation
LNN	Linear, non-Newtonian
LPi	Linear and pseudosteady-state intersection
LRi	Linear and radial flow regime lines intersection
NN	Non-Newtonian
rNN	Radial (any point on radial flow)
sp	Spherical
spNN	Spherical, non-Newtonian
sw	Spherical or hemispherical wellbore
x	Maximum point during wellbore storage period
w	Wellbore
z	Vertical

REFERENCES

Bird, R. B., Stewart, W. E. & Lightfoot, E. N. (2007). *"Transport Phenomena* (Revised Second Edition ed.)." John Wiley & Sons. ISBN 978-0-470-11539-8.

Chatas, A. T. (1966, June 1). "Unsteady Spherical Flow in Petroleum Reservoirs." *Society of Petroleum Engineers.* doi:10.2118/1305-PA

Ci-qun, L. (1988). "Transient Spherical Flow of non-Newtonian Power-law Fluids in Porous Media." *Applied Mathematics and Mechanics, 9* (6), 521-525.

Escobar, F. H., Muñoz, O. F., Sepulveda, J. A. & Montealegre, M. (2005). *"New Finding on Pressure Response In Long, Narrow Reservoirs." CT&F – Ciencia, Tecnología y Futuro.* Vol. *2*, No. 6.

Escobar, F. H., Martinez, J. A. & Montealegre-M., Matilde. (2010). "Pressure and Pressure Derivative Analysis for a Well in a Radial Composite Reservoir with a Non-Newtonian/Newtonian Interface." *CT&F*, Vol. *4*, No. 1. p. 33-42. Dec. 2010.

Escobar, F. H., Martinez, J. A. & Bonilla, L. F. (2012a). *"Transient Pressure Analysis for Vertical Wells With Spherical Power-Law Flow." CT&F.* Vol. 5 No. 1. P. 19-25. Dec.

Escobar, F. H., Bonilla, D. F. & Cicery, Y. Y. (2012b). "Pressure and Pressure Derivative Analysis for Pseudoplastic Fluids in Vertical Fractured Wells." *Journal of Engineering and Applied Sciences.* Vol. 7. Nro. 8. Aug.

Escobar, F. H., Vega, L. J. & Bonilla, L. F. (2012c). "Determination of Well-Drainage Area for Power-Law Fluids by Transient Pressure Analysis." *CT&F.* Vol. 5 No. 1. P. 45-55. Dec.

Escobar, F. H., Ascencio, J. M. & Real, D. F. (2013). "Injection and Fall-off Tests Transient Analysis of non-Newtonian Fluids." *Journal of Engineering and Applied Sciences.* Vol. 8. Nro. 9. P. 703-707. Sep.

Escobar, F. H. (2012). "Transient Pressure and Pressure Derivative Analysis for Non-Newtonian Fluids." *New Technologies in the Oil and Gas Industry*, Dr. Jorge Salgado Gomes (Ed.), ISBN: 978-953-51-0825-2, InTech, DOI: 10.5772/50415. Available from: http://www.intechopen.com/books/new-technologies-in-the-oil-and-gas-industry/transient-pressure-and-pressure-derivative-analysis-for-non-newtonian-fluids.

Fan, Y. (1998). "A New Interpretation Model for Fracture-Calibration Treatments." *SPE Journal.* P. 108-114. June.

Ikoku, C. U. (1978). *"Transient Flow of Non-Newtonian Power-Law Fluids in Porous Media."* Ph.D. Dissertation. Stanford University.

Ikoku, C. U. (1979, January 1). "Practical Application Of Non-Newtonian Transient Flow Analysis." *Society of Petroleum Engineers.* doi:10.2118/8351-MS

Ikoku, C. U. & Ramey H. J., Jr. (1978). "Numerical Solution of the Nonlinear Non-Newtonian Partial Differential Equation." paper SPE 7661. *American Institute of Mining, Metallurgical, and Petroleum Engineers*, 1978.

Ikoku, C. U. & Ramey, H. J. (1979, June 1). "Transient Flow of Non-Newtonian Power-Law Fluids in Porous Media." *Society of Petroleum Engineers.* doi:10.2118/7139-PA

Ikoku, C. U. & Ramey, H. J. (1980, February 1). "Wellbore Storage and Skin Effects During the Transient Flow of Non-Newtonian Power-Law Fluids in Porous Media." *Society of Petroleum Engineers.* doi:10.2118/7449-PA.

Ikoku, C. U., (1982). "Well Test Analysis for Enhanced Oil Recovery Projects." *ASME Journal of Energy Resources Technology*, June 1, 1982.

Katime-Meindl, I. & Tiab, D. (2001, January 1). "Analysis of Pressure Transient Test of Non-Newtonian Fluids in Infinite Reservoir and in the Presence of a Single Linear Boundary by the Direct Synthesis Technique." *Society of Petroleum Engineers*. doi:10.2118/71587-MS

Martinez, J. A., Escobar, F. H. & Cantillo, J. H. (2011). "Application f the TDS Technique to Dilatant Non-Newtonian/Newtonian Fluid Composite Reservoirs." *Ingeniería e Investigación*. Vol. *31*. No. 3. p. 130-134. Aug.

Moncada, K., Tiab, D., Escobar, F. H., Montealegre-M, M., Chacon, A., Zamora, R. A., Nese, S. L. (2005). "Determination of Vertical and Horizontal Permeabilities for Vertical Oil and Gas Wells with Partial Completion and Partial Penetration using Pressure and Pressure Derivative Plots without Type-Curve Matching." *CT&F – Ciencia, Tecnología y Futuro*. Vol. *2*, No. 6. P. 77-95. ISSN 0122-5383. Dec.

Odeh, A. S. & Yang, H. T. (1979, June 1). "Flow of Non-Newtonian Power-Law Fluids Through Porous Media." *Society of Petroleum Engineers*. doi:10.2118/7150-PA.

Tiab, D. (1993, January 1). "Analysis of Pressure and Pressure Derivatives Without Type-Curve Matching: I-Skin and Wellbore Storage." *Society of Petroleum Engineers*. This paper was prepared for presentation at the Production Operations Symposium held in Oklahoma City, OK, U.S. A., March 21-23. doi:10.2118/25426-MS.

Vongvuthipornchai, S. & Raghavan, R. (1987, December 1). "Pressure Falloff Behavior in Vertically Fractured Wells: Non-Newtonian Power-Law Fluids." *Society of Petroleum Engineers*. doi:10.2118/13058-PA.

Chapter 8

NON-NEWTONIAN/NEWTONIAN INTERFACE

BACKGROUND

Non-Newtonian fluids are often used in various drilling, workover and enhanced oil recovery processes. Most of the fracturing fluids injected into reservoir-bearing formations behave non-Newtonianly, and these fluids are often mistakenly approximated by Newtonian fluid flow models. In the field of well testing, several analytical and numerical models taking into account Bingham and pseudoplastic, non-Newtonian behavior have been introduced in the literature to study the transient nature of these fluids in porous media for better reservoir characterization. Most of them deal with fracture wells and homogeneous formations, and well-test interpretation is conducted via straight-line conventional analysis or type-curve matching. Only a few studies consider pressure-derivative analysis. Thus, a more practical and accurate way of characterizing such systems is presented in this chapter.

As noted above, in many activities of the oil and gas industry, engineers have to deal with completion and stimulation treatment fluids such as polymer solutions, drilling muds, gels, cements and some heavy crude oils that display non-Newtonian behavior. When it is necessary to conduct a treatment with a non-Newtonian fluid in an oil-bearing formation, this comes in contact with conventional oil, which may have a Newtonian nature. If this is the case, the definition of two media with entirely different mobilities is developed, as sketched in Figure 8.1 where the invaded zone may have displaced the Newtonian oil and be replaced by either a pseudoplastic, dilatant or Bingham plastic fluid (radial composite reservoir). This zone is labeled as zone 1. The external zone corresponds to the virgin zone containing Newtonian oil and it is labeled as zone 2. If a pressure test is run in such a system, a very particular situation arises and can lead to a huge interpretation mistake if not dealt with appropriately, since it may be confused with something else.

PSEUDOPLASTIC/NEWTONIAN INTERFACE

A partial differential equation for the radial flow of non-Newtonian fluids that follow a power-law relationship through porous media was proposed by Ikoku and Ramey (1979). Coupling the non-Newtonian Darcy's law with the continuity equation, they derived a rigorous partial differential equation, which, in linearized form, is:

$$\frac{1}{r^n}\frac{\partial}{\partial r}\left(r^n \frac{\partial P}{\partial r}\right) = Gr^{1-n}\frac{\partial P}{\partial t} \qquad (8.1)$$

where:

$$G = \frac{3792.188 n\phi\mu_{eff}}{k_1}\left(96681.605\frac{h}{qB}\right)^{1-n} \qquad (7.4)$$

The effective viscosity given in Equation (7.4) is defined in oil-field units by:

$$\mu_{eff} = \frac{H}{12}\left(9+\frac{3}{n}\right)^n \left(1.59344 \times 10^{-12} k_1 \phi\right)^{(1-n)/2} \qquad (7.5)$$

The system under consideration assumes the radial flow of a non-Newtonian fluid and a slightly compressible Newtonian fluid through porous media. It is assumed that the reservoir is homogeneous, isotropic and of a constant thickness. The shape of the reservoir is cylindrical, with a finite outer radius. The non-Newtonian fluid is injected through a well in the center of the field. The fluid is considered to be non-Newtonian and pseudoplastic, and to obey the power law. The Newtonian fluid possesses a constant viscosity, as is usually the case in well-test analysis. It is also assumed piston-like of the Newtonian fluid by the non-Newtonian fluid. Figure 8.1 sketches the composite reservoir under consideration.

Martinez, Escobar and Bonilla (2012) solved Equation (8.1) numerically using finite-difference approximation, with the initial and boundary conditions normally used in well-test analysis and using compatibility and continuity conditions along the interface. Their results agree quite well with those reported by Ikoku and Ramey (1979).

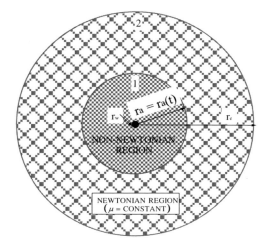

Figure 8.1. Composite non-Newtonian/Newtonian radial reservoir.

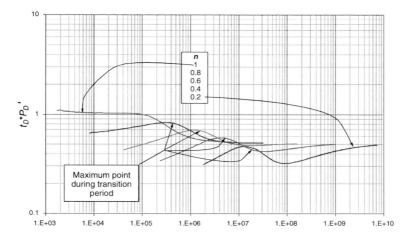

Figure 8.2. Dimensionless pressure-derivative behavior of a non-Newtonian pseudoplastic/Newtonian interface for an invasion radius, $r_a = 200$ ft; adapted from Escobar et al. (2010).

Since the pressure wave is traveling through a porous medium saturated with a non-Newtonian fluid, the pressure derivative has a certain degree of inclination, as the flow behavior index reduces its value. Once the travel wave gets into the Newtonian zone, the pressure derivative forms a plateau, as expected, after an obvious transition period. Figure 8.2 illustrates such a situation for different values of n. Notice than for $n = 1$, two different plateaus are observed, since two different mobilities are dealt with. The smaller the n, the longer the transition time required to get the horizontal radial flow line during the Newtonian zone.

The dimensionless pressure, P_{DNN}, and dimensionless time, t_{DNN}, for region 1 (non-Newtonian) are expressed as:

$$P_{DNN} = \frac{P - P_i}{141.2(96681.605)^{1-n}\left(\dfrac{qB}{h}\right)^n \dfrac{\mu_{eff} r_w^{1-n}}{k_1}} \tag{7.6}$$

$$t_{DNN} = \frac{t}{Gr_w^{3-n}} \tag{7.8}$$

The dimensionless pressure, P_{DN}, and dimensionless time, t_{DN}, for region 2 (Newtonian) are expressed by:

$$P_{DN} = \frac{k_2 h \Delta P}{141.2 q \mu_N B} \tag{8.2}$$

$$t_{DN} = \frac{0.0002637k_2t}{\phi\mu_N c_t r_w^2} \tag{8.3}$$

and the dimensionless radius, r_D, is defined by:

$$r_D = \frac{r}{r_w} \tag{7.7}$$

Notice that permeability is Equations (7.4) through (7.8) has been labeled as k_1, indicating that it is the permeability in the non-Newtonian zone. However, k_1 and k_2 ought to have the same value, since they are the same formation. Skin factor in the Newtonian zone is expected to be less than that in zone 1, since more work is required to displace a non-Newtonian fluid than a Newtonian fluid. Besides, as can be seen in Figures 7.6 and 8.1, as the fluid becomes more non-Newtonian (n further from 1), the shear stress increases for a given shear strain. In linear flow regime behavior, the pressure gradient is about the same, contrary to radial flow, in which the pressure gradient is higher as the wave approaches the wellbore; thus, linear flow behavior keeps the one-half slope on the pressure derivative no matter the n value.

Escobar (2012) and Escobar, Martinez and Montealegre (2010) extended the *Tiab's Direct Synthesis* Technique (*TDS* Technique), introduced by Tiab (1993), using several features found in the pressure and pressure derivative log-log plot for each region, as given below.

The definition of the dimensionless pressure derivative in the non-Newtonian region was given previously:

$$\left(t_D * P_D'\right)_{rNN} = 0.00708\left(96681.605\right)^{n-1}\left(\frac{h}{qB}\right)^n \frac{k_1}{\mu_{eff}} r_w^{n-1}\left(t*\Delta P'\right)_r \tag{7.37}$$

Escobar, Martinez and Montealegre (2010) found that the dimensionless pressure derivative during radial flow regime is governed by:

$$\log\left[t_D * P_D'\right]_r = \alpha\log\left(t_{DNN}\right)_r + \log 0.5 \tag{7.35}$$

Or:

$$\left(t_D * P_D'\right)_{rNN} = 0.5t_{DNN}^\alpha \tag{7.36}$$

The combination of Equations (7.5), (7.35) and (7.37) yields:

$$k_1 = \left(\frac{\left[70.6(96681.605)^{(1-\alpha)(1-n)} \left(\frac{0.0002637 t_r}{n\phi c_t} \right)^{\alpha} \left(\frac{qB}{h} \right)^{n-\alpha(n-1)} \left(\frac{r_w^{\alpha(n-3)+(1-n)}}{(t*\Delta P')_{r1}} \right) \right]^{\frac{1}{1-\alpha}}}{\left(\left(\frac{H}{12} \right) \left(9 + \frac{3}{n} \right)^n (1.59344 \times 10^{-12} \phi)^{(1-n)/2} \right)} \right)^{\frac{2}{1+n}} \tag{7.39}$$

where α is the slope of the pressure-derivative curve on the non-Newtonian region and is defined as:

$$\alpha = \frac{1-n}{3-n} \tag{7.40}$$

Escobar et al. (2010) also found that during the non-Newtonian radial flow regime, the governing well pressure expression is given by:

$$P_{DNNr} = 0.5 \left[\ln t_{DNNr} + 0.80907 + 2s \right] \tag{7.41}$$

The skin factor expression is found by the division of Equation (7.41) by Equation (7.35):

$$s = \frac{1}{2} \left[\frac{\left[\frac{0.0002637 k_1 t_{r1}}{n\phi\mu_{eff} c_t r_w^{3-n}} \left(96681.605 \frac{h}{qB} \right)^{n-1} \right]^{\alpha} \left(\frac{\Delta P}{(t*\Delta P)} \right)_{r1} -}{\ln \left(\frac{k_1 t_{r1}}{n\phi\mu_{eff} c_t r_w^{3-n}} \left(96681.605 \frac{h}{qB} \right)^{n-1} \right) + 7.43} \right] \tag{7.42}$$

The radius of the injected non-Newtonian fluid bank, or the invasion radius, is calculated using the following correlation (not valid for $n=1$), obtained from reading the time at which the pressure derivative has its maximum value during the transition from non-Newtonian to Newtonian behavior:

$$r_a = \left[\frac{G \left(0.2258731 n^3 - 0.2734284 n^2 + 0.5064602 n + 0.5178275 \right)^{1/\alpha}}{t_M} \right]^{1/(n-3)} \tag{8.4}$$

Lund and Ikoku (1981) found that the radius of the non-Newtonian fluid bank can be found using the radius investigation equation proposed by Ikoku and Ramey (1979):

$$r_a = \left[\Gamma \left(\frac{2}{3-n} \right) \right]^{1/(n-1)} \left[\frac{(3-n)^2 t}{G} \right]^{1/(3-n)} \tag{8.5}$$

where t is the end time of the straight line found on a non-Newtonian Cartesian graph of ΔP vs. $t^{1-n/3-n}$. For the Newtonian region, the permeability and skin factor were presented by Tiab (1993) as:

$$k_2 = \frac{70.6 q \mu_N B}{h(t*\Delta P')_r} \quad (1.18)$$

$$s_2 = \frac{1}{2}\left[\left(\frac{\Delta P}{t*\Delta P'}\right)_{r2} - \ln\left(\frac{k_2 t_{r2}}{\phi \mu_N c_t r_w^2}\right) + 7.43\right] \quad (1.21)$$

Another important feature in this system is the time of intercept of the non-Newtonian/Newtonian radial flow lines. The infinite-acting dimensionless pressure derivative is given by:

$$(t_D * P_D')_{rN} = 0.5 \quad (8.6)$$

The combination of Equations (7.36) and (8.6) yields:

$$\frac{(t_D * P_D')_{rNN}}{(t_D * P_D')_{rN}} = t_{DNN}^{\alpha} \quad (8.7)$$

Replacing the dimensionless quantities and solving for k_1 yields:

$$k_1 = \left[\left\langle\left(\frac{H}{12}\right)\left(9+\frac{3}{n}\right)^n (1.59344\times 10^{-12}\phi)^{(1-n)/2}\left(96681.605\frac{hr_w}{qB}\right)^{1-n}\right\rangle^{1-1/\alpha} \frac{\mu_N^{1/\alpha} \phi c_t r_w^2 n}{0.0002637 t_{irN\_NN}}\right]^{1/2} \quad (8.8)$$

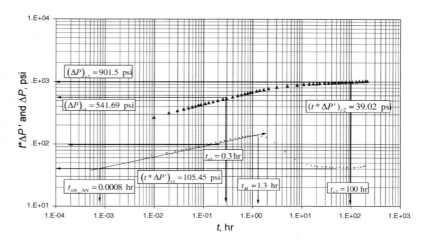

Figure 8.3. Pressure and pressure derivative log-log plot for example 8.1; adapted from Escobar et al. (2010).

Non-Newtonian/Newtonian Interface

where t_{irN_NN} is the time of intersection of the non-Newtonian and Newtonian radial lines.

Example 8.1

An injection test for a well in a closed reservoir was presented by Lund and Ikoku (1981) with the information given below.

$q = 300$ BPD	$k = 100$ md	$\phi = 0.2$
$r_w = 0.33$ ft	$h = 16.4$ cm	$n = 0.6$
$H = 20$ cp*s^{n-1}	$B = 1$ rb/STB	$c_t = 6.89 \times 10^{-6}$ psi^{-1}
$k_{sp} = 15$ md	$h_p = 13$ ft	$P_i = 2500$ psi
$r_e = 2650$ ft	$\mu_N = 3$ cp	

It is necessary to estimate the permeability and the skin factor in each area and the radius of the injected non-Newtonian fluid bank.

Solution by TDS technique

The log-log plot of pressure and pressure derivative against injection time is given in Figure 8.3. From that plot, the following information is read:

$t_{r1} = 0.3$hr	$\Delta P_{r1} = 541.54$ psi	$(t^*\Delta P')_{r1} = 105.45$ psi
$t_{r2} = 120$ hr	$\Delta P_{r2} = 991.5$ psi	$(t^*\Delta P')_{r2} = 39.02$ psi
$t_M = 1.3$hr	$t_{irN\_NN} = 0.0008$ hr	
t_{inj} (before shut-in)= 8.98 days		

First, α is evaluated with Equation (7.40), and the non-Newtonian effective fluid permeability is estimated with Equation (7.39):

$$\alpha = \frac{1-n}{3-n} = \frac{1-0.6}{3-0.6} = 0.1667$$

$$k_1 = \left(\begin{array}{c} \left[70.6\left(96681.605\right)^{(1-0.1667)(1-0.6)} \left(\dfrac{0.0002637(0.3)}{0.6(0.2)(6.89\times10^{-6})} \right)^{0.1667} \right]^{\frac{1}{1-0.1667}} \\ \left[\left(\dfrac{300*1}{16.4} \right)^{0.6-0.1667(0.6-1)} \left(\dfrac{0.33^{0.1667(0.6-3)+(1-0.6)}}{105.45} \right) \right] \\ \left(\left(\dfrac{20}{12} \right)\left(9 + \dfrac{3}{0.6} \right)^{0.6} \left(1.59344\times10^{-12}(0.2) \right)^{(1-0.6)/2} \right) \end{array} \right)^{\frac{2}{1+0.6}} = 100.4 \text{ md}$$

Equation (7.5) is used to find the effective viscosity, and then the skin factor in the non-Newtonian region is found with Equation (7.42):

$$\mu_{\mathit{eff}} = \left(\frac{20}{12}\right)\left(9+\frac{3}{0.6}\right)^{0.6}\left(1.59344\times10^{-12}(100.4)(0.2)\right)^{(1-0.6)/2} = 0.06465 \text{ cp}*(\text{s/ft})^{n-1}$$

$$s_1 = \frac{1}{2}\left[\begin{array}{l}\left[\dfrac{0.0002637(100.4)(0.3)}{(0.6)(0.2)(0.06465)(6.89\times10^{-6})(0.33^{3-0.6})}\left(96681.605\dfrac{16.4}{300(1)}\right)^{0.6-1}\right]^{0.1667}\left(\dfrac{541.54}{105.45}\right) \\ -\ln\left(\dfrac{100.4*0.3}{(0.6)(0.2)(0.06465)(6.89\times10^{-6})(0.33^{3-0.6})}\left(96681.605\dfrac{16.4}{300(1)}\right)^{0.6-1}\right)+7.43\end{array}\right] = 10.47$$

Parameter G is estimated with Equation (7.4). This value is used in Equation (8.4) to find the non-Newtonian fluid bank radius:

$$G = \frac{3792.188(0.6)(0.2)(6.89\times10^{-6})(0.06465)}{100.4}\left(96681.605\frac{16.4}{300(1)}\right)^{1-0.6} = 6.228\times10^{-5} \frac{\text{hr}}{\text{ft}^{3-n}}$$

$$r_a = \left[\frac{6.228\times10^{-5}\left(0.2258731(0.6)^3 - 0.2734284(0.6)^2 + 0.5064602(0.6) + 0.5178275\right)^{1/0.1667}}{1.3}\right]^{1/(0.6-3)}$$

$r_a = 120.42$ ft

Equations (1.18) and (1.21) are used to estimate the permeability and skin factor of the Newtonian zone:

$$k_2 = \frac{70.6(300)(3)(1)}{16.4(39.02)} = 99.29 \text{ md}$$

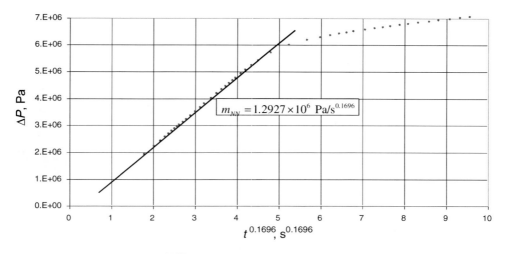

Figure 8.4. Pressure drop versus $t^{0.1696}$; adapted from Lund and Ikoku (1981).

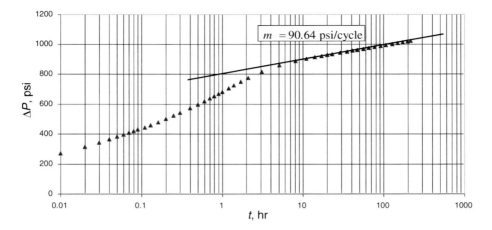

Figure 8.5. Semilog plot of pressure drop versus t.

$$s_2 = \frac{1}{2}\left[\left(\frac{991.5}{39.02}\right)_{r2} - \ln\left(\frac{99.29(100)}{0.2(3)(6.89\times10^{-6})(0.33^2)}\right) + 7.43\right] = 4.51$$

Equation (8.8) is used to verify the non-Newtonian effective fluid permeability:

$$k_1 = \left[\frac{\left\langle\left(\frac{20}{12}\right)\left(9+\frac{3}{0.6}\right)^{0.6}(1.59344\times10^{-12}(0.2))^{0.2}\left(96681.605\frac{16.4(0.33)}{300(1)}\right)^{0.4}\right\rangle^{-5}}{\frac{3^6(0.2)(6.89\times10^{-6})(0.33^2)(0.6)}{0.0002637(0.0008)}}\right]^{1/2} = 95.36 \text{ md}$$

Solution by Conventional Analysis

Lund and Ikoku (1981) built the plot shown in Figure 8.4. From there, a slope value of 1272900 psi/s$^{0.1696}$ was found for the non-Newtonian region. Permeability is found using Equation (7.15):

$$k_1 = \left(\frac{0.00055204}{2\pi(5)}\right)\left\{\frac{\left[(1-0.6)\Gamma\left(\frac{2}{3-0.6}\right)\right]^{n-3}\left[\frac{0.020}{12}\left(9+\frac{3}{0.6}\right)^{0.6}(150)^{\frac{1-0.6}{2}}\right]^2 (3-0.6)^{2(1-0.6)}}{(0.6*1\times10^{-9})^{1-0.6}(1272900)^{3-0.6}}\right\}^{\frac{1}{1+0.6}}$$

$k_1 = 99.24$ md

From the semilog plot given in Figure 8.5, a slope value of 90.64 psi/cycle is found for the Newtonian region. Equation (1.10) is used to find permeability:

$$k_2 = \frac{162.6 q \mu B}{hm} = \frac{162.6(300)(3)(1)}{(5)(90.64)} = 98.44 \text{ md}$$

Effective viscosity is found with Equation (7.44):

$$\mu_{eff} = \frac{0.02}{12}(9+3/0.6)^{0.6}(150*9.87\times10^{-14}*0.2)^{(1-0.6)/2} = 0.0134 \text{ Pa}\cdot\text{s}$$

Parameter G is found with an MKS version of Equation (7.4):

$$G = \frac{n\phi\mu_{eff}}{k_1}\left(\frac{h}{qB}\right)^{1-n} = \frac{(0.6)(0.2)(0.0134)}{100}\left(\frac{5}{500*1}\right)^{1-0.6} = 2.55\times10^{-6} \text{ s/m}^{2.4}$$

$$r_a = \left[\Gamma\left(\frac{2}{3-0.6}\right)\right]^{1/(n-1)}\left[\frac{(3-n)^2\,775800}{2.25\times10^{-6}}\right]^{1/(3-n)} = 39.8 \text{ m (130.6 ft)}$$

DILATANT/NEWTONIAN INTERFACE

Before the work presented by Martinez, Escobar and Cantillo (2011), the unique study on well-test analysis for dilatant Non-Newtonian fluids had been presented by Okpobiri and Ikoku (1983). They studied both the flow behavior index and the fluid consistency in pressure fall-off tests and conducted a reservoir characterization using the conventional straight-line method. Then, Martinez et al. (2011) studied the transient pressure behavior of dilatant fluids, with the purpose of complementing the *TDS* technique for a composite reservoir with dilatant non-Newtonian/Newtonian interface. For this purpose, the models developed previously by Lund and Ikoku (1981) were solved numerically by Martinez et al. (2012) to obtain the pressure and pressure-derivative behavior, and new expressions to characterize the reservoir by means of the *TDS* technique were developed by Martinez et al. (2011) and verified with synthetic data.

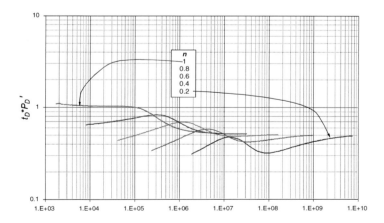

Figure 8.6. Dimensionless pressure-derivative behavior of a non-Newtonian dilatant/Newtonian interface for an invasion radius; adapted from Martinez et al. (2011).

Non-Newtonian/Newtonian Interface

For modeling purposes, consider again the composite reservoir sketched in Figure 8.1. In this case, the non-Newtonian fluid is dilatant, meaning $1 < n < 2$. The system under consideration assumes the radial flow of both a dilatant non-Newtonian and a slightly compressible Newtonian fluid through porous media. It is assumed that the reservoir is homogeneous and isotropic, and that it possesses a constant thickness. The shape of the reservoir is cylindrical with a finite outer radius. The dilatant fluid is injected through a well in the center of the field. The fluid is considered to be non-Newtonian dilatant, and to obey the power law. The Newtonian fluid possesses a constant viscosity, as is usually considered in well-test analysis. It is also assumed a piston-like behavior of the Newtonian fluid by the non-Newtonian fluid.

Since the pressure wave is traveling through a porous medium saturated with a non-Newtonian fluid, the pressure derivative has a certain degree of inclination—negative slope—as the flow behavior index increases its value. Once the travel wave arrives at the Newtonian zone, the pressure derivative forms a plateau, as expected, after an obvious transition period. Figure 8.6 illustrates such a situation for different n values. Notice that for $n = 1$, two different plateaus are observed, since two different mobilities are dealt with. The larger the n, the longer the time required to obtain the horizontal radial-flow line in the Newtonian zone.

As for the case of pseudoplastic behavior, the equations for the estimation of permeability and skin factor, Equations (7.39) and (7.42), are the same, since the dilatant fluid also obeys the power-law behavior. However, since the shape of the curves for the pseudoplastic and dilatant cases differ (see Figures 8.2 and 8.6), the respective expressions to find the invasion radius ought to be different. The radius of the injected non-Newtonian fluid bank is calculated using the following correlation (valid for $n>1$), obtained from reading the time at which the radial flow non-Newtonian ends, t_{e_rNN}:

$$
r_a = \left[\frac{G\left(0.46811e^{0.76241n}\right)^{1/\alpha}}{t_{e\_rNN}} \right]^{\frac{1}{(n-3)}}
\tag{8.9}
$$

Again, the nature of both fluids suggests that Equation (8.8) also applies to the dilatant case, since the governing equations are the same, although the n value takes different ranges.

Table 8.1. Reservoir and fluid data for example 8.2

Parameter	Values	Parameter	Values
P_R, Psi	2500 (1.724×10^7 Pa)	q, bbl/D	250 (4.60×10^{-4} m³/s)
r_e, ft	5000 (1524 m)	B, rb/STB	1.0 (1 m³/m³)
r_w, ft	0.3 (0.09144 m)	c_t, 1/psi	6.89×10^{-6} (1×10^{-9} 1/Pa)
h, ft	25 (7.62 m)	r_a, ft	100 (30.48 m)
ϕ, %	20	H, cp*s^{n-1}	10 (0.01 Ns^n/m²)
k, md	100 (9.87×10^{-14} m²)	μ_N, cp	3 (0.003 Pa.s)
		n	1.4

Example 8.2

An injection test was simulated using the information from Table 8.1. Pressure and pressure-derivative data are reported in Figure 8.7. It is necessary to estimate permeability and skin factor in each area and the radius of the non-Newtonian fluid bank using the *TDS* technique.

Solution by TDS technique

The log-log plot of pressure and pressure derivative against injection time is given in Figure 8.7. From that plot, the following information is read:

t_{r1} = 1.0 hr ΔP_{r1} = 3785.19 psi $(t*\Delta P')_{r1}$ = 165.51 psi
t_{r2} = 190 hr ΔP_{r2} = 4128.33 psi $(t*\Delta P')_{r2}$ = 21.99 psi
t_{e_rNN} = 3.1 hr t_{irN_NN} = 2150 hr

First, α is evaluated with Equation (7.40) and the non-Newtonian effective fluid permeability is estimated with Equation (7.39):

$$\alpha = \frac{1-n}{3-n} = \frac{1-1.4}{3-1.4} = -0.25$$

$$k_1 = \left(\frac{(96681.605)^{(1-1.4)} \left[70.6 \left(\frac{0.0002637(1.0)}{1.4(0.2)(6.89\times 10^{-6})} \right)^{-0.25} \right]^{\frac{1}{1+0.25}}}{\left[\left(\frac{250*1}{25} \right)^{1.4+0.25(1.4-1)} \left(\frac{1}{165.51} \right) \right]} \left(\left(\frac{10}{12} \right) \left(9 + \frac{3}{1.4} \right)^{1.4} \left(1.59344 \times 10^{-12}(0.2) \right)^{(1-1.4)/2} \right) \right)^{\frac{2}{1+1.4}} = 94.16 \text{ md}$$

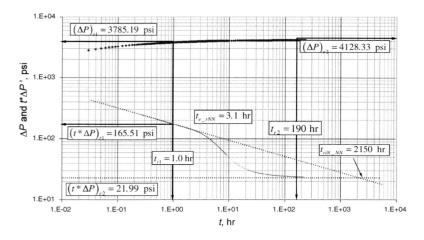

Figure 8.7. Pressure and pressure derivative versus time log-log plot for example 8.2.

Non-Newtonian/Newtonian Interface

Equation (7.5) is used to find the effective viscosity, and then the skin factor in the non-Newtonian region is found with Equation (7.42),

$$\mu_{eff} = \left(\frac{10}{12}\right)\left(9+\frac{3}{1.4}\right)^{1.4}\left(1.59344\times10^{-12}(94.16)(0.2)\right)^{(1-1.4)/2} = 3098.61 \text{ cp*(s/ft)}^{n-1}$$

$$s_1 = \frac{1}{2}\left[\left[\frac{0.0002637(94.16)(1.0)}{(1.4)(0.2)(3098.61)(6.89\times10^{-6})(0.3^{3-1.4})}\left(96681.605\frac{25}{250(1)}\right)^{1.4-1}\right]^{-0.25}\left(\frac{3785.19}{165.51}\right)_{r1}\right.$$
$$\left.-\ln\left(\frac{94.16*1.0}{(1.4)(0.2)(3098.61)(6.89\times10^{-6})(0.3^{3-1.4})}\left(96681.605\frac{25}{250(1)}\right)^{1.4-1}\right)+7.43\right]$$

$$s_1 = -1.94$$

Parameter G is estimated with Equation (7.4). This value is used in Equation (8.9) to find the non-Newtonian fluid bank:

$$G = \frac{3792.188(1.4)(0.2)(6.89\times10^{-6})(3098.61)}{94.16}\left(96681.605\frac{25}{250(1)}\right)^{1-1.4} = 0.00613 \text{ hr / ft}^{3-n}$$

$$r_a = \left[\frac{6.13\times10^{-3}\left(0.46811e^{0.76241(1.4)}\right)^{-1/0.25}}{3.1}\right]^{1/(1.4-3)} = 105.85 \text{ ft}$$

Equations (1.18) and (1.21) are used to estimate the permeability and skin factor of the Newtonian zone:

$$k_2 = \frac{70.6(250)(3)(1)}{25(21.99)} = 96.32 \text{ md}$$

$$s_2 = \frac{1}{2}\left[\left(\frac{4128.33}{21.99}\right)_{r2}-\ln\left(\frac{96.32(190)}{0.2(3)(6.89\times10^{-6})(0.3^2)}\right)+7.43\right] = 85.27$$

Equation (8.8) is used to verify the non-Newtonian effective fluid permeability:

$$k_1 = \left[\frac{\left\langle\left(\frac{10}{12}\right)\left(9+\frac{3}{1.4}\right)^{1.4}\left(1.59344\times10^{-12}(0.2)\right)^{-0.2}\left(96681.605\frac{25(0.3)}{250(1)}\right)^{-0.4}\right\rangle^5}{\frac{3^{-4}(0.2)(6.89\times10^{-6})(0.3^2)(1.4)}{0.0002637(2150)}}\right]^{1/2} = 109.95 \text{ md}$$

Nomenclature

B	Volumetric factor, RB/STB
c_t	System total compressibility, 1/psi
h	Formation thickness, ft
H	Consistency (power-law parameter), $cp*s^{n-1}$
G	Group defined by Equation (7.4)
G	Minimum pressure gradient, psi/ft
G_D	Dimensionless pressure gradient
k	Permeability, md
k	Flow consistency parameter
m	Slope
n	Flow behavior index (power-law parameter)
P	Pressure, psi
q	Flow/injection rate, STB/D
t	Time, hr
t_f	Injection time, hr
r	Radius, ft
r_a	Invasion/penetration radius of the non-Newtonian fluid, ft
$t*\Delta P'$	Pressure derivative, psi
t_D*P_D'	Dimensionless pressure derivative

Greeks

α	Slope of the pressure-derivative curve during radial flow regime
Δ	Change, drop
ϕ	Porosity, Fraction
Γ	Gamma function
μ	Viscosity, cp
μ_{eff}	Effective viscosity for power-law fluids, $cp*(s/ft)^{n-1}$

Suffixes

1	Non-Newtonian zone
2	Newtonian zone
app	Apparent
D	Dimensionless
e_rNN	End of non-Newtonian radial flow regime
eff	Effective
erf	End of radial flow
i	Initial
i	Intersection between early unit-slope and radial lines
irN_NN	Intersection of the radial pressure-derivative lines between the two zones

inv	Investigation
N	Newtonian
NN	Non-Newtonian
r	Radial flow regime
rNN	Radial (any point on radial flow)
w	Wellbore

REFERENCES

Escobar, F. H. (2012). *"Transient Pressure and Pressure Derivative Analysis for Non-Newtonian Fluids." New Technologies in the Oil and Gas Industry*, Dr. Jorge Salgado Gomes (Ed.), ISBN: 978-953-51-0825-2, InTech, DOI: 10.5772/50415. Available from: http://www.intechopen.com/books/new-technologies-in-the-oil-and-gas-industry/transient-pressure-and-pressure-derivative-analysis-for-non-newtonian-fluids.

Escobar, F. H., Martinez, J. A. & Montealegre-M., Matilde. (2010). "Pressure and Pressure Derivative Analysis for a Well in a Radial Composite Reservoir with a Non-Newtonian/Newtonian Interface." *CT&F – Ciencia, TEcnología y Futuro*, Vol. *4*, No. 1. 33-42. Dec.

Ikoku, C. U. & Ramey, H. J. (1979, June 1). "Transient Flow of Non-Newtonian Power-Law Fluids in Porous Media." *Society of Petroleum Engineers*. doi:10.2118/7139-PA

Lund, O. & Ikoku, C. U. (1981). "Pressure Transient Behavior of Non-Newtonian/Newtonian Fluid Composite Reservoirs." *Society of Petroleum Engineers of AIME*. 271-280. April.

Martinez, J. A., Escobar, F. H. & Cantillo, J. H. (2011). "Application f the TDS Technique to Dilatant Non-Newtonian/Newtonian Fluid Composite Reservoirs." *Ingeniería e Investigación*. Vol. *31*. No. 3. 130-134. Aug.

Martinez, J. A., Escobar, F. H. & Bonilla, L. F. (2012). "Numerical Solution for a Radial Composite Reservoir Model with a No-Newtonian/Newtonian Interface. *Journal of Engineering and Applied Sciences*. Vol. 7. Nro. 8. Aug.

Okpobiri, G. A. & Ikoku, C. U. (1983, January 1). "Pressure Transient Behavior of Dilatant Non-Newtonian/Newtonian Fluid Composite Reservoirs." *Society of Petroleum Engineers*. doi:10.2118/12307-MS

Tiab, D. (1993, January 1). "Analysis of Pressure and Pressure Derivatives Without Type-Curve Matching: I-Skin and Wellbore Storage." *Society of Petroleum Engineers*. This paper was prepared for presentation at the Production Operations Symposium held in Oklahoma City, OK, U.S.A., March 21-23. doi:10.2118/25426-MS

Chapter 9

FINITE SYSTEMS

BACKGROUND

Conventional well-test interpretation models do not apply to reservoirs containing non-Newtonian fluids such as some paraffinic and heavy crude oils and some completion and stimulation treatment fluids (polymer solutions, foams, drilling muds, etc.). Non-Newtonian fluids are generally classified as time independent, time dependent and viscoelastic. Examples of the first classification are Bingham, pseudoplastic and dilatant fluids, which are commonly encountered by petroleum engineers. Power-law fluids are also divided into two branches: pseudoplastic, in which the flow index behavior (n) is smaller than the unity, and dilatant, in which the flow index behavior falls between 1 and 2. Many fluids in the oil industry display pseudoplastic behavior. The pressure derivative of a power-law fluid differs from that of a Newtonian one. For a Newtonian fluid, the flow index n is equal to unity, and the pressure derivative has a zero slope during the radial flow regime. For pseudoplastic fluids, the pressure derivative during the radial flow regime is a straight line with an increasing slope as the flow index behavior decreases. For dilatant fluids, the situation is the opposite.

Pseudoplastic and dilatant fluids have no yield point. For pseudoplastic fluids, the slope of shear stress versus shear rate decreases progressively and tends to become constant for high values of shear stress. The simplest model is power law:

$$`\tau = k\gamma^n; \quad n < 1 \tag{7.2}$$

k and n are constants that differ for each particular fluid. k measures the flow consistency and n measures the deviation from the Newtonian behavior, for which $k = \mu$ and $n = 1$ in Equation (7.2). Dilatants fluids are similar to pseudoplastic fluids, except that the apparent viscosity increases as the shear stress increases. The power-law model also describes the behavior of dilatant fluids, but $n > 1$.

For well-test interpretation, several analytical and numerical models taking into account Bingham and pseudoplastic non-Newtonian behavior have been introduced in the literature for a better understanding of reservoir behavior. Most of them dealt with fractured wells and homogeneous formations, and well-test interpretation was conducted via either the classical straight-line conventional analysis or type-curve matching. Only a few studies considered pressure-derivative analysis. However, there exists a need for a more practical and accurate way of characterizing such systems.

Many studies in petroleum engineering, chemical engineering and rheology have focused on non-Newtonian fluid behavior through porous formations; among them are Hirasaki and Pope (1974); Ikoku (1979); Ikoku and Ramey (1979); Odeh and Yang (1979); Savins (1969) and Van Poollen and Jargon (1969). Ikoku (1978) also presented several analytical solutions, including finite systems, which were used by Escobar, Vega and Bonilla (2012), and were also reported by Escobar (2012), for non-Newtonian fluids in homogeneous and heterogeneous reservoirs. Several numerical and analytical models have been proposed to study the transient behavior of non-Newtonian fluids in porous media. Since all were published before the 1980s, when the pressure derivative concept was nonexistent, the interpretation was conducted using either conventional analysis or type-curve matching. The conventional method is poor for the identification of the flow regime and has no verification. On the other hand, type-curve matching has been limited to regular identification and verification. Pressure derivative and deconvolution are very good for both purposes.

Vongvuthipornchai and Raghavan (1987) were the first to use the pressure-derivative concept for well-test analysis of non-Newtonian fluids, and, later on, Katime-Meindl and Tiab (2001) presented the first extension of the *TDS* technique to non-Newtonian fluids. Igbokoyi and Tiab (2007) used type-curve matching for interpretation of the pressure test for non-Newtonian fluids in infinite systems, including skin and wellbore storage effects. Recent applications of the derivative function to non-Newtonian system solutions are presented by Escobar, Martínez and Montealegre (2010) and Martínez, Escobar and Montealegre (2010a), who applied the *TDS* technique to radial composite reservoirs with a non-Newtonian/Newtonian interface for pseudoplastic and dilatant systems, respectively. However, all of the aforementioned cases considered infinite reservoirs, and, therefore, the drainage area was never involved. Therefore, Escobar, Vega and Bonilla (2012) formulated a practical interpretation methodology, based on the models proposed by Ikoku (1978), to find well-drainage area from transient-pressure analysis using the pressure-derivative curve.

BOUNDED RESERVOIRS

The analytical solution in the Laplace space domain for a reservoir with a no-flow (also referred as closed) external boundary and constant-rate production at the well is given as the closed reservoirs under constant-rate case is given by Ikoku (1978), as:

$$\overline{P}(z) = \frac{\left\{ K_{2/(3-n)}\left[\frac{2}{3-n}\sqrt{z}r_{eD}^{(3-n)/2}\right] \cdot I_{\frac{1-n}{3-n}}\left[\frac{2}{3-n}\sqrt{z}\right] + I_{2/(3-n)}\left[\frac{2}{3-n}\sqrt{z}r_{eD}^{(3-n)/2}\right] \cdot K_{\frac{1-n}{3-n}}\left[\frac{2}{3-n}\sqrt{z}\right]\right\}}{\left(z^{3/2}\left\{I_{2/(3-n)}\left[\frac{2}{3-n}\sqrt{z}r_{eD}^{(3-n)/2}\right] \cdot K_{2/(3-n)}\left[\frac{2}{3-n}\sqrt{z}\right] - K_{2/(3-n)}\left[\frac{2}{3-n}\sqrt{z}r_{eD}^{(3-n)/2}\right] \cdot I_{2/(3-n)}\left[\frac{2}{3-n}\sqrt{z}\right]\right\}\right)} \tag{9.1}$$

Using the solutions provided by Ikoku (1978), Escobar et al. (2012) presented pressure and pressure-derivative plots for such behaviors, as shown in Figures 9.1 and 9.2. It is seen in these plots that for closed systems, the late-time pressure-derivative behaviors for both pseudoplastic and dilatant cases always display unit-slope lines equivalent to the case, as for those of Newtonian fluids. As for the Newtonian behavior, the late-time pressure derivative decreases in both dilatant and pseudoplastic cases.

Equation (7.8) is rewritten based on the reservoir drainage area, so that:

$$t_{DA} = \frac{t}{G\pi r_e^{3-n}} \qquad (9.2)$$

Escobar et al. (2012) proposed a combination of Equations (9.2), (7.105) and (7.37) to develop an analytical expression to find the well-drainage area:

$$A = \pi \left[\frac{t_{rpiNN}}{G} \left(\frac{1}{4} \right)^{1/\alpha - 1} \right]^{2/(3-n)} \qquad (9.3)$$

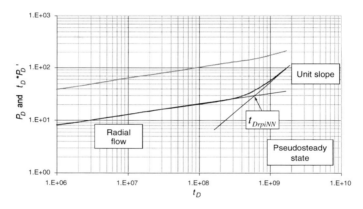

Figure 9.1. Dimensionless pressure and pressure-derivative versus dimensionless time behavior for closed-boundary system for a non-Newtonian pseudoplastic fluid with $n = 0.5$ and $r_e = 2000$ ft; adapted from Escobar et al. (2012).

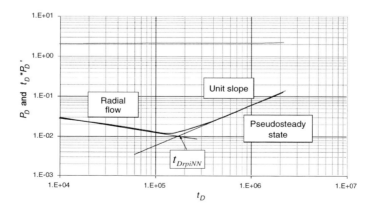

Figure 9.2. Dimensionless pressure and pressure-derivative versus dimensionless time behavior for closed-boundary system for a non-Newtonian dilatant fluid with $n = 1.5$ and $r_e = 2000$ ft; adapted from Escobar et al. (2012).

where t_{rpiNN} is the intersection point formed by the straight lines of the radial and pseudosteady-state flow regimes exhibited by the derivative response. Equation (9.3) was multiplied by the term $(\pi^{(1/\alpha-1)})^{1/3-n}$ to account for correction factor. This is valid for both dilatant and pseudoplastic non-Newtonian fluids. This correction factor was empirically included, since the solution deviates from its expected value as the absolute value of n increases.

CONSTANT-PRESSURE RESERVOIRS

The Laplace-space-domain analytical solution for a constant-pressure boundary system and constant-rate production at the well is given as the closed reservoirs under constant-rate case by Ikoku (1978) as:

$$\overline{P}(z) = \frac{\left\{I_{\frac{1-n}{3-n}}\left[\frac{2}{3-n}\sqrt{z}r_{eD}^{(3-n)/2}\right]\cdot K_{\frac{1-n}{3-n}}\left[\frac{2}{3-n}\sqrt{z}\right] - K_{\frac{1-n}{3-n}}\left[\frac{2}{3-n}\sqrt{z}r_{eD}^{(3-n)/2}\right]\cdot I_{\frac{1-n}{3-n}}\left[\frac{2}{3-n}\sqrt{z}\right]\right\}}{\left(z^{3/2}\left\{I_{2/(3-n)}\left[\frac{2}{3-n}\sqrt{z}\right]\cdot K_{\frac{1-n}{3-n}}\left[\frac{2}{3-n}\sqrt{z}r_{eD}^{(3-n)/2}\right] + K_{2/(3-n)}\left[\frac{2}{3-n}\sqrt{z}\right]\cdot I_{\frac{1-n}{3-n}}\left[\frac{2}{3-n}\sqrt{z}r_{eD}^{(3-n)/2}\right]\right\}\right)} \quad (9.4)$$

Figures 9.3 and 9.4 show the pressure and pressure-derivative behavior for constant-pressure systems. The late behavior is similar to that of Newtonian fluids. There is no pressure-derivative expression for open-boundary systems. Therefore, Escobar, Hernandez and Tiab (2010b) provided an approach by drawing a negative unit-slope line tangentially to the steady-state flow regime. Then, for pseudoplastic fluids, the following correlation, which works for the full range of dilatant fluids (which is $1 < n < 2$), was also developed:

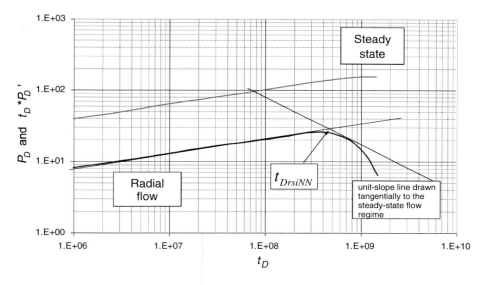

Figure 9.3. Dimensionless pressure and pressure derivative versus dimensionless time behavior for constant-pressure boundary system for a non-Newtonian pseudoplastic fluid with $n = 0.5$ and $r_e = 2000$ ft; adapted from Escobar et al. (2012).

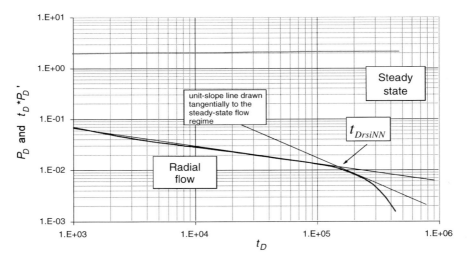

Figure 9.4. Dimensionless pressure and pressure-derivative versus dimensionless time behavior for constant-pressure boundary system for a non-Newtonian dilatant fluid with $n = 1.5$ and $r_e = 2000$ ft; adapted from Escobar et al. (2012).

$$t_{DA_{NN}} = -0.003n^2 + 0.0337n + 0.052 \qquad (9.5)$$

t_{DANN} refers to the dimensionless time based on the reservoir area given for non-Newtonian fluids. Equating Equations (9.5) and (9.3) and solving for reservoir drainage yields:

$$A = \pi \left[\frac{t_{rsiNN}}{G\pi\left(-0.003n^2 + 0.0337n + 0.052\right)} \right]^{2/(3-n)} \qquad (9.6)$$

For dilatant fluids, the correlation found is:

$$t_{DA_{NN}} = 0.9175n^3 - 3.7505n^2 + 5.1777n - 2.2913 \qquad (9.7)$$

By the same token, for the pseudoplastic case, the correlation is:

$$A = \pi \left[\frac{t_{rsiNN}}{G\pi\left(0.9175n^3 - 3.7505n^2 + 5.1777n - 2.2913\right)} \right]^{2/(3-n)} \qquad (9.8)$$

t_{rsiNN} in Equations (9.5) and (9.7) corresponds to the intersection point formed between the straight line of the radial and negative unit-slope line drawn tangentially to the steady-state flow regime exhibited by the derivative response.

Example 9.1

Escobar et al. (2012) presented a synthetic pressure test for a bounded reservoir, in which fluid displays a power-law pseudoplastic behavior. Pressure and pressure-derivative data for this example are reported in Table 9.1 and Figure 9.5, and other relevant information is given below:

$n = 0.5$	$h = 16.4$ ft	$k = 350$ md
$q = 300$ BPD	$\phi = 5\%$	$B_o = 1$ rb/STB
$\mu_{eff} = 20$ cp*s^{n-1}	$c_t = 0.0000689$ psi^{-1}	$r_w = 0.33$ ft
$H = 20$ cp*s^{n-1}	$r_e = 2000$ ft	$P_i = 2500$ psi

Solution

The effective viscosity is found with Equation (7.5) and parameter G with Equation (7.4):

$$\mu_{eff} = \left(\frac{20}{12}\right)\left(9 + \frac{3}{0.5}\right)^{0.5} \left(1.59344 \times 10^{-12}(350)(0.05)\right)^{(1-0.5)/2} = 1.28967 \times 10^{-2}\ \text{cp*(s/ft)}^{-0.5}$$

$$G = \frac{(3792.188)(0.5)(0.05)(0.0000689)(0.0129)}{350}\left(\frac{96681.605(16.4)}{300}\right)^{0.5} = 1.22487 \times 10^{-5}\ \text{hr/ft}^{2.5}$$

α is found with Equation (7.40):

$$\alpha = \frac{1-n}{3-n} = \frac{1-0.5}{3-0.5} = 0.2$$

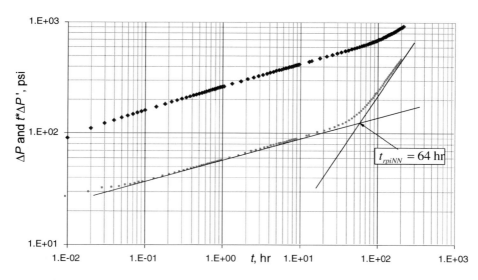

Figure 9.5. Pressure and pressure derivative versus time log-log plot for example 9.1; adapted from Escobar et al. (2012).

Finite Systems

Table 9.1. Pressure and pressure-derivative data for example 9.1

t, hr	ΔP, psi	$t*\Delta P'$, psi	t, hr	ΔP, psi	$t*\Delta P'$, psi	t, hr	ΔP, psi	$t*\Delta P'$, psi	t, hr	ΔP, psi	$t*\Delta P'$, psi
0	90.05	25.88	0.7	243.16	50.54	8.5	407.17	83.45	90	673.05	193.81
0	109.87	28.51	0.7	246.00	51.13	8.9	411.01	83.87	94	681.21	201.22
0	121.77	30.80	0.8	248.71	51.82	9.3	414.71	84.42	98	689.33	208.74
0	130.49	31.41	0.8	251.31	52.31	9.7	418.29	85.10	102	697.42	216.33
0.1	137.47	31.85	0.9	253.82	52.63	13	443.89	90.62	104	701.45	220.18
0.1	143.33	32.91	0.9	256.23	53.00	14	450.66	91.91	109	711.51	229.85
0.1	148.41	33.16	0.9	258.56	53.47	18	474.37	97.56	114	721.55	239.65
0.1	152.91	33.98	1	260.81	54.01	22	494.17	100.93	119	731.56	249.53
0.1	156.96	34.75	1	264.05	54.93	26	511.28	105.30	124	741.56	259.48
0.1	160.65	35.22	1.3	277.44	57.50	30	526.42	108.49	129	751.55	269.49
0.1	172.89	37.49	1.7	293.06	60.60	34	540.06	112.51	134	761.53	279.56
0.2	182.52	39.56	2.1	305.99	63.29	38	552.54	116.38	139	771.50	289.68
0.2	190.53	40.90	2.5	317.09	65.84	42	564.10	120.53	144	781.47	299.84
0.3	197.43	42.25	2.9	326.86	67.48	46	574.94	125.19	149	791.43	310.03
0.3	203.51	43.19	3.3	335.61	69.31	50	585.20	129.94	154	801.39	320.25
0.3	208.97	44.33	3.7	343.55	70.90	54	595.01	135.25	159	811.35	330.49
0.4	213.93	45.25	4.1	350.83	72.19	58	604.44	140.77	164	821.30	340.75
0.4	218.49	46.08	4.5	357.57	73.65	62	613.58	146.56	174	841.21	361.33
0.5	222.70	46.93	4.9	363.85	74.84	66	622.49	152.73	184	861.12	381.95
0.5	226.64	47.63	5.3	369.73	76.01	70	631.20	159.03	194	881.02	402.60
0.5	230.32	48.38	6.1	380.50	77.79	74	639.77	165.66	204	900.92	423.28
0.6	233.79	49.04	6.9	390.20	79.34	78	648.22	172.45	215	922.82	446.05
0.6	237.07	49.65	7.3	394.72	80.40	82	656.57	179.39			
0.7	240.19	50.11	8.1	403.19	82.59	86	664.84	186.56			

From Figure 9.5, the intercept point, t_{rpiNN}, of the radial and pseudosteady-state straight lines is 64 hr, which is used in Equation (9.3) to find the reservoir area:

$$A = \pi \left[\frac{64}{1.225 \times 10^{-5}} * \left(\frac{1}{4\pi} \right)^{1/0.2-1} \right]^{3/(3-0.5)} = 3.124 \times 10^{6} \, \text{ft}^2 = 71.72 \, \text{Acres}$$

Notice that this reservoir has an external radius of 2000 ft, which represents an area of 288.5 acres. This yields an absolute error of 0.224%.

Example 9.2

Escobar et al. (2012) also provided a synthetic pressure for a dilatant fluid, for which the pressure and pressure derivative versus time log-log plot is given in Figure 9.6 and Table 9.2. Other relevant data are given in example 9.1, except that $n = 1.5$.

Table 9.2. Pressure and pressure-derivative data for example 9.2

t, hr	ΔP, psi	t*ΔP', psi	t, hr	ΔP, psi	t*ΔP', psi	t, hr	ΔP, psi	t*ΔP', psi	t, hr	ΔP, psi	t*ΔP', psi
0.01	7215.73	3545.88	0.7	10492.14	284.40	13	11020.74	106.79	90	11170.50	45.85
0.03	8767.72	1105.56	0.74	10507.82	279.35	14	11028.54	104.09	98	11174.35	42.14
0.05	9224.10	797.57	0.78	10522.40	275.41	18	11053.59	97.43	102	11176.04	40.31
0.07	9468.87	677.75	0.82	10536.01	270.54	22	11072.14	89.72	104	11176.83	39.46
0.08	9556.33	640.87	0.86	10548.76	265.20	26	11086.66	85.88	109	11178.66	37.30
0.09	9629.47	613.10	0.9	10560.74	260.37	30	11098.47	80.98	114	11180.31	35.28
0.1	9691.94	581.97	0.98	10582.68	253.55	34	11108.35	77.97	119	11181.79	33.32
0.14	9874.65	512.20	1.04	10597.62	250.14	38	11116.78	74.78	124	11183.11	31.44
0.18	9996.39	473.03	1.32	10654.60	230.27	42	11124.11	71.92	129	11184.31	29.62
0.22	10085.68	436.37	2.12	10755.06	197.06	46	11130.54	69.49	134	11185.38	27.95
0.26	10155.16	411.04	2.92	10814.42	176.99	50	11136.25	66.90	139	11186.34	26.41
0.3	10211.43	386.08	3.72	10855.25	163.51	54	11141.36	64.70	144	11187.20	24.98
0.34	10258.34	371.18	4.52	10885.78	152.96	58	11145.96	62.42	149	11187.98	23.65
0.38	10298.31	355.93	4.92	10898.46	148.37	62	11150.12	60.17	159	11189.30	21.34
0.42	10332.95	342.92	5.72	10920.11	141.13	66	11153.89	58.09	169	11190.37	19.36
0.46	10363.41	332.64	6.92	10945.97	130.71	70	11157.32	55.88	179	11191.24	17.73
0.5	10390.49	322.14	7.32	10953.29	128.59	74	11160.45	53.84	189	11191.93	16.36
0.54	10414.79	314.13	8.12	10966.45	125.16	78	11163.31	51.80	199	11192.49	15.25
0.58	10436.79	306.24	8.92	10977.98	120.82	82	11165.92	49.73	204	11192.73	14.76
0.66	10475.20	291.52	9.72	10988.21	117.00	86	11168.31	47.81	215	11193.18	13.89

Solution

As for the former example, the effective viscosity is found with Equation (7.5) and parameter *G* with Equation (7.4):

$$\mu_{eff} = \left(\frac{20}{12}\right)\left(9+\frac{3}{1.5}\right)^{1.5}\left(1.59344\times10^{-12}(350)(0.05)\right)^{(1-1.5)/2} = 27499.9\,\text{cp*(s/ft)}^{0.5}$$

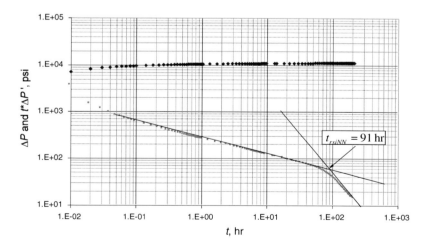

Figure 9.6. Pressure and pressure derivative versus time log-log plot for example 9.2; adapted from Escobar et al. (2012).

$$G = \frac{(3792.188)(1.5)(0.05)(0.0000689)(27499.9)}{350}\left(\frac{96681.605(16.4)}{300}\right)^{-0.5}$$

$G = 0.00247 \text{ hr/ft}^{2.5}$

α is found with Equation (7.40):

$$\alpha = \frac{1-n}{3-n} = \frac{1-1.5}{3-1.5} = -0.33334$$

For dilatant fluids, the correlation is found with Equation (9.7);

$$t_{DA_{NN}} = 0.9175(1.5)^3 - 3.7505(1.5)^2 + 5.1777(1.5) - 2.2913 = 0.1315$$

From Figure 9.4, the intercept point, t_{rsiNN}, of the radial and steady-state straight lines is 91 hr. This value is used in Equation (9.8) to find the well-drainage area:

$$A = \frac{\pi}{43560}\left[\frac{91}{(0.00247)\pi(0.1315)}\right]^{2/(3-1.5)} = 287.22 \text{ Acres}$$

Compared to the actual area, there is an absolute error of 0.219%.

Nomenclature

B	Volumetric factor, RB/STB
c_t	System total compressibility, 1/psi
h	Formation thickness, ft
H	Consistency (power-law parameter), $cp*s^{n-1}$
G	Group defined by Equation (7.4)
G	Minimum pressure gradient, psi/ft
G_D	Dimensionless pressure gradient
k	Permeability, md
k	Flow consistency parameter
m	Slope
n	Flow behavior index (power-law parameter)
P	Pressure, psi
q	Flow/injection rate, STB/D
t	Time, hr
r	Radius, ft
r_e	Reservoir external radius, ft
$t*\Delta P'$	Pressure derivative, psi

| t_D*P_D' | Dimensionless pressure derivative |
| z | Laplace parameter |

Greeks

α	Slope of the pressure-derivative curve during radial flow regime
Δ	Change, drop
ϕ	Porosity, Fraction
μ_{eff}	Effective viscosity for power-law fluids, $cp*(s/ft)^{n-1}$

Suffixes

app	Apparent
D	Dimensionless
DA	Dimensionless with respect to area
e	External
eff	Effective
i	Initial
NN	Non-Newtonian
r	Radial flow regime
rNN	Radial (any point on radial flow)
rpiNN	Intersect of radial and pseudosteady-state lines
rsiNN	Intersect of radial line and negative-unit slope tangent to steady state period
w	Wellbore

REFERENCES

Escobar, F. H. (2012). *"Transient Pressure and Pressure Derivative Analysis for Non-Newtonian Fluids." New Technologies in the Oil and Gas Industry*, Dr. Jorge Salgado Gomes (Ed.), ISBN: 978-953-51-0825-2, InTech, DOI: 10.5772/50415. Available from: http://www.intechopen.com/books/new-technologies-in-the-oil-and-gas-industry/transient-pressure-and-pressure-derivative-analysis-for-non-newtonian-fluids.

Escobar, F. H., Martínez, J. A. & Montealegre, M. (2010a). "Pressure and pressure derivative analysis for a well in a radial composite reservoir with a Non-Newtonian/Newtonian interface." *CT&F – Ciencia, Tecnología y Futuro, 4* (2),33-42.

Escobar, F. H., Hernandez, Y. A. & Tiab, D. (2010b). "Determination of reservoir drainage area for constant-pressure systems using well test data." *CT&F – Ciencia, Tecnología y Futuro*. Vol. *4*, No. 1. p. 51-72. June.

Escobar, F. H., Vega, L. J. & Bonilla, L. F. (2012). "Determination of Well-Drainage Area for Power-Law Fluids by Transient Pressure Analysis." *CT&F*. Vol. 5 No. 1. P. 45-55. Dec.

Hirasaki, G. J. & Pope, G. A. (1974). "Analysis of factors influencing mobility and adsorption in the flow of polymer solutions through porous media." *SPE Journal, 14*(4), 337-346.

Igbokoyi, A. & Tiab, D. (2007). "New type curves for the analysis of pressure transient data dominated by skin and wellbore storage: Non-Newtonian fluid." *SPE Production and Operations Symposium*. Oklahoma City, U.S.A. SPE 106997.

Ikoku, C. U. (1978). Transient flow of Non-Newtonian Power-Law fluids in porous media. Ph. D. Thesis dissertation, Petroleum Engineering Department, Stanford University, Stanford, U.S.A., 257pp.

Ikoku, C. U. (1979). "*Practical application of Non-Newtonian transient flow analysis.*" *SPE* 64th *Annual Technical Conference and Exhibition*, Las Vegas, U.S.A., SPE 8351.

Ikoku, C. U. & Ramey, H. J. Jr. (1979). "Transient flow of Non-Newtonian Power-law fluids through in porous media." *SPE Journal*. 164-174.

Katime-Meindl, I. & Tiab, D. (2001). "Analysis of pressure transient test of non-newtonian fluids in infinite reservoir and in the presence of a single linear boundary by the direct synthesis technique." *SPE Annual Technical Conference and Exhibition*, New Orleans, U.S.A. SPE 71587.

Martínez, J. A., Escobar, F. H. & Montealegre, M. (2011). "Vertical well pressure and pressure derivative analysis for Bingham fluids in a homogeneous reservoirs." *Dyna, 78*(166), 21-28.

Odeh, A. S. & Yang, H. T. (1979). "Flow of non-Newtonian Power-Law fluids through in porous media." *SPE Journal. 19*(3), 155-163.

Savins, J. G. (1969). Non-Newtonian flow Through in Porous Media. *Industrial and Engineering Chemistry. 61*(10),18-47.

van Poollen, H. K. & Jargon, J. R. (1969). "Steady-State and Unsteady-State Flow of Non-Newtonian Fluids Through Porous Media." *SPE Journal. 9*(1), 80-88.

Vongvuthipornchai, S. (1985). Well Test analysis for Non-Newtonian Fluid Flow. PhD. Dissertation, Petroleum Engineering Department, University of Tulsa, Tulsa, U.S.A. 376pp.

Vongvuthipornchai, S. & Raghavan, R. (1987). "Well Test Analysis of Data Dominated by Storage and Skin: Non-Newtonian Power-Law Fluids." *SPE Formation Evaluation. 2*(4),618-628.

Chapter 10

NATURALLY FRACTURED RESERVOIRS AND BINGHAM PLASTIC FLUIDS

BACKGROUND

Taking into account that more than half of the world's reserves are contained in naturally fractured occurring formations, it may be possible to find in them some cases of a heavy oil that could behave non-Newtonianly, in need of an appropriate characterization. It has recently been found that porous material and natural fractures may have a fractal distribution. Among others, Camacho, Fuentes-Cruz and Vazquez (2008) have devoted their attention to the study of naturally fractured fractal reservoirs. Escobar, Salcedo and Pinzon (2015a) conducted a study on fractal reservoirs with Newtonian power-law fluids. However, they did not elaborate on the estimation of the flow behavior index. This is because the pressure derivative slope during the radial flow regime changes as a result of the combined effect of the flow behavior index and the fractal dimension. Therefore, the flow behavior index, n, must be known. For this reason, Escobar et al. (2015)'s work will not be included in this book. Instead, it will be included in a later work by Escobar, Morales-Lopez and Gomez (2015b) on a methodology for interpretation in naturally fractured fractal reservoirs.

Laboratory experiments and field tests indicate that certain fluids exhibit Bingham-type, non-Newtonian behavior in porous media. In these cases, flow takes place only once the applied pressure gradient exceeds a certain minimum value, known as the threshold pressure gradient. The flow of oil in many heavy oil reservoirs does not follow Darcy's law but may be approximated by a Bingham fluid. This chapter will include an interpretation technique for Bingham fluids in vertical wells. Although a work by Escobar, Zhao and Zhang (2014) on the threshold gradient of horizontal wells may fit into this chapter, it will not be included here.

NON-NEWTONIAN FLOW IN NATURALLY FRACTURED DOUBLE-POROSITY SYSTEMS

Before the recent work of Escobar et al. (2011), which provided an interpretation technique based on features found on the pressure and pressure-derivative plot, and the work of Escobar, Martinez and Silva (2013), which implemented conventional analysis, only the work presented by Olarewaju (1992) is reported in the literature as far as non-Newtonian fluid flow through naturally fractured reservoirs is concerned. Olarewaju presented an analytical solution for the transient behavior of double-porosity infinite formations, which bear non-Newtonian pseudoplastic fluids, and his analytical solution also considered wellbore storage effects and skin factor. However, he presented neither well-test data interpretation techniques nor examples, and his application was focused only on a homogeneous case.

The viscosity of a non-Newtonian fluid varies with pressure and temperature. For a pseudoplastic fluid, the slope of the shear stress versus shear rate decreases progressively and tends to become constant for high values of shear stress. Its most simplistic model is the power law, defined by:

$$\tau = k\gamma^n; \; n < 1 \tag{7.2}$$

k measures the fluid consistency, and n measures the deviation from the Newtonian behavior. Both parameters differ for each fluid. The governing well-pressure solution in the Laplacian domain for a double-porosity system with a non-Newtonian fluid, including wellbore storage and skin effects, was provided by Olarewaju (1992) as:

$$\overline{P}_{DNN} = \frac{K_{\frac{1-n}{3-n}}\left(\frac{2}{3-n}\sqrt{zf(z)}\right) + s\sqrt{zf(z)}K_{\frac{2}{3-n}}\left(\frac{2}{3-n}\sqrt{zf(z)}\right)}{z\left((zf(z))^{1/2}K_{\frac{2}{3-n}}\left(\frac{2}{3-n}\sqrt{zf(z)}\right) + \tilde{s}C_D\left[K_{\frac{1-n}{3-n}}\left(\frac{2}{3-n}\sqrt{zf(z)}\right) + s\sqrt{zf(z)}K_{\frac{2}{3-n}}\left(\frac{2}{3-n}\sqrt{zf(z)}\right)\right]\right)} \tag{10.1}$$

Neglecting skin and wellbore storage effects, Equation (10.1) becomes:

$$\overline{P}_{DNN} = \frac{K_{\frac{1-n}{3-n}}\left(\frac{2}{3-n}\sqrt{zf(z)}\right)}{z\left(\sqrt{zf(z)}K_{\frac{2}{3-n}}\left(\frac{2}{3-n}\sqrt{zf(z)}\right)\right)} \tag{10.2}$$

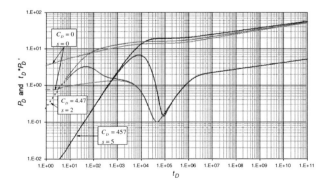

Figure 10.1. Effects of wellbore storage and skin on the dimensionless pressure and pressure derivative for a double-porosity system bearing a Newtonian fluid. $\omega = 0.01$ and $\lambda = 1 \times 10^{-6}$; adapted from Escobar et al. (2011).

The Laplacian parameter, $f(z)$, is a function of the model type and fracture-system geometry and is given by:

$$f(z) = \frac{\omega(1-\omega)z + \lambda}{(1-\omega)z + \lambda} \tag{10.3}$$

TDS TECHNIQUE FOR NATURALLY FRACTURED RESERVOIRS

Equation (7.22) is rewritten as:

$$C = \frac{k_{fb}ht_i}{10605.44e^{-1.85n}\mu_{eff}\left(96681.605\dfrac{hr_w}{qB}\right)^{1-n}} \tag{10.4}$$

Escobar et al. (2010) provided equations for reservoir permeability and skin factor for homogeneous systems, which can be extended to double-porosity systems as:

$$\frac{k_{fb}}{\mu_{eff}} = \left[70.6(96681.605)^{(1-\alpha)(1-n)}\left(\frac{0.0002637t_{r1NN}}{n\omega\phi c_t}\right)^{\alpha}\left(\frac{qB}{h}\right)^{n-\alpha(n-1)}\left(\frac{r_w^{\alpha(n-3)+(1-n)}}{(t*\Delta P')_{r1NN}}\right)\right]^{\frac{1}{1-\alpha}} \tag{10.5}$$

$$\frac{k_{fb}}{\mu_{eff}} = \left[70.6(96681.605)^{(1-\alpha)(1-n)}\left(\frac{0.0002637t_{r2NN}}{n\phi c_t}\right)^{\alpha}\left(\frac{qB}{h}\right)^{n-\alpha(n-1)}\left(\frac{r_w^{\alpha(n-3)+(1-n)}}{(t*\Delta P')_{r2NN}}\right)\right]^{\frac{1}{1-\alpha}} \tag{10.6}$$

$$s = \frac{1}{2}\left[\left[\frac{0.0002637k_{fb}t_{r1NN}}{n\omega\phi\mu_{eff}c_t r_w^{3-n}}\left(96681.605\frac{h}{qB}\right)^{n-1}\right]^{\alpha}\left(\frac{\Delta P}{(t*\Delta P)}\right)_{r1NN} - \ln\left(\frac{k_{fb}t_{r1NN}}{n\omega\phi\mu_{eff}c_t r_w^{3-n}}\left(96681.605\frac{h}{qB}\right)^{n-1}\right) + 7.43\right] \tag{10.7}$$

$$s = \frac{1}{2}\left[\left[\frac{0.0002637k_{fb}t_{r2NN}}{n\phi\mu_{eff}c_t r_w^{3-n}}\left(96681.605\frac{h}{qB}\right)^{n-1}\right]^{\alpha}\left(\frac{\Delta P}{(t*\Delta P)}\right)_{r2NN} - \ln\left(\frac{k_{fb}t_{r2NN}}{n\phi\mu_{eff}c_t r_w^{3-n}}\left(96681.605\frac{h}{qB}\right)^{n-1}\right) + 7.43\right] \tag{10.8}$$

Equations (10.5) and (10.7) use the radial flow regime before the transition period, while Equations (10.6) and (10.8) use the radial flow (second) during homogeneous behavior. However, the first radial flow (heterogeneous behavior) may be masked by wellbore storage.

For practical purposes, instead of Equations (10.7) and (10.8), it is better to use the skin factor equation introduced by Escobar, Martinez and Bonilla (2012):

$$s = \frac{1}{2}\left(\frac{(\Delta P)_{r1NN}}{(t*\Delta P')_{r1NN}} - \frac{1}{\alpha}\right)\left(\frac{t_{r1NN}}{\omega G\, r_w^{3-n}}\right)^{\alpha} \tag{10.9}$$

$$s = \frac{1}{2}\left(\frac{(\Delta P)_{r2NN}}{(t*\Delta P')_{r2NN}} - \frac{1}{\alpha}\right)\left(\frac{t_{r2NN}}{G\,r_w^{3-n}}\right)^\alpha \qquad (10.10)$$

Figure 10.1 displays the dimensionless pressure and pressure derivative versus time log-log plot behavior for a non-Newtonian fluid in a double-porosity system with wellbore storage and skin effects. Estimations of the wellbore storage coefficient and skin factors have already been studied in Chapter 7 for homogeneous reservoirs, and their development can also be applied for naturally fractured systems. Notice in Figure 10.1 that wellbore storage affects the minimum of the pressure derivative (trough), impacting the estimation of the dimensionless storativity ratio and much less impact on the interporosity flow parameter as found by Tiab Igbokoyi and Restrepo (2007). This correction was first presented for Newtonian fluids by Engler and Tiab (1996) and was later improved by Tiab et al. (2007). Both cases can be extended to non-Newtonian fluids.

Figure 10.2 contains an ideal pressure and pressure derivative log-log plot for a Newtonian double-porosity systems. In this plot, the infinite-acting radial flow regime is represented by a horizontal straight line on the pressure-derivative curve. The first segment corresponds to pressure depletion in the fracture network, while the second portion is due to the pressure response of an equivalent homogeneous reservoir. On the other hand, the transition period, which displays a trough on the pressure-derivative curve during the transition period only, depends on the dimensionless fracture storativity ratio, ω. The characteristic features depicted in Figure 10.2 are used to adequately characterize the reservoir system, as described in detail by Engler and Tiab (1996). Escobar et al. (2011) took the same characteristic points into account to generate relationships to estimate the flow behavior index, n, and the interporosity flow coefficient and the storativity ratio. For completeness, the reader should refer to Chapter 7, where expressions to determine permeability, skin factor and wellbore storage are provided. In the examples used in this chapter, however, these properties are calculated.

As seen in Figure 10.2, the radial flow for the Newtonian case ($n = 1$) is characterized by a zero-slope straight line on the pressure-derivative curve. As the value of n decreases, the slope of the derivative during radial flow increases, as shown in Figures 10.3 through 10.5. The dimensionless pressure and dimensionless pressure derivative log-log plot in Figure 10.3 is given for a constant interporosity flow parameter, a constant n value and a variable dimensionless fracture storativity ratio. It is observed in this plot that the higher the ω, the less pronounced the trough. In Figure 10.4, the effect of the variable of the interporosity flow parameter for constant values of dimensionless storativity ratio and flow behavior index is shown. Notice in that plot that as the value of λ decreases, the transition period shows up later. Finally, Figure 10.5 shows the effect of changing the value of the flow behavior index for constant values of λ and ω. The effect of the increasing the slope of the pressure derivative curve is observed as the value of n decreases. Needless to say, neither wellbore storage nor skin effects are considered in Figure 10.5.

The infinite-acting radial flow regime is identified by a straight line for which the slope increases as the value of the flow behavior index decreases (see Figure 10.5). The first segment of this line corresponds to the fracture-network-dominated period, and the second one (once the transition effects are no longer present) responds for an equivalent homogeneous reservoir. The slope of the pressure-derivative curve is also defined by:

$$\alpha = \frac{1-n}{3-n} \tag{7.40}$$

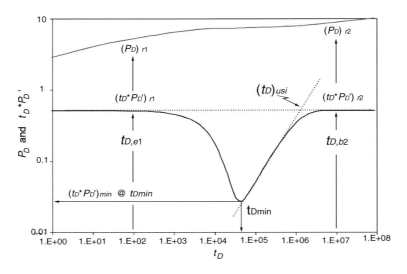

Figure 10.2. Characteristic points and lines for a naturally fractured reservoir bearing a Newtonian fluid. $\omega = 0.1$ and $\lambda = 1 \times 10^{-6}$; adapted from Engler and Tiab (1996).

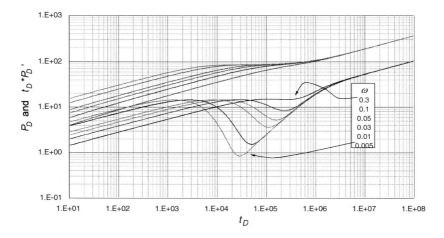

Figure 10.3. Dimensionless pressure and pressure derivative log-log plot for variable dimensionless storage coefficient, $\lambda = 1 \times 10^{-6}$ and $n = 0.2$ for a heterogeneous reservoir; adapted from Escobar et al. (2011).

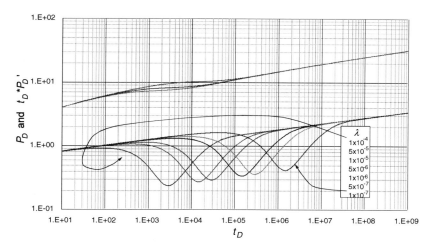

Figure 10.4. Dimensionless pressure and pressure derivative log-log plot for variable interporosity flow parameter, $\omega=0.05$ and $n=0.8$ for a heterogeneous reservoir; adapted from Escobar et al. (2011).

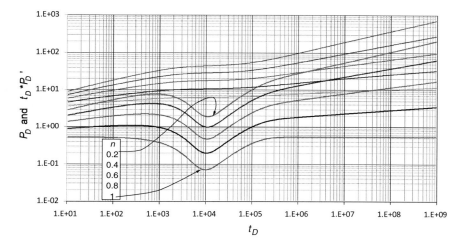

Figure 10.5. Dimensionless pressure and pressure derivative log-log plot for variable flow behavior index, $\omega=0.03$ and $\lambda=1\times10^{-5}$ for a heterogeneous reservoir; adapted from Escobar et al. (2011).

Also, the slope of the pressure derivative during the radial flow regime is related to the flow behavior index by:

$$n = -1.8783425 - 7.8618321m^3 + 0.19406557m^{0.5} + 2.8783425e^{-\alpha} \quad (10.11)$$

As observed in Figure 10.3, as the dimensionless storativity ratio decreases, the transition period is more pronounced, no matter the value of the interporosity flow parameter. Therefore, a correlation for ω as a function of the minimum time value of the pressure derivative at the transition point (trough), the flow behavior index and the beginning of the second infinite-acting radial flow regime was developed by Escobar et al. (2011) and also presented by Escobar (2012):

$$\frac{1}{\omega} = \left| \frac{3180.6369 + 551.0582\left(\ln \frac{t_{min}}{t_{b2}} \right)^2 - \frac{2053.5888}{x^{0.5}} + \frac{75.337547}{x}}{-\frac{1.4787073}{x^{1.5}} - \frac{910.05377}{n^{0.5}} + \frac{988.80592}{n} - \frac{459.61296}{n^{1.5}} + \frac{73.93695}{n^2}} \right| \quad (10.12)$$

The above correlation is valid for $0 \le \omega \le 1$, with an error lower than 3%.

Another way to estimate ω uses a correlation that is a function of the intersection time between the unit-slope pseudosteady-state straight line developed during the transition period and the time at the trough. This correlation is also valid for $0 \le \omega \le 1$, with an error lower than 0.7%.

$$\omega = 0.019884508 - \frac{1.153351}{y} + \frac{43.428536}{y^2} - \frac{555.85387}{y^3} +$$
$$\frac{3232.6805}{y^4} - \frac{6716.9801}{y^5} - \frac{0.0093613189}{n} + \frac{0.0042870178}{n^2} + \quad (10.13)$$
$$\frac{0.00027356586}{n^3} - \frac{0.0005221335}{n^4} + \frac{0.000072466135}{n^5}$$

A final correlation to estimate ω, valid in the range $0 \le \omega \le 1$, with an error lower than 0.4%, is given as follows:

$$\omega = \frac{\begin{bmatrix} -0.098427346 + 0.00046337048y + 0.000025063353y^2 - \\ 0.00000050316996y^3 + 0.0036057682n - 0.0073959605n^2 \end{bmatrix}}{1 - 0.36468068y - 0.064934748n - 0.047596083n^2} \quad (10.14)$$

The interporosity flow parameter also plays an important role in the characterization of double-porosity systems. From Figure 10.4, it is observed that the smaller the value of λ, the later the transition period will occur. Based on this observation, a correlation is obtained using the time at the trough and the dimensionless storativity ratio, as presented by the following expression:

$$\lambda = \frac{\begin{pmatrix} 6.9690127 \times 10^{-7} + 3.4893658 \times 10^{-8}n - 3.2315082 \times 10^{-8}n^2 - \\ 5.9013807w + 21571690w^2 + 3.6102987 \times 10^{12}w^3 \end{pmatrix}}{1 + 0.0099353372n - 3740035.1w + 6.7143604 \times 10^{12}w^2} \quad (10.15)$$

Equation 10.15 is valid for $1 \times 10^{-4} < \lambda < 9 \times 10^{-7}$, with an error lower than 4%. A correlation involving the coordinates of the trough is given as:

$$\lambda = -0.00082917155 - 0.0014247498n - 0.00028717451\,Z - 0.00077173053n^2$$
$$- 3.2538271 \times 10^{-5} Z^2 - 0.0003203949nz - 0.0001423889n^3 - 1.212213 \times 10^{-6} Z^3 \quad (10.16)$$
$$- 1.7831692 \times 10^{-5} nZ^2 - 8.6457217 \times 10^{-5} n^2 Z$$

which is valid for $1 \times 10^{-4} < \lambda < 9 \times 10^{-7}$, with an error lower than 3.7%. Another expression for λ within the same range, involving minimum time at the trough, is given for an error lower than 1.3%:

$$\ln \lambda = -2.1223034 - 0.09473309n + 0.077489686n^{0.5} \ln(n) - \frac{0.010651118}{n^{0.5}} - \frac{0.043958503}{w^{0.5}}$$
$$+ \frac{1.5653137 \times 10^{-5} \ln w}{w} + \frac{0.00024143014}{w} + \frac{8.7148736 \times 10^{-9}}{w^{1.5}} - \frac{4.0331364 \times 10^{-13}}{w^2} \quad (10.17)$$

Example 10.1

Figure 10.6 and Table 10.1 contain pressure and pressure derivative versus time data from a pressure test simulated by Escobar et al. (2011), with the information given below:

q = 1000 BPD $\qquad k$ = 1000 md $\qquad \phi$ = 0.05
r_w = 0.25 ft $\qquad \mu_{eff}$ = 5 cp $\qquad h$ = 50 ft
n = 0.76 $\qquad B$ = 1.2 rb/STB $\qquad \lambda$ = 4.5x10^{-6}
c_t = 1x10^{-6} psi^{-1} $\qquad \omega$ = 0.05

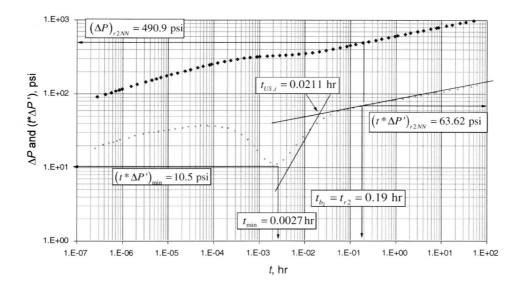

Figure 10.6. Pressure and pressure derivative log-log plot for synthetic example 10.1.

Naturally Fractured Reservoirs and Bingham Plastic Fluids

Table 10.1. Pressure data for example 10.1

t, hr	ΔP, psi	$t*\Delta P'$, psi	t, hr	ΔP, psi	$t*\Delta P'$, psi	t, hr	ΔP, psi	$t*\Delta P'$, psi
2.83E-07	91.19	17.11	1.75E-04	272.83	33.30	0.190	490.88	63.62
4.25E-07	98.43	19.08	2.38E-04	283.07	31.98	0.255	509.99	66.33
5.67E-07	104.03	20.30	3.33E-04	293.52	29.40	0.362	533.56	68.55
7.08E-07	108.63	20.61	4.60E-04	302.62	26.25	0.492	555.03	71.23
8.50E-07	112.54	21.55	6.33E-04	310.42	22.80	0.687	579.22	74.19
9.92E-07	115.95	22.34	8.86E-04	317.24	18.38	0.947	603.33	77.07
1.56E-06	126.47	23.64	1.03E-03	319.79	15.98	1.096	614.56	77.33
2.12E-06	134.12	25.50	1.48E-03	324.76	12.19	1.597	644.43	79.81
3.26E-06	145.21	26.77	1.99E-03	327.88	10.85	2.117	667.58	83.80
4.39E-06	153.32	28.03	2.83E-03	331.17	10.38	3.045	698.51	86.28
5.52E-06	159.75	28.44	3.84E-03	334.29	11.79	4.085	724.39	89.78
6.94E-06	166.32	29.17	5.37E-03	338.49	14.79	5.793	756.26	92.55
9.21E-06	174.68	29.84	7.40E-03	343.76	18.94	7.873	785.24	97.11
1.18E-05	182.11	30.32	1.04E-02	351.15	24.45	8.616	793.94	97.16
1.60E-05	191.75	31.53	1.48E-02	360.75	30.77	12.775	832.96	99.57
2.34E-05	203.95	32.53	2.09E-02	372.58	37.32	17.232	863.72	105.22
3.07E-05	213.02	33.55	2.96E-02	386.90	43.65	25.550	905.75	107.99
4.43E-05	225.42	34.01	4.18E-02	403.19	49.17	34.463	938.89	113.34
5.91E-05	235.30	34.77	5.92E-02	421.39	54.36	51.100	984.16	116.32
8.17E-05	246.60	34.61	8.36E-02	440.73	57.74	68.926	1019.85	122.75
9.08E-05	250.27	34.40	9.98E-02	451.11	58.30			
1.23E-04	260.69	34.34	0.132	468.09	60.95			

The dimensionless storativity ratio and the interporosity flow parameter are estimated from these data.

Solution

From Figure 10.6, the following characteristic points are read:

t_{min} = 0.0027 hr $\qquad$ t_{b2} = 0.19 hr

$t_{US,i}$ =0.0211 hr $\qquad$ $(t*\Delta P')_{min}$ = 10.5 psi

Using Equations (7.8) and (7.37), the above data are transformed into dimensionless quantities as follows:

t_{Dmin} =27619 $\qquad$ t_{Db2} = 851939

$t_{DUS,i}$ =300000 $\qquad$ $(t_D*P_D')_{min}$ =0.357

A slope of α = 0.108 is estimated with two points on the second straight-line portion of the plot. Equation (7.40) yields a flow behavior index of 0.759989.

During the infinite-acting radial flow regime, the following points are arbitrarily read:

$(t)_{r2NN}$ =0.19 hr $(\Delta P)_{r2NN}$ = 490.9 psi $(t*\Delta P')_{r2NN}$ = 63.62 psi

Permeability and skin factors are found, using Equations (7.6) and (7.8), at a value of 1026 md and a skin factor of 3.56, respectively. The results agree with the input value of the simulation.

$$k_{fb} = 5\left[\frac{70.6(96681.605)^{(1-0.108)0.76}\left(\dfrac{0.0002637(0.19)}{0.76(0.05)(1\times10^{-6})}\right)^{0.108}}{\left(\dfrac{1000(1.2)}{50}\right)^{0.76-0.108(0.76-1)}\left(\dfrac{0.25^{0.108(0.76-3)+(1-0.76)}}{63.62}\right)}\right]^{\frac{1}{1-0.108}} = 1026 \text{ md}$$

$$s = \frac{1}{2}\left[\left[\frac{0.0002637(1026)(0.19)}{(0.76)(0.05)(5)(1\times10^{-6})(0.25)^{3-0.76}}\left(96681.605\frac{50}{1000(1.2)}\right)^{0.76-1}\right]^{0.108} \left(\frac{490.9}{63.62}\right) - \ln\left(\frac{1026(0.19)}{(0.76)(0.05)(5)(1\times10^{-6})(0.25)^{3-0.76}}\left(96681.605\frac{50}{1000(1.2)}\right)^{n-1}\right) + 7.43\right] = 3.56$$

The remaining calculations are reported in Table 10.2.

Table 10.2. Summary of results for example 10.1

Equation	ω	Equation	λ
10.12	0.0299	10.15	5.010×10^{-6}
10.13	0.0303	10.16	5.036×10^{-6}
10.14	0.0304	10.17	3.656×10^{-6}

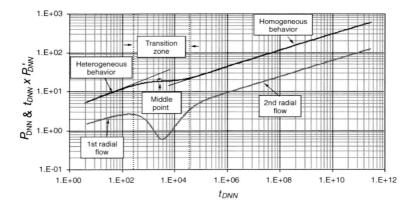

Figure 10.7. Dimensionless pressure and pressure derivative log-log plot displaying the typical behavior of a dual-porosity reservoir for ω=0.01, λ=1x10^{-7} and n=0.25; adapted from Escobar et al. (2013).

CONVENTIONAL ANALYSIS FOR NATURALLY FRACTURED RESERVOIRS

Escobar, Martinez and Silva (2013) presented expressions to complement the conventional straight-line method, so that pressure tests in naturally fractured reservoirs with non-Newtonian power laws can be interpreted. This is accomplished mainly by estimating the interporosity flow parameter and dimensionless storage coefficient. The developed equations are successfully tested on well-pressure tests reported in the literature.

Escobar et al. (2013) based their work on observations of the pressure behavior against time to estimate both the dimensionless storage coefficient, ω, and the interporosity flow parameter, λ. Estimations of permeability and skin factor were presented in Chapter 7. Although conventional analysis is not based on the pressure-derivative plot, it is not unacceptable to use it as reference. Therefore, Figure 10.7 illustrates some characteristic points found on the dimensionless pressure and pressure derivative versus time log-log plot for a naturally fractured reservoir with a power-law fluid with a flow behavior index, n, of 0.25. Notice that when radial flow occurs in both systems, fractures and matrix-fracture set, two straight lines are seen in that plot, and they are interrupted by a depression caused by the transition from heterogeneous to equivalent homogeneous behavior. The first straight line corresponds to an early time when fractures dominate the flow, and the second line is for a late time, corresponding to the response of an equivalent homogeneous reservoir. Note that the transition middle point of the pressure curve corresponds approximately to the minimum point of the pressure-derivative curve. As far as pressure derivative is concerned, during the radial flow regime, for the case in which the flow behavior index is less than one ($0 < n < 1$, unconventional pseudoplastic fluid), the developed slope decreases as the n value increases until becoming fully flat (zero slope), as happens for the conventional-fluid case, which means n equals unity, making it a Newtonian fluid.

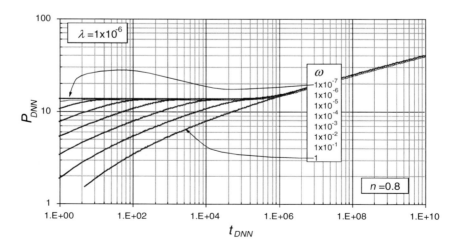

Figure 10.8. Effect of the dimensionless storage coefficient on the pressure response of a heterogeneous reservoir; adapted from Escobar et al. (2013).

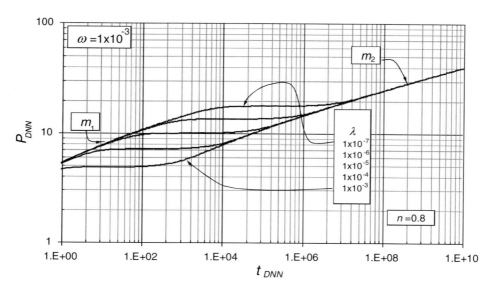

Figure 10.9. Effect of the interporosity flow parameter on the pressure response in a heterogeneous reservoir; adapted from Escobar et al. (2013).

The emphasis of Escobar et al. (2013)'s work was to develop expressions for the application of the straight-line, conventional-analysis interpretation methodology for estimating naturally fractured (dual-porosity) reservoir parameters. The idea was to use the slope, the intercept and the relationship of logarithmic cycles of a pressure vs. time behavior log-log plot, which are used in the calculation of the parameters. Figure 10.8 shows a log-log plot of dimensionless pressure vs. dimensionless time, with the dimensionless storage coefficient varying in the range of 1×10^{-7} to 1 and with constant values of λ and n. Neither wellbore storage nor skin effects are taken into account, as governed by Equation 10.2. Observe that the impact of the dimensionless storage coefficient is high. Although not shown here, the lower its value, the more pronounced is the pressure derivative during the transition from heterogeneous to equivalent homogeneous behavior.

Figure 10.9 shows the effect of the variation of the interporosity flow parameter, keeping constant both the dimensionless storage coefficient and the flow behavior index. In all cases, the typical s-shape behavior of double porosity is observed. Besides, Figure 10.10 shows the effect of the flow index behavior on the pressure behavior. For this case, both the dimensionless storage coefficient and the interporosity flow parameter are kept constant. This plot shows that as the flow behavior index becomes higher, the slope decreases; thus, an inverse relationship is established. It is worth noting that, although for values of $n \leq 0.6$, the slopes (m'_1 and m'_2) are the same, for values of flow behavior index close to unity ($n > 0.7$), the radial flow corresponding to the equivalent-homogeneous system is smaller than the first one; this decreases the parallelism between the two straight lines.

Naturally Fractured Reservoirs and Bingham Plastic Fluids 227

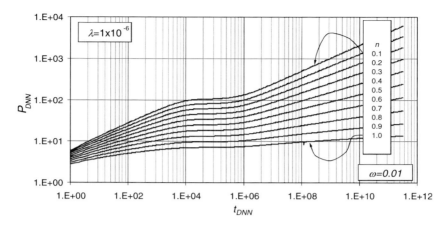

Figure 10.10. Effect of flow behavior index on pressure response in a heterogeneous reservoir; adapted from Escobar et al. (2013).

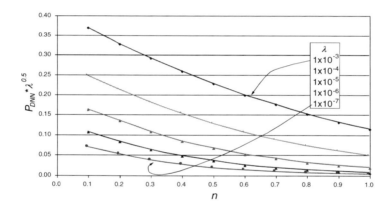

Figure 10.11. Effect of the interporosity flow parameter on the dimensionless pressure multiplied by the square root of the interporosity flow parameter versus the flow behavior index; adapted from Escobar et al. (2013).

Table 10.3. Correlations for the determination of ω; adapted from Escobar et al. (2013)

Equation No.	Correlations
(10.22)	$\ln(\omega) = -2.3025851(Cycles)$
(10.23)	$\ln(\omega) = -2.3025721 - 2.3516663(Cycles)^{0.5}\ln(Cycles)$
(10.24)	$\ln(\omega) = -1.0518712 - 1.2508026(Cycles)^{1.5}$
(10.25)	$\ln(\omega) = -2.302705 - 1.6461502(Cycles)\ln(Cycles)$
(10.26)	$\omega^{0.5} = -0.023404325 + 0.9230909 e^{-Cycles}$
(10.27)	$\ln \omega = \dfrac{0.99684106 + 5.4967445 e^{-(cycles \times n)} - \ln P_{DNN}}{0.29403111}$

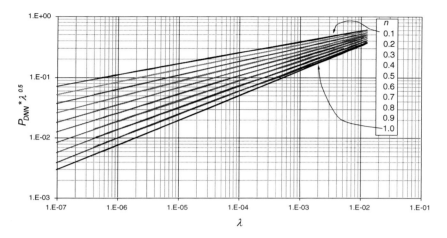

Figure 10.12. Effect of the interporosity flow parameter on the dimensionless pressure multiplied by the square root of the interporosity flow parameter versus the interporosity flow parameter for several values of the flow behavior index; adapted from Escobar et al. (2013).

Equation (7.40) is rewritten using m' instead of α:

$$n = \frac{3m'-1}{m'-1} \qquad (10.18)$$

Since the first semilog line may be affected by wellbore storage, using the second semilog slope is recommended. Escobar et al. (2013) studied the behavior of the dimensionless pressure times the square root of the interporosity flow parameter versus the dimensionless time multiplied by (λ/ω); a summary of the observations is reported in Figure 10.11. From this behavior, the following averaged equation was obtained:

$$\ln(P_{DNN}\lambda^{0.5}) = 0.16412\ln(\lambda) + 0.2484 + [0.14644\ln(\lambda) + 0.09761]n + [0.11021\ln(\lambda) + 0.45275]n^{3/2} \qquad (10.19)$$

from which the interporosity flow parameter was solved as a function of the flow behavior index and the dimensionless pressure value read at the midpoint of a log-log plot of pressure versus time, which was proved for λ values between 1×10^{-3} and 1×10^{-7} and is valid for $0.1 \leq n < 1$, as follows:

$$\lambda = \frac{(0.2484) + n(0.09761) + n^{1.5}(0.45275) - \ln(P_{DNN})}{(0.33588) - n(0.14644) - n^{1.5}(0.11021)} \qquad (10.20)$$

A stricter regression analysis was achieved, with the purpose of minimizing the reading errors as much as possible. For this particular case, 10 different values of λ (1×10^{-2} and 5.5×10^{-7}) were employed to generate many scenarios in the log-log plot of $P_{DNN}\lambda^{0.5}$ vs. $t_D\lambda/\omega$. Again, the range of the flow index behavior was set between 0.1 and 1, and a range of ω

values between 1x10^{-6} and 1 was used in attempt to cover the most common values found in heterogeneous reservoirs. These relationships are graphically shown in Figure 10.12 which yields the correlation given by Equation (10.21):

$$P_{DNN}\lambda^{0.5} = \left[1.2087 e^{0.5315n}\right]\lambda^{0.2568n+0.1515} \tag{10.21}$$

As far as the determination of the dimensionless storage coefficient, ω, is concerned, Escobar et al. (2013) made a further observation of the pressure behavior, as shown in Figure 10.13. In this case, a plot of dimensionless pressure times the dimensionless storage coefficient against dimensionless time was built. Notice that pressure curves with the same value of ω and different values of λ coincided for $t_D \geq 1\times10^8$; therefore, characteristic points were read at that dimensionless time, as reported in Figure 10.14. These observations led to the estimation of the dimensionless storage coefficient, ω, from a reading of the relationship between the distance equivalent to the transition zone and one log cycle measurement. The correlations developed by Escobar et al. (2013) are reported in Table 10.3. Equations were proved for ω values ranging from 1x10^{-6} to 1. The study found that Equation (10.27) presents very good results, but the exponential included in it approaches unity as n increases; at that point, accuracy is lost. Therefore, this equation is only recommended for $n < 0.6$.

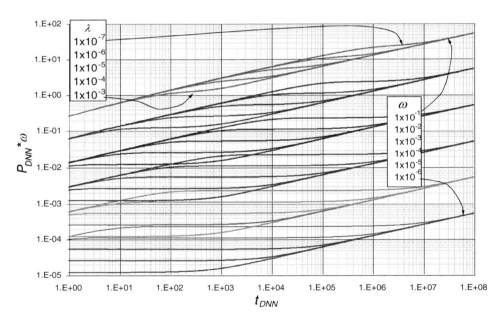

Figure 10.13. Effect of λ and ω on the pressure response for a dual-porosity reservoir, $n = 0.1$; adapted from Escobar et al. (2013).

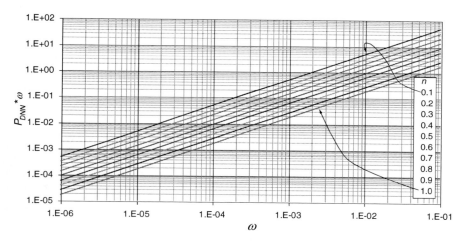

Figure 10.14. Values of the dimensionless pressure times the dimensionless storage coefficient against storage coefficient for several values of the flow behavior index; adapted from Escobar et al. (2013).

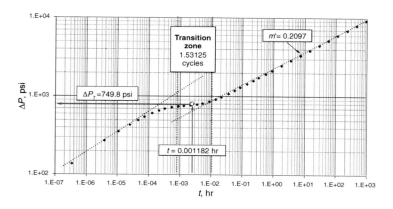

Figure 10.15. Log-log plot of ΔP vs. t for example 10.2; adapted from Escobar et al. (2013).

Equation (10.27) presents very good results, however, the exponential included in it approaches unity as n increases; then, accuracy is lost. Thereby, this equation is only recommended for $n < 0.6$.

The values of permeability and skin factor can be estimated from the Cartesian plot of either ΔP or P_{wf} vs. $t^{(1-n)/(3-n)}$ or ΔP or P_{ws} versus $(t_p+\Delta t)^{(1-n)/(3-n)} - \Delta t^{(1-n)/(3-n)}$, according to the expressions presented in Chapter 7.

$$k = \frac{\mu_{eff}[q/(2\pi k)]^{(n+1)/2}[(3-n)^2/(n\phi c_t)]^{(1-n)/2}}{[m_{NN}(1-n)\Gamma(2/(3-n))]^{(3-n)/2}} \quad (7.12)$$

$$s = \left(\frac{\Delta P_o}{r_w^{1-n}}\right)\left(\frac{2\pi k}{q}\right)^n\left(\frac{k}{\mu_{eff}}\right) + \left(\frac{1}{1-n}\right) \quad (7.13)$$

Naturally Fractured Reservoirs and Bingham Plastic Fluids

Table 10.4. Pressure drop-time data from example 10.2

t, hr	ΔP, psi	t, hr	ΔP, psi	t, hr	ΔP, psi
3.515E-07	136.48	2.239E-03	766.55	0.5758	1926.75
3.866E-06	271.31	3.589E-03	784.72	0.9213	2130.83
1.090E-05	353.68	5.388E-03	807.74	1.84	2470.72
2.495E-05	431.45	8.987E-03	850.18	3.69	2863.15
4.604E-05	494.97	0.0144	906.00	7.37	3316.08
7.416E-05	546.60	0.0216	968.96	14.74	3838.76
1.304E-04	607.35	0.0360	1067.89	29.48	4441.85
2.147E-04	656.82	0.0576	1176.40	58.97	5137.70
3.272E-04	692.06	0.0864	1282.02	117.93	5940.58
5.522E-04	724.10	0.1440	1430.35	235.87	6866.92
8.896E-04	742.23	0.2303	1582.45	471.73	7935.70
1.339E-03	752.90	0.3455	1726.60	943.46	9168.82

Example 10.2

Escobar et al. (2013) reported a synthetic example that was generated with the information given below, and the pressure-time data is given in both Table 10.4 and the log-log plot of Figure 10.15.

$q = 500$ BPD (9.2×10^{-4} m³/s) $k = 2000$ BPD (1.97×10^{-12} m²)
$\phi = 0.05$ $r_w = 0.25$ ft (0.0762 m)
$\mu_{eff} = 1.5$ cp (0.0015 Pa.s) $h = 120$ ft (36.576 m)
$n = 0.48$ $B = 1.0$ rb/STB
$\lambda = 3 \times 10^{-5}$ $c_t = 1 \times 10^{-6}$ psi^{-1} (1.45×10^{-10} 1/Pa)
$\omega = 0.03$

It is necessary to find the naturally fractured reservoir parameters.

Solution

In the plot given in Figure 10.15, the transition zone has a length of 1.5312 log cycles. Other important information read from this plot is as follows:

$t = 0.01182$ hr $\Delta P_T = 749.844$ psi $m' = 0.2097$

Equation (10.18) is used to find index flow behavior:

$$n = \frac{3m'-1}{m'-1} = \frac{3(0.2097)-1}{0.2097-1} = 0.4693$$

The ΔP_T is entered into Equation (7.6) to convert it to dimensionless form, P_{DNN}:

$$P_{DNN} = \frac{749.844}{141.2(96681.605)^{1-0.48}\left(\dfrac{500\times1.2}{120}\right)^{0.48}\left(\dfrac{1.5\times0.25^{1-0.48}}{2000}\right)} = 17.194$$

λ is calculated using Equation (10.20):

$$\ln\lambda = \frac{(0.2484)+(0.48\times0.09761)+\left(0.48^{1.5}\times0.45275\right)-\ln(17.194)}{(0.33588)-(0.48\times0.14644)-\left(0.48^{1.5}\times0.11021\right)}$$

$\lambda = 2.82\times10^{-5}$
λ is estimated using Equation (10.21):

$$P_{DNN}\lambda^{0.5} = \left[1.2087e^{0.5315n}\right]\lambda^{0.2568n+0.1515}$$

$\lambda = 2.36\times10^{-5}$
The G parameter is found using Equation (7.4):

$$G = \frac{3792.188(0.48)(0.05)(1\times10^{-6})(1.5)}{2000}\left(96681.605\frac{120}{500\times1.2}\right)^{1-0.48} = 1.1563\times10^{-5}\frac{\text{hr}}{\text{ft}^{3-n}}$$

The time t is calculated using Equation (7.8), at which point an approximated ΔP is read, taking into account that $t_{DNN}=1\times10^{8}$:

$$t=(1\times10^{8})(1.1563\times10^{-5})(0.25^{3-0.48})=36.146 \text{ hr}$$

The corresponding ΔP_o is read for the above time value. The closest time point is:

$t = 35.379$ hr $\qquad\qquad \Delta P_o = 4615.345$ psi

The above pressure drop is converted to dimensionless form using Equation (7.6):

$$P_{DNN} = \frac{4615.345}{141.2(96681.605)^{1-0.48}\left(\dfrac{500\times1.2}{120}\right)^{0.48}\left(\dfrac{1.5\times0.25^{1-0.48}}{2000}\right)} = 105.804$$

According to Figure 10.15, the length of the transition zone corresponds to:
Cycles $=1.53125$
ω is found using Equations (10.22) through (10.26):

$$\ln(\omega) = -2.3025851(1.53125)$$
$$\omega = 0.0294$$

Naturally Fractured Reservoirs and Bingham Plastic Fluids

$$\ln(\omega) = -2.3025721 - 2.3516663(1.53125)^{0.5}\ln(1.53125)$$
$$\omega = 0.0289$$

$$\ln(\omega) = -1.0518712 - 1.2508026(1.53125)^{1.5}$$
$$\omega = 0.03265$$

$$\ln(\omega) = -2.302705 - 1.6461502(1.53125)\ln(1.53125)$$
$$\omega = 0.0341$$

$$\omega^{0.5} = -0.023404325 + 0.9230909e^{-1.53125}$$
$$\omega = 0.031$$

Since $n<0.6$, ω can be estimated using Equation (10.27) with the number of cycles P_{DNN}:

$$\ln \omega = \frac{0.99684106 + 5.4967445e^{-(1.53125 \times 0.48)} - \ln(105.804)}{0.29403111}$$
$$\omega = 0.0302$$

A summary of results is provided in Table 10.5:

Table 10.5. Summary of results for example 10.2

Parameter	Equation	Value
ω	Actual	0.03
	10.22	0.0294
	10.23	0.0289
	10.24	0.0326
	10.25	0.0341
	10.26	0.031
	10.27	0.0302
	Escobar et al. (2011)	0.03115
		0.03366
		0.03413
λ	Actual	3×10^{-5}
	10.20	2.816×10^{-5}
	10.21	2.358×10^{-5}
	Escobar et al. (2011)	7.678×10^{-6}
		2.755×10^{-6}
		3.380×10^{-5}

BINGHAM FLUIDS IN HOMOGENEOUS RESERVOIRS

Martinez, Escobar and Montealegre (2011) presented a technique for interpreting the behavior of pressure and pressure derivative for a Bingham-type fluid in a homogeneous reservoir drained by a vertical well using the *TDS* technique. The proposed method involves observing the influence of the minimum pressure gradient, which characterizes this behavior, and characteristic points, which are used for estimating formation permeability, drainage area and skin factor. Martinez et al. (2011) presented in the literature for the first time the pressure derivative for Bingham non-Newtonian fluids. The higher the minimum pressure gradient, the more asymmetrically concave the pressure derivative becomes. Additionally, it was observed in closed systems that the late unit-slope pressure derivative coincides with the same one for Newtonian fluids. For the case of horizontal wells with this type of fluid, the reader is referred to the work of Owayed and Tiab (2008).

As a special kind of non-Newtonian fluid, Bingham fluids (or Bingham plastics) exhibit a finite yield stress at zero shear rates. There is no gross movement of fluids until the yield stress, τ_y, is exceeded. Once this is accomplished, it is also necessary to cut efforts to increase the shear rate—i.e., they behave as Newtonian fluids. These fluids behave as a straight line crossing the *y* axis in $\tau = \tau_y$, when the shear stress, τ, is plotted against the shear rate, $\dot{\gamma}$, in Cartesian coordinates. The characteristics of these fluids are defined by two constants: the yield, τ_y, which is the stress that must be exceeded for flow to begin, and the Bingham plastic coefficient, μ_B. According to Bear (1972), the rheological equation for a Bingham plastic is:

$$\tau = \tau_y + \mu_B \dot{\gamma} \qquad (10.28)$$

The Bingham plastic concept has been found to closely approximate many real fluids existing in porous media, such as paraffinic oils, heavy oils, drilling muds and fracturing fluids, which are suspensions of finely divided solids in liquids. Laboratory investigations have indicated that the flow of heavy oil in some fields has non-Newtonian behavior and approaches the Bingham type.

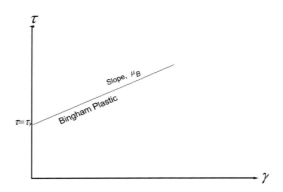

Figure 10.16. Graphic representation of Bingham fluid; adapted from Bear (1972).

For a phenomenological description of flow in porous media, some equivalent or apparent viscosities for non-Newtonian fluid flow are needed in Darcy's equation. Therefore, many experimental and theoretical studies have investigated rheological models or correlations of apparent viscosities and flow properties for a given non-Newtonian fluid or porous material. For flow problems in porous media involving non-Newtonian Bingham fluids, the formulation of Darcy's law has been modified to:

$$\vec{u} = -\frac{k}{\mu_B}\left(1 - \frac{G}{|\nabla P|}\right)\nabla p \quad \text{for} \quad |\nabla P| > G \tag{10.29}$$

and:

$$\vec{u} = 0 \quad \text{for} \quad |\nabla P| \le G \tag{10.30}$$

where G is the pressure gradient corresponding to the yield stress in a porous medium. The above conditions show that in this type of fluid, there is no flow until $|\nabla P|$ exceeds the minimum pressure gradient, G. The two Bingham-fluid parameters, G and μ_B, should be determined by laboratory experiments or by a well test for a porous medium flow problem. For heavy oils, a reasonable value of G is in the order of 10^4 Pa/m (0.44 psi/ft).

Wu (1990) presented the governing equation for this problem. He also provided a complex analytical integral solution, which requires numerical integration. Wu (1990) interpreted the pressure tests by numerical solutions and regression analysis, which means matching the well-pressure response to the simulator response.

The problem considered by Martinez et al. (2011), which was presented by Wu (1990), involves the production of a Bingham fluid from a fully penetrating vertical well in a horizontal reservoir of constant thickness; the formation is saturated only with the Bingham fluid. The basic assumptions are:

1. Isothermal, isotropic, and homogeneous formation
2. Single-phase horizontal flow without gravity effects
3. Darcy's law applies (Equation (10.29))
4. Constant fluid properties and formation permeability

The governing flow equation can be derived by combining the modified Darcy's law with the continuity equation and is expressed in a radial coordinate system as:

$$\frac{1}{r}\frac{\partial}{\partial r}\left[r\left(\frac{\partial P}{\partial r} - G\right)\right] = \frac{\phi\mu_B c_t}{k}\frac{\partial P}{\partial t} \tag{10.31}$$

The initial condition is:

$$P(r, t = 0) = P_i, \qquad r \ge r_w \tag{10.32}$$

At the wellbore inner boundary, $r = r_w$, the fluid is produced at a given production rate, q, so the inner boundary condition is:

$$q = 2\pi r h \frac{k}{\mu_B}\left(\frac{\partial P}{\partial r} - G\right)_{r=r_w} \tag{10.33}$$

The dimensionless pressure, P_D, the dimensionless time, t_D, the dimensionless radius, r_D, and the dimensionless pressure gradient, G_D (conveniently introduced here), are expressed as:

$$P_D = \frac{kh\Delta P}{141.2 q \mu_B B} \tag{10.34}$$

$$t_D = \frac{0.0002637 kt}{\phi \mu_B c_t r_w^2} \tag{10.35}$$

$$r_D = \frac{r}{r_w}, \quad r_{eD} = \frac{r_e}{r_w} \tag{10.36}$$

$$G_D = \frac{G r_w kh}{141.2 q \mu_B B} \tag{10.37}$$

For radial flow and Newtonian fluids, the dimensionless pressure derivative is:

$$(t_D * P_D')_r = \frac{kh(t*\Delta P')_r}{141.2 q \mu_B B} = 0.5 \tag{10.38}$$

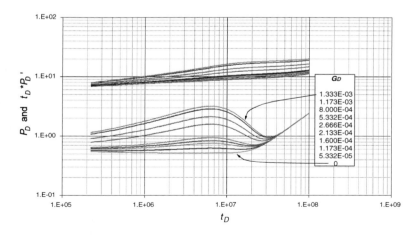

Figure 10.17. Dimensionless pressure and derivative pressure for $r_{eD} = 9375$; adapted from Escobar et al. (2013).

Naturally Fractured Reservoirs and Bingham Plastic Fluids

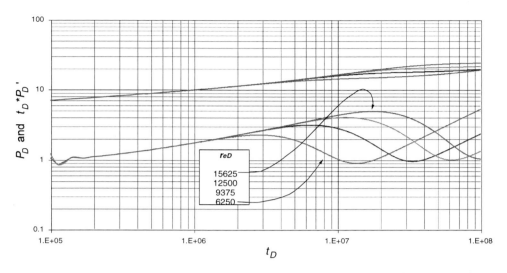

Figure 10.18. Dimensionless pressure and derivative pressure for $G_D = 1.333 \times 10^{-3}$.

Martinez et al. (2011) solved Equation (10.31) numerically. For a Bingham-type, non-Newtonian fluid, this behavior changes by observing that there is a point where the dimensionless pressure derivative is high. This increases with an increase of G_D and the reservoir radius, as shown in Figures 10.17 and 10.18. Figure 10.19 shows the trend between the dimensionless outer radius and the dimensionless derivative pressure maximum for various G_D. The slope of each line is the product $0.20536G_D$. Thus, grouping all of the straight lines into one yields the following relationship:

$$\left(t_D * P_D'\right)_{r,max} = 0.5 + 0.20536 G_D r_{eD} \tag{10.39}$$

$$\left(t_D * P_D'\right)_{r,max} - 0.20536 G_D r_{eD} = 0.5 \tag{10.40}$$

Plugging the dimensionless quantities into the above expressions and solving for permeability yields, respectively:

$$k = \frac{70.6 q \mu_B B}{h\left[(t * \Delta P')_{r,max} - 0.20536 G r_e\right]} \tag{10.41}$$

The behavior of the dimensionless pressure is added to the equation for radial flow and Newtonian fluid to produce an additional quantity that we call the "Bingham effect," which does not depend upon reservoir size, as shown in Figure 10.20.

$$P_{Dr} = 0.5\left[ln\, t_D + 0.80907 + 2s\right] + B_{eff} \tag{10.42}$$

where:

$$B_{eff} = 1.69602 G_D t_D^{0.50304} \quad (10.43)$$

The skin factor, s, can be obtained by dividing Equation (10.42) with Equation (10.40):

$$s = \frac{1}{2}\left[\frac{\Delta P_{r,max}}{(t*\Delta P')_{r,max} - 0.20536 Gr_e} - ln\left(\frac{kt_{r,max}}{\phi\mu_B c_t r_w^2}\right) + 7.43\right] - B_{eff} \quad (10.44)$$

In $G = 0$, the fluid is Newtonian, which leads to the normal equations for obtaining permeability and skin factor, as presented by Tiab (1993).

As observed in Figure 10.17, the late pressure derivative coincides with that of a Newtonian fluid. Thus, using Equation (7.105), the reservoir drainage area can be estimated from any convenient point during the late pseudosteady-state pressure derivative.

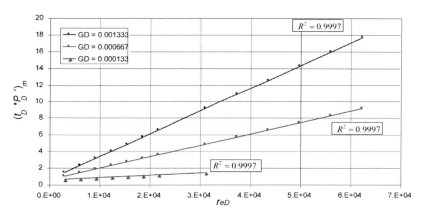

Figure 10.19. Relationship between the dimensionless radius and the dimensionless derivative pressure at its peak; adapted from Escobar et al. (2013).

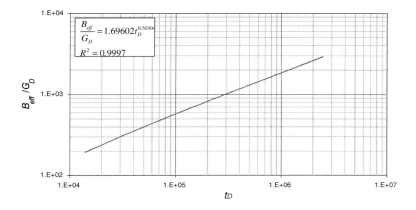

Figure 10.20. Correlation for the "Bingham effect"; adapted from Escobar et al. (2013).

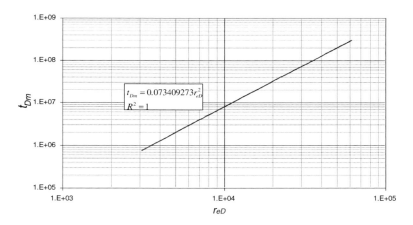

Figure 10.21. Relationship between the dimensionless outer radius and the maximum dimensionless time; adapted from Escobar et al. (2013).

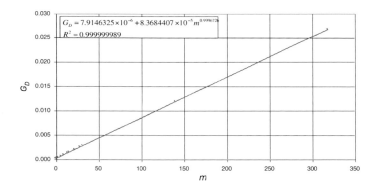

Figure 10.22. Relationship between Cartesian slope from the pressure derivative during radial flow and dimensionless pressure gradient; adapted from Escobar et al. (2013).

$$A = \frac{0.234qBt_{pss}}{\phi c_t h (t*\Delta P')_{pss}} \tag{10.45}$$

Permeability can also be determined by relating the dimensionless outer radius with the maximum dimensionless time. This relationship works for any G_D, as shown in Figure 10.21. The resulting equation is:

$$k = \frac{278.3817710 r_e^{2.0} \phi \mu_B c_t}{t_{r,max}} \tag{10.46}$$

Equations (10.41) and (10.46) are functions of the external reservoir radius. When the late pseudosteady-state flow is not developed, permeability is obtained by equating Equations (10.46) and (10.41). This yields:

$$k = \frac{0.20536k^{1.5}t_{r,max}^{0.5}G}{\left(t*\Delta P'\right)_{r,max}\left(278.3817710\phi\mu_B c_t\right)^{0.5}} + \frac{70.6q\mu B}{h\left(t*\Delta P'\right)_{r,max}} \tag{10.47}$$

Equation (10.47) is solved iteratively using the Newton-Raphson method (or any other) by choosing an initial value of permeability, until the difference between the new and previous value is less than 0.001.

$$f(k) = \frac{0.20536k^{1.5}t_{r,max}^{0.5}G}{\left(t*\Delta P'\right)_{r,max}\left(278.3817710\phi\mu_B c_t\right)^{0.5}} + \frac{70.6q\mu B}{h\left(t*\Delta P'\right)_{r,max}} - k \tag{10.48}$$

$$f'(k) = \frac{0.30804k^{0.5}t_{r,max}^{0.5}G}{\left(t*\Delta P'\right)_{r,max}\left(278.3817710\phi\mu_B c_t\right)^{0.5}} - 1 \tag{10.49}$$

$$k_{n+1} = k_n - \frac{f(k)}{f'(k)} \tag{10.50}$$

Figure 10.22 shows a relation between the dimensionless minimum pressure gradient and the Cartesian slope of the pressure-derivative values during the radial flow regime. If there is no peak in the derivative pressure, obtaining the Cartesian slope of the derivative pressure against time, we can obtain the permeability:

$$k = \frac{141.2q\mu B}{Gr_w h}\left(7.9146325\times10^{-6} + 8.3684407\times10^{-5}m^{0.9996726}\right) \tag{10.51}$$

Example 10.3

Adapted from Escobar et al. (2013). Using information provided by Wu, Pruess, and Witherspoon (1992), the formation permeability and the skin factor are obtained from a reservoir that produces a Bingham-type fluid with a $G = 0.0044$ psi/ft (100 Pa/m). Pressure and pressure-derivative data are reported in Table 10.6 and Figures 10.23 and 10.24.

$P_i = 1450$ psi $\qquad q = 272$ STB/D $\qquad h = 3.2$ ft
$\phi = 20\%$ $\qquad k = 1000$ md $\qquad \mu_B = 1$ cp
$r_w = 0.32$ ft $\qquad B = 1$ rb/STB $\qquad c_t = 4.52\times10^{-6}$ psi^{-1}

Solution

From Figure 10.24, the following information is read:

$\Delta P_{r,max} = 110.5$ psi $\qquad (t*\Delta P')_{r,max} = 9.7$ psi

Naturally Fractured Reservoirs and Bingham Plastic Fluids

$(t*\Delta P')_{p1} = 0.43$ psi $\qquad t_{r,max} = 4.01$ hr

$(t*\Delta P')_r = 6.96$ psi

The drainage area is obtained from Equation (10.45) using information from the late pseudosteady-state regime:

$$A = \frac{0.234(272)}{0.2(4.52\times10^{-6})(3.2)(0.43)} = 51167925.8 \text{ ft}^2$$

Assuming a circular reservoir shape, the reservoir radius, r_e, is 4035.75 ft. Formation permeability is estimated from Equation (10.41):

$$k = \frac{70.6(272)(1)(1)}{3.2*[9.7-0.20536(0.0044)(4035.75)]} = 991.35 \text{ md}$$

The dimensionless minimum pressure gradient, Equation (10.37), is:

$$G_D = \frac{0.0044(0.32)(991.35)(3.2)}{141.2(272)(1)(1)} = 1.163\times10^{-4}$$

Table 10.6. Pressure and pressure-derivative data for example 10.3

t, hr	ΔP, psi	$t*\Delta P'$, psi	t, hr	ΔP, psi	$t*\Delta P'$, psi	t, hr	ΔP, psi	$t*\Delta P'$, psi
0.01	63.34	8.89	0.7	95.40	7.78	8.69	117.80	8.51
0.02	69.52	8.64	0.75	95.94	7.82	9.25	118.32	8.43
0.03	72.67	7.89	0.8	96.45	7.90	9.81	118.80	8.35
0.04	74.78	7.46	0.85	96.94	8.00	10.37	119.25	8.31
0.05	76.39	7.16	0.9	97.39	8.01	11.8	120.28	8.27
0.06	77.68	7.16	0.95	97.83	8.03	13.42	121.30	8.37
0.07	78.76	6.96	1	98.25	8.06	15.04	122.20	8.61
0.08	79.70	6.92	1.05	98.64	8.12	16.66	123.04	8.95
0.09	80.52	7.03	1.41	101.11	8.49	18.28	123.83	9.40
0.1	81.26	6.95	1.97	104.03	9.00	19.9	124.59	9.89
0.15	84.10	7.04	2.53	106.31	9.30	21.52	125.34	10.44
0.2	86.13	7.24	3.09	108.19	9.43	23.14	126.07	11.04
0.25	87.72	7.20	3.65	109.78	9.51	24.76	126.79	11.66
0.3	89.04	7.34	4.21	111.15	9.45	26.38	127.51	12.32
0.35	90.16	7.38	4.77	112.35	9.41	28	128.23	12.99
0.4	91.15	7.45	5.33	113.41	9.29	29.62	128.94	13.68
0.45	92.03	7.53	5.89	114.34	9.16	31.24	129.65	14.39
0.5	92.82	7.58	6.45	115.18	9.02	32.86	130.36	15.10
0.55	93.54	7.66	7.01	115.93	8.86	35.99	131.73	16.49
0.6	94.21	7.72	7.57	116.61	8.74			
0.65	94.83	7.76	8.13	117.23	8.63			

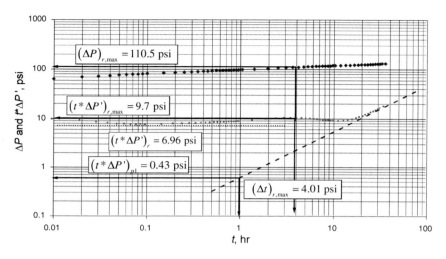

Figure 10.23. Pressure and pressure derivative for example 10.3.

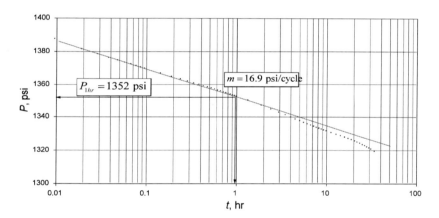

Figure 10.24. Semilog plot of pressure vs. time for example 10.3.

$t_{r,max}$ is converted to dimensionless form using Equation (10.35):

$$t_D = \frac{0.0002637(991.35)(4.01)}{(0.2)(1)(4.52\times 10^{-6})(0.32^2)} = 11324345.9$$

The correlation given in Figure 10.20 is used to find the "Bingham effect":

$$B_{eff} = G_D 1.69602 t_D^{0.50304} = (1.163\times 10^{-4})1.69602(11324345.9)^{0.50304} = 0.4112$$

Skin factor is estimated using Equation (10.44):

$$s = \frac{1}{2}\left[\frac{\dfrac{110.5}{9.7 - 0.205(0.0044)(4035.75)} -}{\ln\left(\dfrac{991.35(4.01)}{0.2(1)(4.52\times10^{-6})(0.32^2)}\right)}\right]$$

$$-1.7(3.634\times10^{-5})\left(\frac{0.0002637(991.35)(4.01)}{(0.2)(1)(4.52\times10^{-6})(0.32^2)}\right)^{0.50304} - 0.4112 = -0.5082$$

Equation (10.46) is used to estimate formation permeability as follows:

$$k = \frac{278.4(4035.75^{2.0})(0.2)(1)(4.52\times10^{-6})}{4.01} = 1022.15 \text{ md}$$

Since G_D appears to be small enough for the application of the semilog (conventional) straight-line method, the semilog slope is determined using Figure 10.24 to be 16.9 psi/cycle. Then, using Equation (1.10), the following is found:

$$k = \frac{162.6q\mu B}{mh} = \frac{162.6(272)(1)(1)}{16.9(3.2)} = 817.8 \text{ md}$$

Equation (1.11) is used to find the skin factor:

$$s = 1.1513\left[\frac{1352-1450}{-14} - \log\left(\frac{987.2}{0.2(1)(4.52\times10^{-6})0.32^2}\right) + 3.23\right] = 0.24$$

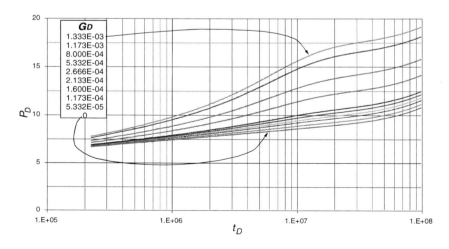

Figure 10.25. Dimensionless semilog plot.

Permeability is found again using Equation (1.18):

$$k = \frac{70.6q\mu B}{h(t*\Delta P')_r} = \frac{70.6(272)(1)(1)}{(3.2)6.96} = 862.2 \text{ md}$$

Even though the G_D is small for the application of the conventional method, Equation (1.10) and TDS technique for Newtonian fluids, if Equation (1.18) (the estimation of permeability) differs by about 20%, then these methods may be applied if $G_D < 5.33 \times 10^{-4}$, since no straight line is formed during radial flow, as seen in Figure 10.25.

Nomenclature

B	Volumetric factor, RB/STB
c_t	System total compressibility, 1/psi
h	Formation thickness, ft
H	Consistency (power-law parameter), $cp*s^{n-1}$
G	Group defined by Equation (7.4)
G	Minimum pressure gradient, psi/ft
G_D	Dimensionless pressure gradient
k	Permeability, md
k	Flow consistency parameter
m	Slope
n	Flow behavior index (power-law parameter)
P	Pressure, psi
q	Flow/injection rate, STB/D
t	Time, hr
r	Radius, ft
r_e	Reservoir external radius, ft
$t*\Delta P'$	Pressure derivative, psi
t_D*P_D'	Dimensionless pressure derivative
y	t_{USi}/t_{min}
w	ω/t_{Dmin}
z	Laplace parameter
Z	$\ln[(t_D*P_D')_{min}/t_{Dmin}]$

Greeks

α	Slope of the pressure-derivative curve during radial flow regime
Δ	Change, drop
ϕ	Porosity, Fraction
μ_B	Bingham plastic coefficient, cp
μ_{eff}	Effective viscosity for power-law fluids, $cp*(s/ft)^{n-1}$

Suffixes

app	Apparent
D	Dimensionless
DA	Dimensionless with respect to area
e	External
eff	Effective
i	Initial
NN	Non-Newtonian
pss	Any point during late pseudosteady state regime
*p*1	Pressure derivative on the pseudosteady-state line read at 1 hr
r	Radial flow regime
r,max	Maximum during radial flow regime
rNN	Radial (any point on radial flow)
rpiNN	Intersect of radial and pseudosteady-state lines
rsiNN	Intersect of radial line and negative-unit slope tangent to steady state period
w	Wellbore

REFERENCES

Bear, J. (1972). *"Dynamics of Fluids in Porous Media."* Elsevier Science Publishers, New York City.

Camacho. R. Fuentes-Cruz. G. & Vasquez. M. A. (2008). "Decline-Curve Analysis of Fractured Reservoirs With Fractal Geometry." *Society of Petroleum Engineers.* doi:10.2118/104009-PA. June 1.

Engler, T. W. & Tiab, D. (1996). "Analysis of Pressure and Pressure Derivate without type curve matching, 4. Naturally Fractured Reservoir," *Journal of petroleum Science and Engineering 15*, p. 127-138.

Escobar, F. H. (2012). *"Transient Pressure and Pressure Derivative Analysis for Non-Newtonian Fluids."* New Technologies in the Oil and Gas Industry, Dr. Jorge Salgado Gomes (Ed.), ISBN: 978-953-51-0825-2, InTech, DOI: 10.5772/50415. Available from: http://www.intechopen.com/books/new-technologies-in-the-oil-and-gas-industry/transient-pressure-and-pressure-derivative-analysis-for-non-newtonian-fluids.

Escobar, F. H., Martinez, J. A. & Montealegre-M., Matilde. (2010). "Pressure and Pressure Derivative Analysis for a Well in a Radial Composite Reservoir with a Non-Newtonian/Newtonian Interface." *CT&F – Ciencia, TEcnología y Futuro*, Vol. *4*, No. 1. 33-42. Dec.

Escobar, F. H., Zambrano, A. P., Giraldo, D. V. & Cantillo, J. H. (2011). "Pressure and Pressure Derivative Analysis for Non-Newtonian Pseudoplastic Fluids in Double-Porosity Formations." *CT&F*, Vol. *5*, No. 3. p. 47-59. ISSN 0122-5383. June.

Escobar, F. H., Martinez, J. A. & Bonilla, L. F. (2012). "Transient Pressure Analysis for Vertical Wells With Spherical Power-Law Flow." *CT&F*. Vol. 5 No. 1. P. 19-25. Dec.

Escobar, F. H., Martinez, A. & Silva, D. M. (2013). "Conventional Pressure Analysis for Naturally-Fractured Reservoirs with Non-Newtonian Pseudoplastic Fluids." *Revista Fuentes: El Reventón Energético*. Vol. *11*. Nro. 1. p. 27-34. ISSN 1657-6527. Ene/Jun.

Escobar, F. H., Zhao, Y. L. & Zhang L. H. (2014). "Interpretation of Pressure Tests in Horizontal Wells in Homogeneous and Heterogeneous Reservoirs with Threshold Pressure Gradient." *Journal of Engineering and Applied Sciences*. ISSN 1819-6608. Vol. *9*. Nro. 11. p. 2220-2228.

Escobar, F. H., Salcedo, L. N. & Pinzon, C. (2015a). *"Pressure and pressure derivative analysis for fractal homogeneous reservoirs with power-law fluids." Paper accepted for publication in Journal of Engineering and Applied Sciences.*

Escobar, F. H., Lopez-Morales, L. & Gomez, K. T. (2015b). "Pressure and Pressure Derivative Analysis for Naturally-Fractured Fractal Reservoirs." *Journal of Engineering and Applied Sciences*. Vol. *10*. Nro. 2. P. 915-923. 2015.

Martinez, J. A., Escobar, F. H. & Montealegre-M, M. (2011). "Vertical Well Pressure and Pressure Derivative Analysis for Bingham Fluids in a Homogeneous Reservoirs." *Dyna*, Year 78, Nro. 166, 21-28. Dyna. ISSN 0012-7353.

Owayed, J. F. & Tiab. D. (2008). "Transient pressure behavior of Bingham non-Newtonian fluids for horizontal wells." *Journal of Petroleum Science and Engineering*. Volume *61*, Issue 1, April, Pages 21-32.

Olarewaju, J. S. (1992). "A Reservoir Model of Non-Newtonian Fluid Flow." *Society of Petroleum Engineers*. Paper SPE 25301.

Tiab, D. (1993, January 1). "Analysis of Pressure and Pressure Derivatives Without Type-Curve Matching: I-Skin and Wellbore Storage." *Society of Petroleum Engineers*. This paper was prepared for presentation at the Production Operations Symposium held in Oklahoma City, OK, U.S.A., March 21-23. doi:10.2118/25426-MS.

Tiab, D., Igbokoyi, A. & Restrepo, D. P., 2007. *"Fracture Porosity From Pressure Transient Data."* Paper IPTC 11164 presented at the *International Petroleum Technology Conference* held in Dubai, U. A.E., 4–6 December.

Wu, Y. S. (1990). *"Theoretical Studies of Non-Newtonian and Newtonian Fluid Flow Through Porous Media."* Ph.D. dissertation, U. of California, Berkeley.

Wu, Y. S., Pruess, K. & Witherspoon, P. A. (1992). "Flow and Displacement of Bingham Non-Newtonian Fluids in Porous Media." *SPE Reservoir Engineering*. August 1992, p. 369-376

PART THREE: DIVERSE MODERN TOPICS

Chapter 11

TRANSIENT-RATE ANALYSIS

BACKGROUND

Most well-test analysis methods have a constant-rate internal boundary condition while well-flowing pressure varies with time; this is also known as pressure-transient analysis (PTA). However, several common reservoir production conditions result in flow at a constant pressure and flow-rate changes. This is better known as transient-rate analysis (RTA). Wells in low-permeability reservoirs are often produced at constant wellbore-flowing pressure by necessity. Thus, well testing could be eliminated in many cases as being of little value or economically unjustifiable, due to the resulting production loss compared with what can be obtained from constant wellbore pressure production data. Other examples of such a production mode include production into a constant production separator or pipe line and open flow to the atmosphere.

As observed in Figure 11.1, the behavior of a well producing at constant bottom-hole pressure is similar to one operating at a constant flow rate. In a constant-pressure flow test, the well produces at a constant bottom-hole pressure, and flow rate is recorded with time. Since rate solutions are found on basic flow principles, flow-rate data can be used for reservoir characterization and different property estimations. Thus, this technique can be considered as an alternative to conventional well-testing techniques.

Most common rate-analysis methods involve type-curve matching. The pioneer in this area was Fetkovich (1980). This technique assumes a circular, homogeneous reservoir and is not applicable to heterogeneous systems.

Arab (2003) performed a detailed study to apply the *TDS* technique to a well producing under constant-pressure conditions. Escobar, Sanchez and Cantillo (2008) complemented Arab's work for gas wells.

CONVENTIONAL ANALYSIS FOR HOMOGENEOUS RESERVOIRS

The dimensionless quantities in oil-field units are defined by:

$$t_D = \frac{0.0002637kt}{\phi \mu c_t r_w^2} \tag{1.2}$$

$$q_D = \frac{141.2q\mu B}{kh\Delta P} \tag{11.1}$$

$$r_D = \frac{r}{r_w} \tag{1.4}$$

Van Everdingen and Hurst (1949) provided the analytical Laplace-space solution for well production at constant bottom-hole pressure:

$$\overline{q}_D = \frac{K_1(\sqrt{u})}{\sqrt{u}\left[K_0(\sqrt{u}) + s\sqrt{u}K_1(\sqrt{u})\right]} \tag{11.1}$$

As for the case of Equation (1.7), at the wellbore radius, when $t_D \geq 100$, it yields:

$$\frac{1}{q_D} = \frac{1}{2}\left[\ln(t_D) + 0.80907 + 2s\right] \tag{11.2}$$

In oil-field units:

$$\frac{1}{q} = \frac{162.6\mu B}{kh(P_i - P_{wf})}\left[\log\left(\frac{kt}{\phi \mu c_t r_w^2}\right) - 3.227 + 0.8686s\right] \tag{11.3}$$

The original solution for pressure behavior in finite systems was introduced by Muskat (1934). According to van Everdingen and Hurst (1949), the constant bottom-hole pressure solution is related to the constant terminal flow-rate solution by the following equation:

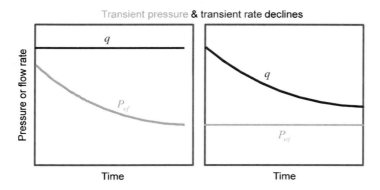

Figure 11.1. Transient-pressure and transient-rate declines.

$$\overline{P}_D * \overline{q}_D = \frac{1}{u^2} \tag{11.3}$$

Thus, the constant-pressure solution is given by:

$$\overline{q}_D = \frac{I_1(r_{eD}\sqrt{u})K_1(\sqrt{u}) - K_1(r_{eD}\sqrt{u})I_1(\sqrt{u})}{\sqrt{z}\left[I_0(\sqrt{u})K_1(r_{eD}\sqrt{u}) + K_0(\sqrt{u})I_1(r_{eD}\sqrt{u})\right]} \tag{11.5}$$

For late times, the above equation leads to:

$$q_D = \frac{1}{\ln(r_{eD}) - 0.75} \exp\left[\frac{-2t_D}{r_{eD}^2\left[\ln(r_{eD}) - 0.75\right]}\right] \tag{11.6}$$

For $t_D \geq t_{Dpss}$, this flow period is known as exponential decline. For non-radial systems, the shape factor is included:

$$q_D = \frac{2}{\ln(4A_D/\gamma C_A)} \exp\left[\frac{-4\pi t_D}{A_D \ln(4A_D/\gamma C_A)}\right] \tag{11.7}$$

where dimensionless area and radius are defined by:

$$A_D = \frac{A}{r_w^2} \tag{11.8}$$

$$r_{eD} = \frac{r_e}{r_w e^{-s}} \tag{11.9}$$

As for the PTA, transient-rate data during transient flow allow for the estimation of permeability and skin factor by semilog analysis, while data recorded under pseudosteady-state conditions allow for the determination of the area. Equation (11.2) leads to:

$$\frac{1}{q} = \frac{162.6\mu B}{kh(P_i - P_{wf})}\left[\log\left(\frac{kt}{\phi\mu c_t r_w^2}\right) - 3.227 + 0.8686s\right] \tag{11.10}$$

Hence, a plot of $(1/q)$ versus $\log t$ yields a straight line for which the slope, m, and intercept, $(1/q)_{1hr}$, reveal the permeability and skin factor, respectively:

$$k = \frac{162.6\mu B}{mh(P_i - P_{wf})} \tag{11.11}$$

$$s = 1.1513\left[\frac{(1/q)_{1hr}}{m} - \log\left(\frac{k}{\phi\mu c_t r_w^2}\right) + 3.227\right] \qquad (11.12)$$

During boundary-dominated flow, Equation (11.7) indicates the predominance of an exponential behavior. Thus, a plot log q versus t will provide a negative-slope straight line for which both slope, $M_{decline}$, and intercept, q_{int}, yield the reservoir area:

$$M_{decline} = -\frac{2(0.0002637)k}{r_{eD}^2\left[\ln(r_{eD}) - 0.75\right]\phi\mu c_t r_w^2} \qquad (11.13)$$

$$q_{int} = \frac{kh\Delta P}{141.2B\mu\left[\ln(r_{eD}) - 0.75\right]} \qquad (11.14)$$

Solving for the dimensionless external radius:

$$r_{eD} = \exp\left(\frac{141.2B\mu}{kh\Delta P}q_{int} + 0.75\right) \qquad (11.15)$$

TDS TECHNIQUE FOR HOMOGENEOUS RESERVOIRS

As performed by Arab (2003), the natural log derivative of Equation (11.2) leads to:

$$t_D(1/q_D)' = \frac{d(1/q_D)}{d\ln(t_D)} = \frac{d\left[\frac{1}{2}\left[\ln(t_D) + 0.80907 + 2s\right]\right]}{d\ln(t_D)} = \frac{1}{2} \qquad (11.16)$$

The dimensionless reciprocal rate derivative is given by:

$$t_D(1/q_D)' = \frac{kh\Delta P}{141.2\mu B}[t*(1/q)'] \qquad (11.17)$$

The combination of Equations (11.6) and (11.7) yields an expression for permeability determination:

$$k = \frac{70.6\mu B}{h\Delta P[t*(1/q)']_r} \qquad (11.18)$$

where $[t*(1/q)']_r$ is the reciprocal rate derivative—horizontal line—during the radial flow regime. Skin factor is estimated from the division of Equation (11.2) by Equation (11.16):

Transient-Rate Analysis

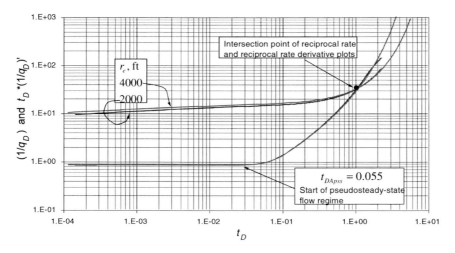

Figure 11.2. Reciprocal rate and reciprocal-rate derivative for closed reservoirs (skin factor=0); adapted from Arab (2003).

$$s = 0.5\left\{\frac{(1/q)_r}{[t*(1/q)']_r} - \ln\left(\frac{kt_r}{\phi\mu c_t r_w^2}\right) + 7.43\right\} \qquad (11.19)$$

Equation (11.19) is analogous to Equation (1.21) for the PTA case. $(1/q)_r$ is the reciprocal rate read at an arbitrary time, t_r, during the radial flow regime.

Taking the derivative of Equation (11.6) yields:

$$t_D(1/q_D)' = \frac{2t_D}{r_{eD}^2}\exp\left[\frac{2t_D}{r_{eD}^2\left[\ln(r_{eD}) - 0.75\right]}\right] \qquad (11.20)$$

Arab (2003) found a very important feature in the reciprocal-rate derivative plot. As seen in Figure 11.2, the pseudosteady state always starts at t_{DApss} = 0.0554. By plugging this value into Equation (11.20), the reservoir radius can be found:

$$r_e = \sqrt{\frac{0.0015 kt_{sp}}{\phi\mu c_t}} \qquad (11.21)$$

The intercept of the reciprocal rate and the reciprocal-rate derivative is governed by:

$$t_{Drpi} = \frac{1}{2}r_{eD}^2\left(\ln(r_{eD}) - 0.75\right) \qquad (11.22)$$

for which the solution reported by Arab (2003) is:

$$r_{eD} = 1.0292 t_{Drpi}^{0.4627} \tag{11.23}$$

In oil-field units:

$$r_e = 22.727 \times 10^{-3} r_{weff} \left(\frac{k}{\phi c_t \mu r_{weff}^2} \right)^{0.4627} t_{rpi}^{0.4627} \tag{11.24}$$

If the intersection point is unseen, according to Arab (2003), the ratio between Equations (11.6) and (11.20) yields the area:

$$\frac{0.000527k}{\phi c_t \mu r_{weff}^2} t_{pss} \frac{(1/q)_{pss}}{[t*(1/q)']_{pss}} = \left(\frac{r_e}{r_{weff}} \right)^2 \left[\ln \left(\frac{r_e}{r_{weff}} \right) - 0.75 \right] \tag{11.25}$$

CONVENTIONAL ANALYSIS FOR FINITE-CONDUCTIVITY FRACTURED WELLS

Arab (2003) also reported the governing equation for the reciprocal rate in both finite- and infinite-conductivity fractured well, respectively, as follows:

$$\frac{1}{q_D} = \frac{2.7222}{\sqrt{C_{fD}}} t_{Dxf}^{1/4} \tag{11.25}$$

$$\frac{1}{q_D} = 2.7842 t_{Dxf}^{0.5} \tag{11.26}$$

Plugging the dimensionless parameters into Equations (11.25) and (11.26) leads to:

$$\frac{1}{q} = \frac{48.968 \mu B}{h \Delta P \sqrt{k_f w_f} \sqrt[4]{\phi \mu c_t k}} t^{0.25} \tag{11.27}$$

$$\frac{1}{q} = 6.384 \frac{B}{h \Delta P x_f} \sqrt{\frac{\mu t}{\phi c_t k}} \tag{11.28}$$

The respective Cartesian plots of $1/q$ versus $t^{1/4}$ or $t^{1/2}$, depending on whether one is dealing with the bilinear or linear flow regime, yield either fracture conductivity or half-fracture length from their respective slopes, m_{BL} or m_L:

$$k_f w_f = \left(\frac{48.968 \mu B}{m_{BL} h \Delta P \sqrt[4]{\phi \mu c_t k}} \right)^2 \qquad (11.29)$$

$$x_f = 6.384 \frac{B}{h \Delta P^* m_L} \sqrt{\frac{\mu}{\phi c_t k}} \qquad (11.30)$$

TDS Technique for Finite-Conductivity Fractured Wells

Following the works of Tiab (1994) and Tiab et. al. (1999) for PTA, Arab (2003) also reported the reciprocal-rate derivative of Equations (11.27) and (11.28) as:

$$t_{Dxf} * (1/q)' = \frac{0.6805}{\sqrt{C_{fD}}} t_{Dxf}^{0.25} \qquad (11.31)$$

$$t_{Dxf} * (1/q_D)' = 1.392 t_{Dxf}^{0.5} \qquad (11.32)$$

Again, plugging in the dimensionless terms will provide, respectively:

$$t * (1/q)' = \frac{12.242 \mu B}{h \Delta P \sqrt{k_f w_f} \left(\phi \mu c_t k \right)^{1/4}} t^{0.25} \qquad (11.33)$$

$$t * (1/q)' = \frac{3.19 B}{h \Delta P x_f} \sqrt{\frac{\mu t}{\phi c_t k}} \qquad (11.34)$$

Solving for fracture conductivity and half-fracture length, respectively, yields:

$$k_f w_f = \frac{149.866}{\left(\phi \mu c_t k \right)^{1/2}} \left\{ \frac{\mu B}{h \Delta P \left[t * (1/q)' \right]_{BL1}} \right\}^2 \qquad (11.35)$$

$$x_f = 3.192 \left(\frac{\mu}{\phi c_t k} \right)^{1/2} \left[\frac{B}{h \Delta P * \left[t * (1/q)' \right]_{L1}} \right] \qquad (11.36)$$

In each case, the value of the reciprocal rate derivative is read at the time of 1 hr and extrapolated if necessary.

Arab (2003) also found the point of intersection between the linear flow and the bilinear, radial and linear flow regime derivatives:

$$t'_{BLLi} = 216.35 \frac{\phi \mu c_t}{k} \left(\frac{x_f^2 k}{k_f w_f} \right)^2 \tag{11.37}$$

$$t_{RBLi} = 1106 \frac{\phi \mu c_t}{k^3} (k_f w_f)^2 \tag{11.38}$$

$$t_{RLi} = 489.2 \frac{\phi \mu c_t x_f^2}{k} \tag{11.39}$$

CONVENTIONAL ANALYSIS FOR DOUBLE-POROSITY SYSTEMS

DaPrat (1981) presented the constant-pressure solution for naturally fractured reservoirs:

$$\overline{q}_D(u) = \frac{\sqrt{uf(u)}K_1\left(\sqrt{uf(u)}\right)}{u\left[K_0\left(\sqrt{uf(u)}\right) + u\sqrt{uf(u)}K_1\left(\sqrt{uf(u)}\right)\right]} \tag{11.40}$$

$f(u)$ is given by Equation (2.5).
Equations (1.2) and (11.1) are rewritten as:

$$t_D = \frac{0.0002637 k_2 t}{\mu(\phi c_t)_t r_w^2} \tag{1.2}$$

$$\frac{1}{q_D} = \frac{k_2 h(P_i - P_{wf})}{141.2 \mu B} \frac{1}{q} \tag{11.1}$$

Arab (2003) concluded that the solutions for pressure and rate are similar (see Figure 11.3), so they are taken equal. This means that the analogous form of Equation (2.16) for RTA is:

$$\frac{1}{q_D} = \frac{1}{2}\left[\ln t_D + 0.80907 + E_i\left(-\frac{\lambda t_D}{\omega(1-\omega)} \right) - E_i\left(-\frac{\lambda t_D}{(1-\omega)} \right) \right] \tag{11.41}$$

For early time and also including skin factor, Equation (11.41) becomes, dimensionally:

$$\frac{1}{q} = \frac{162.6 \mu B}{k_2 h(P_i - P_{wf})}\left[\log\left(\frac{k_2 t}{(\phi c_t)_t \mu r_w^2} \right) - \log(\omega) - 3.227 + 0.8686 s \right] \tag{11.42}$$

As for the case of constant-pressure rate, the semilog plot provides an s-shaped plot with two straight lines and one transition period. The slopes of the straight lines and the intercepts are used to find fracture-bulk permeability and skin factor:

$$k_2 = \frac{162.6\mu B}{mh(P_i - P_{wf})} \quad (11.43)$$

For the first straight line:

$$s = 1.1513\left[\frac{(1/q)_{1hr}}{m} - \log\left(\frac{k_2}{\omega[\phi(\mu c_t)_t r_w^2]}\right) + 3.227\right] \quad (11.44)$$

For the second straight line:

$$s = 1.1513\left[\frac{(1/q)_{1hr}}{m} - \log\left(\frac{k_2}{[\phi(\mu c_t)_t r_w^2]}\right) + 3.227\right] \quad (11.45)$$

Similarly to Equation (2.12), according to Kazemi (1969), ω can be determined from the vertical distance between the two semilogs, identified as $[(1/q)_1 - (1/q)_2]$ (see Figure 11.4):

$$\omega = \text{anti}\log\left\{-\frac{[(1/q)_1 - (1/q)_2]}{m}\right\} \quad (11.46)$$

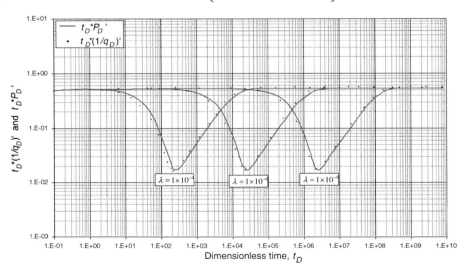

Figure 11.3. Comparison of constant-rate and constant-pressure solutions of naturally fractured systems for $\omega = 0.005$, adapted from Arab (2003).

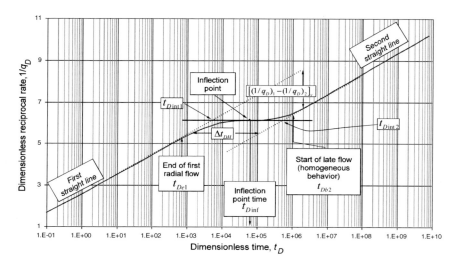

Figure 11.4. Dimensionless semilog plot of reciprocal rate versus time for $\omega = 0.01$, $\lambda = 1\times10^{-6}$, $s=0$; adapted from Warren and Root (1963).

For the PTA case, Samaniego and Cinco-Ley (2209) provided the following expressions for the estimation of the capacity ratio and the interporosity flow parameter using the time values labeled in Figure 11.4:

$$\omega = t_{int1} / t_{int2} \tag{11.47}$$

$$\lambda = \frac{2129.172\omega\phi_{fb}c_{tf}\mu r_w^2}{k_2 t_{int1}} \tag{11.48}$$

$$\lambda = \frac{2129.172(\phi c_t)_t \mu r_w^2}{k_2 t_{int2}} \tag{11.49}$$

Mavor and Cinco-Ley (1979) provided correlations using the starting and ending times of the semilog lines (see Figure 11.4):

$$\lambda = \frac{\omega(1-\omega)}{3.6 t_{De1}} = \frac{(1-\omega)}{1.3 t_{Db2}} \tag{11.50}$$

Tiab and Escobar (2003) developed an expression to estimate the inflection point, t_{inf} or Δt_{inf}, which is pointed out in Figure 11.4. This expression was originally introduced for PTA and can be extended to the semilog plot of semilog $(1/q)$ versus log (t), as pointed out in Figure 2.4:

Transient-Rate Analysis

$$\lambda = \frac{3792\left(\phi c_t\right)_t \mu r_w^2}{k_2 t_{inf}}\left[\omega \ln\left(\frac{1}{\omega}\right)\right] \tag{2.13}$$

Finally, Arab (2003) uses the ordinate, $(1/q)_{inf}$, of the inflection point:

$$\lambda = \exp\left\{-2\left[\frac{k_2 h\left(P_i - P_{wf}\right)}{141.2\mu B}\left(\frac{1}{q}\right)_{inf} - s\right] + 0.2318\right\} \tag{11.51}$$

An analogous equation for PTA is also included:

$$\lambda = \exp\left\{-2\left[\frac{k_2 h}{141.2\mu q B}(P_i - P_{wf})_{inf} - s\right] + 0.2318\right\} \tag{11.52}$$

TDS TECHNIQUE FOR DOUBLE-POROSITY SYSTEMS

Arab (2003) also developed the TDS technique for constant-pressure testing in naturally fractured reservoirs. He started by taking the derivative of Equation (11.41):

$$t_D *(1/q_D)' = \frac{1}{2}\left[1 - \exp\left(-\frac{\lambda t_D}{1-\omega}\right) + \exp\left(-\frac{\lambda t_D}{\omega(1-\omega)}\right)\right] \tag{11.53}$$

Since functions P_D and $(1/q_D)$ show similar behavior, as suggested by Figure 11.3, the relationships given in Chapters 1 and 2 for describing P_D and t_D*P_D' were directly applied for interpretation of the $(1/q_D)$ and $t_D*(1/q_D)'$ curves. From there, Arab (2003) conducted the following analysis:

$$t_D *(1/q_D)' = 0.5 \tag{11.54}$$

Plugging the dimensionless quantities into the above expression and solving for the fracture-bulk permeability yields:

$$k_2 = \frac{70.6\mu B}{h\Delta P\left[t*(1/q)'\right]_{r1,r2}} \tag{11.55}$$

Similarly to the procedure proposed by Engler and Tiab (1996), the log-log plot $[t_D*(1/q_D)']_{min}$ vs. $(\lambda t_D)_{min}$ results in a unit-slope straight line, the equation of which is given by:

$$\ln\left[t_D \times (1/q_D)'\right]_{\min} = \ln\left[\lambda t_{D\min}\right] + \ln(0.63) \qquad (11.56)$$

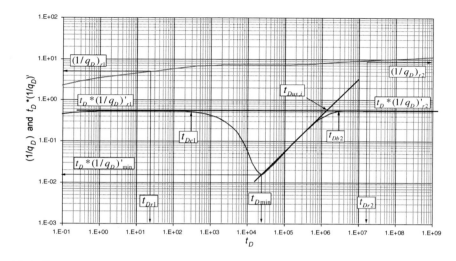

Figure 11.5. Characteristic points and lines found in the log-log plot of reciprocal rate and reciprocal-rate derivative against time; adapted from Arab (2003).

After replacing the dimensionless expressions and solving for the interporosity flow parameter, the following is obtained:

$$\lambda = 42.63 \frac{(\phi c_t)_t \mu r_w^2 \Delta m(P)}{T} \left[\frac{t*(1/q)'}{t}\right]_{\min} \qquad (11.57)$$

Like Engler and Tiab (1996), Arab (2003) also found expressions relating the minimum coordinates, the ending and starting of the radial flow regimes, the pressure derivative during the radial flow regime and the intersection of the unit-slope line formed during the transition period with the radial flow regime, as follows:

$$\omega = 0.15866 \frac{[t_D*(1/q_D)']_{\min}}{[t_D*(1/q_D)']_r} + 0.54653 \left(\frac{[t_D*(1/q_D)']_{\min}}{[t_D*(1/q_D)']_r}\right)^2 \qquad (11.58)$$

$$\frac{50 t_{e1}}{\omega(1-\omega)} = \frac{t_{e2}}{5(1-\omega)} = \frac{t_{\min}}{\omega \ln(1/\omega)} \qquad (11.59)$$

$$\lambda = \frac{(\phi c_t)_t \mu r_w^2}{0.0002637 k_2}\left(\frac{1}{t_{us,i}}\right) \qquad (11.60)$$

Characteristic points and lines are described in Figure 11.5. Arab (2003) also developed another expression for the intercept of the unit-slope line and the radial flow regime:

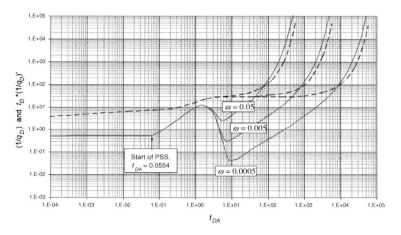

Figure 11.6. Effect of the storativity ratio on the reciprocal rate for a bounded reservoir; adapted from Arab (2003).

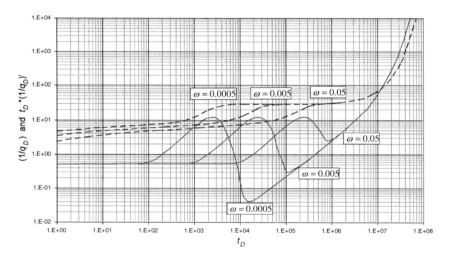

Figure 11.7. Effect of the storativity ratio on the reciprocal rate for a bounded reservoir, $\lambda=1\times10^{-7}$, $r_{eD}=1000$, $s=0$; adapted from Arab (2003).

$$t_{Dus,i} = -\frac{2}{\lambda}\left[\ln\left(\frac{\sqrt{\lambda}}{2}\right) + 0.57721 - s\right] \qquad (11.61)$$

which in oil-field units yields:

$$t_{us,i} = -\frac{(\phi c_t)_t \mu r_w^2}{0.000132k} \frac{1}{\lambda}\left[\ln\frac{\sqrt{\lambda}}{2} + 0.57721 - s\right] \quad (11.62)$$

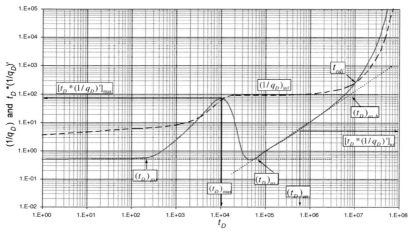

Figure 11.8. Characteristic points and lines found in the log-log plot of reciprocal rate and reciprocal-rate derivative against time for a bounded, naturally fractured reservoir with $\lambda = 1 \times 10^{-7}$, $\omega = 0.0005$, $r_{eD} = 500$, $s = 0$; adapted from Arab (2003).

Similarly to the PTA case, there are expressions for the determination of the skin factor:

$$s = 0.5\left\{\frac{(1/q)_{r1}}{[t*(1/q)']_{r1}} - \ln\left(\frac{k_2 t_{r1}}{(\phi c_t)_t \mu r_w^2}\frac{1}{\omega}\right) + 7.43\right\} \quad (11.63)$$

$$s = 0.5\left\{\frac{(1/q)_{r2}}{[t*(1/q)']_{r2}} - \ln\left(\frac{k_2 t_{r2}}{(\phi c_t)_t \mu r_w^2}\right) + 7.43\right\} \quad (11.64)$$

Arab (2003) also found in his study that the start of the boundary-dominated flow or pseudosteady state is not only a function of reservoir radius, r_e, but also of dimensionless storativity ratio, ω. The dimensionless time with respect to area, t_{DA}, is given by:

$$t_{DA} = \frac{t_D}{\pi r_{eD}^2 \omega} \quad (11.65)$$

$$t_{DApss} \approx 0.0554 \quad (11.66)$$

The combination of Equations (11.65) and (11.66) leads to:

$$\omega = \frac{47.59 \times 10^{-4} k_2 t_{pss}}{\pi r_e^2 (\phi c_t)_t \mu} \tag{11.67}$$

As observed in Figure 11.7, the maximum point ordinate, $[t_D*(1/q_D)']_{max}$, is independent of ω. From this observation, Arab (2003) obtained the following correlation:

$$\lambda = \exp\left[-\ln\left(60.36 \times 10^{-4} \frac{k_2 h (P_i - P_{wf})}{B\mu} [t*(1/q)']_{max} \left(\frac{r_e}{r_w e^{-s}}\right)^{2.4815}\right) / 1.2017\right] \tag{11.68}$$

The characteristic features are given in Figure 11.8. The equation for the late straight line is:

$$\ln\left(t_D*(1/q_D)'\right) = \ln\left(\frac{6.79}{\pi r_{eD}^2} t_{Dus}\right) \tag{11.69}$$

The reservoir external radius can be obtained from any arbitrary point, t_{us} and $[t*(1/q)']_{us}$, on that straight line (Figure 11.8):

$$r_e = \sqrt{\frac{0.0804 B}{(\phi c_t)_t h\Delta P} \frac{t_{us}}{[t*(1/q)']_{us}}} \tag{11.70}$$

The time of intersection formed by the unit-slope line during the transition period and the radial flow regime is given by:

$$t_{Dus,i} = \frac{\pi r_{eD}^2}{13.58} \tag{11.71}$$

Solving for the reservoir external radius,

$$r_e = \sqrt{\frac{k_2}{877.3 (\phi c_t)_t \mu} t_{us,i}} \tag{11.72}$$

The combination of Equations (11.67) and (11.71) leads to:

$$\omega = 1.329 \left(\frac{t_{pss}}{t_{us,i}}\right) \tag{11.73}$$

As seen in Figure 11.8, during the transition period, the $1/q_D$ curve is characterized by a horizontal line, which ordinate yields λ:

$$\lambda = \left[\frac{1}{2} r_{eD}^2 \left((1/q_D)_{inf} - \ln(r_{eD}) + 0.75 \right) \right]^{-1} \qquad (11.74)$$

The substitution of the dimensionless parameters in Equation (11.74) leads to:

$$\lambda = \left[\frac{r_e^2}{2(r_w e^{-s})^2} \left\{ \left(\frac{k_2 h \Delta P(1/q)_{inf}}{141.2 \mu B} \right) - \ln\left(\frac{r_e}{r_w e^{-s}} \right) + 0.75 \right\} \right]^{-1} \qquad (11.75)$$

The intersection of the unit-slope line with the horizontal line of the $1/q_D$ curve also yields an estimation of λ:

$$\lambda = \left[\frac{1}{2} r_{eD}^2 \left(\frac{6.79}{\pi r_{eD}^2} t_{D_{us,h}} - \ln(r_{eD}) + 0.75 \right) \right]^{-1} \qquad (11.76)$$

$$\lambda = \frac{r_e^2}{2(r_w e^{-s})^2} \left(\frac{177.9 \times 10^{-5} k_2}{\pi (\phi c_t)_t \mu r_e^2} t_{us,h} - \ln\left(\frac{r_e}{r_w e^{-s}} \right) + 0.75 \right)^{-1} \qquad (11.77)$$

As can be seen in Figure 11.8, if the test is run long enough, it is possible to observe the intersection of curves $(1/q_D)$ and $t_D*(1/q_D)'$ during the pseudosteady-state period. Arab (2003) found an expression for this intersection point:

$$t_{Drdi} = \frac{1}{2} \omega r_{eD}^2 \left(\ln(r_{eD}) - 0.75 \right) \qquad (11.78)$$

The suffix "*rdi*" means the intercept of reciprocal rate and reciprocal-rate derivative. Combining this expression with Equation (11.67) and Equation (11.71), respectively, provides:

$$r_{eD} = \exp\left(\frac{0.348 t_{rdi}}{t_{pss}} + 0.75 \right) \qquad (11.79)$$

$$\omega = \frac{2.161}{\left(\ln(r_{eD}) - 0.75 \right)} \frac{t_{rdi}}{t_{us,i}} \qquad (11.80)$$

Example 11.1

Arab (2003) reports a synthetic constant-pressure test run in a fractured well in a homogeneous reservoir. Rate data are given in Figure 11.9. Other relevant information is given below:

$P_i = 5200$ psi $B = 1.05$ rb/STB $\mu = 0.85$ cp
$h = 30$ ft $\phi = 20\%$ $c_t = 3.1 \times 10^{-5}$ psi^{-1}
$r_w = 0.29$ ft $x_f = 110$ ft $C_{fD} = 10$
$k = 15$ md $P_{wf} = 3500$ psi

The *TDS* technique is used to characterize this test.

Solution

The following characteristic points are read from Figure 11.9.

$t_r = 40.7$ hr $(1/q)_r = 3.39 \times 10^{-4}$ day/bbl
$[t^*(1/q)']_r = 0.000077$ day/STB $[t^*(1/q)']_{BL1} = 0.0000173$ day/STB
$[t^*(1/q)']_{L1} = 0.000054$ day/STB $t'_{BLLi} = 0.0124$ hr
$t'_{RBLi} = 390$ hr $t'_{RLi} = 2$ hr

Permeability and skin factor are determined with Equations (11.18) and (11.19), respectively:

$$k = \frac{70.6 \mu B}{h \Delta P [t^*(1/q)']_r} = \frac{70.6(0.85)(1.05)}{(30)(1700)(0.000077)} = 16 \text{ md}$$

$$s = 0.5 \left\{ \frac{3.39 \times 10^{-4}}{0.000077} - \ln \left(\frac{(16)(40.7)}{(0.2)(0.85)(3.1 \times 10^{-5})(0.29)^2} \right) + 7.43 \right\} = -4.64$$

Fracture conductivity and half-fractured length are calculated with Equations (11.35) and (11.36), respectively:

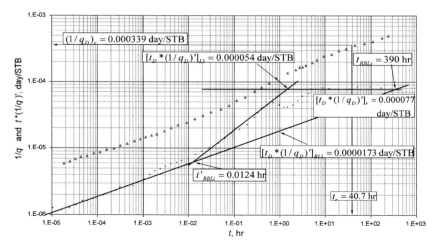

Figure 11.9. Reciprocal rate and reciprocal derivative versus time log-log plot for example 11.1; adapted from Arab (2003).

$$x_f = 3.192\sqrt{\frac{0.85}{0.2(3.1\times 10^{-5})(16)}\left[\frac{1.05}{30(1700)(0.000054)}\right]} = 112 \text{ ft}$$

$$k_f w_f = \frac{149.866}{\left[(0.2)(0.85)(3.1\times 10^{-5})(16)\right]^{1/2}}\left\{\frac{(0.85)(1.05)}{(30)(1700)0.0000173}\right\}^2 = 1680 \text{ md-ft}$$

The estimated parameters are verified with Equation (11.37):

$$t'_{BLLi} = 216.35\frac{\phi\mu c_t}{k}\left(\frac{x_f^2 k}{k_f w_f}\right)^2 = \frac{216.35(0.2)(0.85)(3.1\times 10^{-5})}{16.5}\left(\frac{(111^2)(16.5)}{1660}\right)^2 = 0.011 \text{ hr}$$

which matches the read value of 0.0124 hr.

The time of intersection between the bilinear and radial flow regimes is given by Equation (11.38):

$$t'_{RBLi} = 1106\frac{\phi\mu c_t}{k^3}(k_f w_f)^2 = 1106\frac{(0.2)(0.85)(3.1\times 10^{-5})}{16^3}1680^2 = 397 \text{ hr}$$

The estimated time also matches the time of 390 hr, read from Figure 11.9. The dimensionless fracture conductivity is found with Equation (11.81):

$$C_{fD} = \frac{k_f w_f}{k x_f} \qquad (11.81)$$

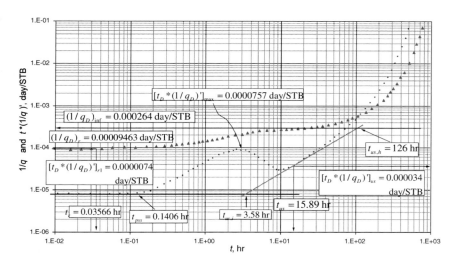

Figure 11.10. Reciprocal rate and reciprocal derivative versus time log-log plot for example 11.2; adapted from Arab (2003).

$$C_{fD} = \frac{1660}{(111)(16.5)} = 9.1$$

Example 11.2

A characterization of a constant-pressure test reported by Arab (2003) for a heterogeneous reservoir is performed using the *TDS* technique. The reciprocal rate and reciprocal-rate derivative data are reported in Figure 11.10. Other important information is given below.

$P_i = 4000$ psi	$B = 1.05$ rb/STB	$\mu = 1.2$ cp
$h = 150$ ft	$\phi = 20\%$	$c_t = 1.1\times10^{-5}$ psi^{-1}
$r_w = 0.29$ ft	$\omega = 0.05$	$r_e = 550$ ft
$k = 200$ md	$P_{wf} = 3600$ psi	$\lambda = 5\times10^{-8}$

Solution
The following information is read from Figure 11.10:

$t_r = 0.03566$ hr	$(1/q)_r = 9.643\times10^{-5}$ day/bbl
$[t*(1/q)']_r = 0.0000074$ day/STB	$[t*(1/q)']_{max} = 0.0000757$ day/STB
$[t*(1/q)']_{us} = 0.000034$ day/STB	$t_{us} = 15.89$ hr
$t_{pss} = 0.1406$ hr	$t_{us,i} = 3.58$ hr
$(1/q)_{inf} = 0.000264$ day/bbl	$t_{us,h} = 126$ hr

Permeability is calculated with Equation (11.57):

$$k_2 = \frac{70.6\mu B}{h\Delta P[t*(1/q)']_{r1,r2}} = \frac{70.6(1.2)(1.05)}{(150)(400)(7.4\times10^{-6})} = 200 \text{ md}$$

Equations (11.70) and (11.72) are used to find the external reservoir radius:

$$r_e = \sqrt{\frac{0.0804B}{(\phi c_t)_t} \frac{t_{us}}{h\Delta P[t*(1/q)']_{us}}} = \sqrt{\frac{0.0804(1.05)}{(0.2)(1.1\times10^5)(150)(400)} \frac{15.89}{3.4\times10^{-5}}} = 552 \text{ ft}$$

$$r_e = \sqrt{\frac{k_2 t_{us,i}}{877.3(\phi c_t)_t \mu}} = \sqrt{\frac{200(3.58)}{877.3(0.2)(1.1\times10^{-5})(1.2)}} = 556 \text{ ft}$$

The capacity ratio is found with Equations (11.67) and (11.73):

$$\omega = \frac{47.59\times10^{-4} k_2 t_{pss}}{\pi r_e^2 (\phi c_t)_t \mu} = \frac{47.59\times10^{-4}(200)0.14}{\pi\times552^2(0.2)(1.1\times10^{-5})(1.2)} = 0.053$$

$$\omega = 1.329 \left(\frac{t_{pss}}{t_{us,i}} \right) = 1.329 \left(\frac{0.1406}{3.58} \right) = 0.052$$

The skin factor is found with Equation (11.63):

$$s = 0.5 \left\{ \frac{9.643 \times 10^{-5}}{7.4 \times 10^{-6}} - \ln \left(\frac{(200)(0.03566)}{(0.2)(1.2)(1.1 \times 10^{-5})0.29^2 (0.052)} \right) + 7.43 \right\} \approx 0$$

The interporosity flow parameter is found with Equations (11.70), (11.75) and (11.77):

$$\lambda = \exp \left[-\ln \left(\frac{60.36 \times 10^{-4} (200)(150)(400)7.57 \times 10^{-5}}{(1.05)(1.2)} \left(\frac{552}{0.29} \right)^{2.4815} \right) / 1.2017 \right] = 4.99 \times 10^{-8}$$

$$\lambda = \left[\frac{552^2}{2(0.29)^2} \left\{ \left(\frac{(200)(150)(400)}{141.2(1.2)(1.05)} \frac{1}{2.64 \times 10^{-4}} \right) - \ln \left(\frac{552}{0.29} \right) + 0.75 \right\} \right]^{-1} = 5.03 \times 10^{-8}$$

$$\lambda = \frac{552^2}{2(0.29)^2} \left(\frac{177.9 \times 10^{-5} (200)}{\pi (0.2)(1.1 \times 10^{-6})(1.2)(552)^2} 126 - \ln \left(\frac{552}{0.29} \right) + 0.75 \right)^{-1} = 5.01 \times 10^{-8}$$

TDS TECHNIQUE FOR GAS WELLS IN HOMOGENEOUS RESERVOIRS

Escobar, Sanchez and Cantillo (2009) presented the *TDS* technique in gas equations for gas reservoirs using the pseudopressure and pseudopressure-derivative functions given by Equations (3.92) and (3.93). The equations will be presented in rigorous time but can be easily converted to pseudotime by means of the pseudotime function given in Equation (3.97), as performed by Escobar, Lopez and Cantillo (2007). The equations are summarized as follows:

$$C = 0.4196 \frac{Tq t_N}{\mu \Delta m(P)_N} = 0.4198 \frac{T}{\mu \Delta m(P)} \left[\frac{t}{t \times (1/q)'} \right]_N \tag{11.82}$$

$$k = 711.5817 \frac{T}{h \Delta m(P) \left[t \times (1/q)' \right]_r} \tag{11.83}$$

$$t_{sr} = 147218.4684 \frac{\mu C}{kh} \left[\ln \left(\frac{0.8935C}{\phi c_t h r_w^2} \right) + 2s \right] \tag{11.84}$$

Transient-Rate Analysis

$$C = 0.056\phi c_t h r_w^2 \left[\frac{t_{Dsr}}{2s + \ln(t_{Dsr})} \right] \tag{11.85}$$

t_{Dsr} is found from Equation (1.2) for $t = t_{Dsr}$. The intercept between the early unit-slope and radial lines results in an expression that can be used to verify permeability:

$$t_i = 1695 \frac{\mu C}{kh} \tag{11.86}$$

The correlations for using the peak during wellbore storage, as originally introduced by Tiab (1993) for gas systems, are:

$$k = 597.7286 \frac{T}{h\Delta m(P)} \Big/ \left(0.1511 \frac{Tt_x}{\mu C \Delta m(P)} - \left[t \times (1/q)' \right]_x \right) \tag{11.87}$$

$$s' = 0.1703 \left(\frac{t_x}{t_i} \right)^{1.24} - 0.5\ln\left(\frac{0.8935C}{\phi c_t h r_w^2} \right) \tag{11.88}$$

$$s' = 0.9184 \left\{ \frac{\left[t \times (1/q)' \right]_x}{\left[t \times (1/q)' \right]_r} \right\}^{1.1} - 0.5\ln\left(\frac{0.8935C}{\phi c_t h r_w^2} \right) \tag{11.89}$$

$$C = \frac{0.1511 Tt_x}{\left\{ \left[t \times (1/q)' \right]_x + 0.84 \left[t \times (1/q)' \right]_r \right\} \mu \Delta m(P)} \tag{11.90}$$

$$k = 9416.1958 \frac{\mu C}{ht_x} \left\{ 0.5 \frac{\left[t \times (1/q)' \right]_x}{\left[t \times (1/q)' \right]_r} + 0.42 \right\} \tag{11.91}$$

For constant pressure production during radial, $t_D > 8000$, the reciprocal rate behavior, including skin factor, displays the following behavior:

$$1/q_{Dr} = 0.5 \left[\ln(t_D) + 0.80907 + 2s' \right] \tag{11.2}$$

Dividing Equation (11.2) by Equation (11.16), plugging in the dimensionless quantities and solving for the apparent skin factor yields:

$$s' = 0.5\left\{\frac{(1/q)_r}{\left[t\times(1/q)'\right]_r} - \ln\left(\frac{kt_r}{\phi\mu c_t r_w^2}\right) + 7.43\right\} \qquad (11.19)$$

For the estimation of the reservoir external radius, Equations (11.21) through (11.25) also apply for gas wells.

TDS TECHNIQUE FOR GAS WELLS IN HETEROGENEOUS RESERVOIRS

The expressions reported here were also introduced by Escobar et al. (2008), following the work performed by Arab (2003). All the equations given for naturally fractured oil reservoirs apply to gas reservoirs, except Equations (11.57) and (11.70), for which equivalents for gas wells are given, respectively, below:

$$\lambda = 4.2314\frac{(\phi c_t)_t\,\mu\,r_w^2\Delta m(P)}{T}\left[\frac{t\times(1/q)'}{t}\right]_{min} \qquad (11.92)$$

$$r_e = \sqrt{0.8107\frac{T}{(\phi c)_t\,\mu h\Delta m(P)}\frac{t_{us}}{\left[t\times(1/q)'\right]_{us}}} \qquad (11.93)$$

FINITE-CONDUCTIVITY FRACTURED GAS WELLS

The dimensionless time based on half-fracture length, x_f, is as follows:

$$t_{Dxf} = \frac{0.0002637kt}{\phi(\mu c_t)_i\,x_f^2} \qquad (11.94)$$

According to Agarwal (1979), the dimensionless pseudotime, based on half-fracture length, x_f, is as follows:

$$t_{aDxf}(P) = \frac{0.0002637kt_a(P)}{\phi x_f^2} \qquad (11.95)$$

Escobar, Castro and Mosquera (2014) extended the solution to the oil diffusivity equation for a well intercepted by a finite-conductivity fracture. Cinco-Ley, Samaniego and Dominguez (1978) extended it to the TRA of gas wells:

$$\frac{1}{q_D} = \left[\frac{2.722}{\sqrt{C_{fD}}}\right] t_{Dxf}^{0.25} \tag{11.96}$$

the derivative of which is given by:

$$t_{Dxf} * (1/q_D)' = \frac{0.6805}{\sqrt{C_{fD}}} t_{Dxf}^{0.25} \tag{11.97}$$

Plugging the dimensionless quantities given by Equations (3.92) and (11.94) into Equation (11.96) yields:

$$\frac{1}{q} = \frac{493.94T}{h[\Delta m(P)]\sqrt{k_f w_f} \sqrt[4]{\phi(\mu c_t)_i k}} t^{0.25} \tag{11.98}$$

Equation (11.97) suggests that the slope m_{BL} from a Cartesian plot of the one-fourth root of time versus the reciprocal rate can be used to estimate fracture conductivity:

$$k_f w_f = \left(\frac{493.94T}{m_{BL} h[\Delta m(P)]\sqrt{k_f w_f} \sqrt[4]{\phi(\mu c_t)_i k}}\right)^2 \tag{11.99}$$

Using the *TDS* technique, the reciprocal rate derivative can also be expressed in oil-field units by:

$$t * (1/q)' = \frac{123.49T}{h\Delta P \sqrt{k_f w_f} \sqrt[4]{\phi(\mu c_t)_i k}} t^{0.25} \tag{11.100}$$

Solving for the apparent fracture conductivity from the above expression yields:

$$(k_f w_f)_{app} = \frac{15242.372}{(\phi \mu c_t k)^{0.5}} \left\{\frac{T}{h[\Delta m(P)][t * (1/q)']_{BL1}}\right\}^2 \tag{11.101}$$

If using pseudotime, as given by Equation (11.95), the above expression becomes:

$$(k_f w_f)_{app} = \frac{15242.372}{(\phi k)^{0.5}} \left\{\frac{T}{h[\Delta m(P)][t * (1/q)']_{BL1}}\right\}^2 \tag{11.102}$$

Equations (11.101) and (11.102) are used to estimate the apparent fracture conductivity from the reciprocal-rate derivative read at either the time of 1 hr or the pseudotime of 1 hr-psi/cp.

INFINITE-CONDUCTIVITY FRACTURED GAS WELLS

The dimensionless reciprocal rate governing equation for linear flow, according to Arab (2003), is given as follows:

$$\frac{1}{q_D} = 2.7842 t_{Dxf}^{0.5} \tag{11.103}$$

the derivative of which yields:

$$t_{Dxf} * (1/q_D)' = 2.3921 t_{Dxf}^{0.5} \tag{11.104}$$

Plugging the dimensionless quantities defined by Equations (3.92) and (11.94) into Equation (11.103) yields:

$$\frac{1}{q} = \frac{64.379T}{x_f h[\Delta m(P)]\sqrt{k\phi(\mu c_t)_i}} t^{0.5} \tag{11.105}$$

The above equation indicates that the slope m_L from a Cartesian plot of the square root of time versus the reciprocal rate can be used to estimate half-fracture length:

$$x_f = \frac{64.379T}{m_L h[\Delta m(P)]\sqrt{k\phi(\mu c_t)_i}} t^{0.5} \tag{11.106}$$

The derivative of Equation (11.104) is:

$$t * (1/q)' = \frac{32.1895T}{x_f h[\Delta m(P)]\sqrt{\phi(\mu c_t)_i k}} t^{0.5} \tag{11.107}$$

Solving for the half-fracture length yields:

$$x_f = \frac{32.1895T}{h[\Delta m(P)][t*(1/q)']_{L1} (\phi[\mu c_t]_i k)^{0.5}} \tag{11.108}$$

As noted previously, the value of the reciprocal-rate derivative is read at the time of 1 hr and extrapolated if necessary. Using the pseudotime function, Equation (11.109) becomes:

$$x_f = \frac{32.1895T}{h[\Delta m(P)][t*(1/q)']_{L1}(\phi k)^{0.5}} \qquad (11.109)$$

Tiab (1994) introduced the concept and definition of the biradial (or elliptical) flow regime. The definition of pressure-derivative behavior is adapted by Escobar et al. (2014) for transient-rate analysis, such that:

$$\frac{1}{q_D} = 2.1361\left(\frac{x_e}{x_f}\right)^{0.72} t_{DA}^{0.36} \qquad (11.110)$$

$$t_{DA}*(1/q_D)' = 0.769\left(\frac{x_e}{x_f}\right)^{0.72} t_{DA}^{0.36} \qquad (11.111)$$

For oil reservoirs, the dimensional form of Equation (11.110) is:

$$\frac{1}{q} = \frac{9.426B}{h\Delta P\left(\phi c_t x_f^2\right)^{0.36}}\left(\frac{\mu}{k}\right)^{0.64} t^{0.36} \qquad (11.112)$$

Notice that Equation (11.112) suggests that a Cartesian plot of either time to the power 9/25 (or 0.36) will provide a slope, m_{BR}, which yields the half-fracture length:

$$x_f = \left(\frac{95.052T}{m_{BR}hk^{0.64}[\Delta m(P)][\phi(\mu c_t)_i]^{0.36}}\right)^{50/36} \qquad (11.113)$$

The dimensional form of Equation (11.111) is given by:

$$[t*(1/q)'] = \frac{3.93313\mu^{0.64}}{h\Delta P\left(\phi c_t x_f^2\right)^{0.36}}\left(\frac{\mu}{k}\right)^{0.64} t^{0.36} \qquad (11.114)$$

The *TDS* technique, solving for x_f when the reciprocal derivative is read at a time of 1 hr, yields:

$$x_f = \left[\frac{3.39313B}{[t*(1/q)']_{BR1}h\Delta P(\phi c_t)^{0.36}}\left(\frac{\mu}{k}\right)^{16/25}\right]^{50/36} \qquad (11.115)$$

For gas reservoirs, the dimensional form of Equation (11.110) is:

$$1/q = \frac{95.052T}{hk^{0.64}\left[\Delta m(P)\right]\left[\phi(\mu c_t)_i\right]^{0.36} x_f^{0.72}} t^{0.36} \tag{11.116}$$

Notice that a Cartesian plot of either time to the power 9/25 (or 0.36) will provide a slope, m_{BR}, which yields the half-fracture length:

$$x_f = \left(\frac{95.052T}{m_{BR} hk^{0.64}\left[\Delta m(P)\right]\left[\phi(\mu c_t)_i\right]^{0.36}}\right)^{50/36} \tag{11.117}$$

The dimensional form of Equation (11.111) for gas wells is given by:

$$[t*(1/q)']_{BR} = \frac{34.219T}{hk^{0.64}\left[\Delta m(P)\right]\left[\phi(\mu c_t)_i\right]^{0.36} x_f^{0.72}} t_{BR}^{0.36} \tag{11.118}$$

Solving for the half-fracture length yields time and pseudotime, respectively:

$$x_f = \left(\frac{34.219T}{hk^{0.64}\left[\Delta m(P)\right]\left[\phi(\mu c_t)_i\right]^{0.36}[t*(1/q)']_{BR1}}\right)^{50/36} \tag{11.119}$$

$$x_f = \left(\frac{34.219T}{hk^{0.64}\left[\Delta m(P)\right]\phi^{0.36}[t*(1/q)']_{BR1}}\right)^{50/36} \tag{11.120}$$

INTERSECTION POINTS IN FRACTURED WELLS

Escobar et al. (2014) found that the governing equation for the reciprocal rate during the pseudosteady-state regime is given by:

$$[t*(1/q_D)']_{pss} = 3\pi(t_{DA})_{pss} \tag{11.121}$$

Changing the reservoir area for half-fracture length in Equation (11.94) and plugging it and Equation (3.94) into Equation (11.121) yields for oil wells:

$$A = \frac{B}{2.85h\Delta P\phi c_t[t*(1/q)']_{PSS1}} \tag{11.122}$$

For the case of gas wells, the resulting equation is given for actual time by:

$$A = \frac{3.54kT}{h\Delta m(P)\phi(\mu c_t)_i[t*(1/q)']_{PSS1}} \tag{11.123}$$

and for pseudotime by:

$$A = \frac{3.54kT}{h\Delta m(P)\phi[t*(1/q)']_{PSS1}} \qquad (11.124)$$

Equations (11.122) through (11.124) are also used by reading the reciprocal-rate derivative at a value of 1 hr or 1 hr-psi/cp. The point of intersection formed by the pseudosteady-state, reciprocal-derivative straight line—Equation (11.121)—with the bilinear flow regime derivative—Equation (11.104)—yields expressions for estimating the drainage area:

$$A = \frac{\sqrt{k_f w_f}}{34.892} \left(\frac{t_{BLPSSi}}{\phi\mu c_t} \right)^{0.75} k^{0.25} \qquad (11.125)$$

The point of intercept between the biradial—Equation (11.114)—and linear—Equation (11.105)—with pseudosteady-state derivative—Equation (11.121)—lines yields useful equations to estimate the drainage area in oil and gas reservoirs:

$$A = \left(\frac{kt_{BRPSSi} x_f^{1.125}}{34.649\phi\mu c_t} \right)^{16/25} \qquad (11.126)$$

$$A = \frac{x_f}{5.79} \sqrt{\frac{kt_{LPSSi}}{\phi\mu c_t}} \qquad (11.127)$$

Other useful intersection derivative points are bilinear–linear, bilinear–biradial, linear–biradial, radial–bilinear, linear–radial, radial–bilinear, radial–biradial and radial–pseudosteady-state, according to Escobar et al. (2014):

$$k = \left(\frac{(k_f w_f)_{app}}{x_f^2} \right)^2 \frac{t'_{BLLi}}{869.375\phi\mu c_t} \qquad (11.128)$$

$$(k_f w_f)_{app} = 12.759 k^{0.7714} x_f^{1.429} \left(\frac{\phi\mu c_t}{t_{BLBRi}} \right)^{0.2143} \qquad (11.129)$$

$$\frac{x_f^2}{k} = \frac{t_{LBRi}}{39\phi\mu c_t} \qquad (11.130)$$

$$t_{RBLi} = 1677 \frac{\phi\mu c_t}{k^3} (k_f w_f)_{app}^2 \qquad (11.131)$$

$$\frac{x_f^2}{k} = \frac{t_{LRi}}{1207\phi\mu c_t} \qquad (11.132)$$

$$\frac{x_f^2}{k} = \frac{t_{RBRi}}{4587\phi\mu c_t} \qquad (11.133)$$

$$A = \frac{kt_{RPPSi}}{201.2\phi\mu c_t} \qquad (11.134)$$

Equations (11.128) through (11.134) apply to both oil and gas using real time. The viscosity and total compressibility ought to be evaluated at initial conditions. For pseudotime, the product μc_t should be dropped and t should be replaced with $t_a(P)$.

LONG HOMOGENEOUS RESERVOIRS—CONVENTIONAL ANALYSIS AND *TDS* TECHNIQUE

Normally, production data are analyzed by decline-curve fitting. However, analogous to transient-pressure analysis, the reciprocal flow rate and its derivative may be analyzed and interpreted for reservoir characterization purposes. In some cases, the formation linear flow regime may be seen once the radial flow regime vanishes. For instance, Figure 11.11 presents reciprocal rate versus time data for a well in Central America in which the formation linear flow regime is seen for at least 20 months. This flow regime is very important and can be presented in fractured wells, horizontal wells and long reservoirs.

Escobar, Rojas and Cantillo (2012a) and Escobar, Rojas and Bonilla (2012b) presented the constant-pressure mathematical formulation for both homogeneous and heterogeneous long formations. They developed both conventional and *TDS* techniques for the previously mentioned cases. They used the dimensionless terms given by Equations (1.35) through (1.39). The dimensionless governing equations for reciprocal rate and its derivative during a single linear (hemilinear) flow regime are given below:

$$\frac{1}{q_D} = \frac{4\pi\sqrt{t_D}}{W_D} + s_L' + s_{DL}' \qquad (11.135)$$

$$\left[t_D * (1/q_D)'\right] = \frac{2\pi\sqrt{t_D}}{W_D} \qquad (11.136)$$

Replacing the dimensionless terms in Equation (11.135) with those given by Equations (1.35) through (1.39) and by (11.17) leads to:

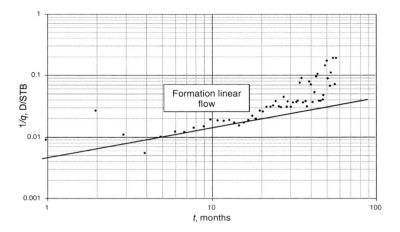

Figure 11.11. Evidence of formation linear flow in a constant-pressure test found in a well in Central America.

$$\frac{1}{q} = \frac{28.8137B}{h\Delta P}\sqrt{\frac{\mu t}{\phi k c_t Y_E^2}} + \frac{141.2\mu B}{kh\Delta P}(s_L' + s_{DL}') \qquad (11.137)$$

This implies that a Cartesian plot of $1/q$ vs. $t^{0.5}$ will produce a straight line with slope m_{LF} and intercept b_{LF}, which leads to:

$$Y_E = \frac{28.8137B}{h\Delta P m_{LF}}\sqrt{\frac{\mu}{\phi k c_t}} \qquad (11.138)$$

$$s_L' + s_{DL}' = \frac{kh\Delta P b_{LF}}{141.2\mu B} \qquad (11.139)$$

The application of the *TDS* technique requires replacing terms in Equation (11.136) with those from Equations (1.35) through (1.39) and (11.17) and solving for the half-fracture length, leading to:

$$Y_E\sqrt{k} = \frac{14.4068B}{h\Delta P[t*(1/q)']_{L1}}\sqrt{\frac{\mu}{\phi c_t}} \qquad (11.140)$$

$[t*(1/q)']_{L1}$ is the value of the reciprocal rate derivative during the linear flow regime, read at a time of 1 hr and extrapolated if needed.

Similar to the process in Equation (1.49), the skin factor caused by convergence from a linear to a dual-linear flow regime is obtained by dividing Equation (11.135) by Equation (11.136) and plugging in the dimensionless terms:

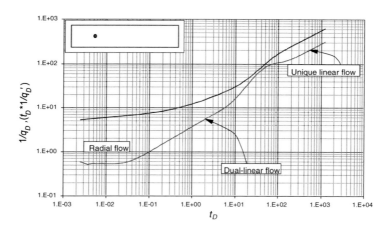

Figure 11.12. Linear flow regimes for a well off-centered inside a closed-boundary rectangular reservoir; adapted from Escobar et al. (2012b).

$$s_L' + s_{DL}' = \left(\frac{1/q}{[t*(1/q)']_L} - 2\right)\frac{1}{9.8008 Y_E}\sqrt{\frac{k\, t_L}{\phi\mu\, c_t}} \qquad (11.141)$$

where, t_L is any convenient time during the linear flow regime, and $(1/q)_L$ and $[t*(1/q)']_L$ are the reciprocal rate and its derivative corresponding to t_L, respectively.

The dual-linear flow regime is presented for all the cases of closed, mixed and constant-pressure boundaries. It takes place when the well is located at any appropriate distance from the lateral boundaries. This behavior is shown in Figure 11.12. The transition period between the linear flow regimes is longer in TRA than in PTA, as can be compared to Figure 1.3. The governing dimensionless reciprocal rate and its derivative, presented by Escobar et al. (2012a, 2012b), are:

$$\frac{1}{q_D} = \frac{5}{2}\frac{\sqrt{\pi t_D}}{W_D} + (s_{DL}' + s) \qquad (11.142)$$

$$[t_D *(1/q_D)'] = \frac{5}{4}\frac{\sqrt{\pi t_D}}{W_D} \qquad (11.143)$$

Once the dimensionless quantities are replaced, Equation (11.142) yields:

$$\frac{1}{q} = \frac{10.1602 B}{h\Delta P}\sqrt{\frac{\mu t}{\phi k c_t Y_E^2}} + \frac{141.2 \mu B}{k h \Delta P}s_{DL} \qquad (11.144)$$

As for the case of Equation (11.137), the above expression also indicates that a Cartesian plot of $1/q$ vs. $t^{0.5}$ will produce a straight line with slope m_{DLF} and intercept b_{DLF}, yielding:

$$Y_E = \frac{10.1602B}{h\Delta P m_{DLF}} \sqrt{\frac{\mu}{\phi k c_t}} \tag{11.145}$$

$$s_{DL}' + s = \frac{kh\Delta P b_{DLF}}{141.2\mu B} \tag{11.146}$$

Application of the *TDS* technique also needs the dimensionless quantities to be replaced in Equation (11.143) to provide:

$$Y_E \sqrt{k} = \frac{5.0801B}{h\Delta P \left[t*(1/q)'\right]_{DL1}} \sqrt{\frac{\mu}{\phi c_t}} \tag{11.147}$$

$[t*(1/q)']_{DL1}$ is the value of the reciprocal rate derivative during the dual-linear flow regime, read at a time of 1 hr and extrapolated if needed. By the same token as in Equation (11.141):

$$s_{DL}' + s = \left(\frac{1/q}{\left[t*(1/q)'\right]_{DL}} - 2\right)\frac{1}{27.7945Y_E}\sqrt{\frac{k\,t_{DL}}{\phi\mu\,c_t}} \tag{11.148}$$

where $t_{DL\,is}$ any convenient time during the dual-linear flow regime, and $(1/q)_{DL}$ and $[t*(1/q)']_{DL}$ are the reciprocal rate and its derivative corresponding to t_{DL}, respectively.

Escobar et al. (2012a, 2012b) also found that parabolic flow is developed under constant-pressure conditions. See Figure 11.13. The governing equations, similar to Equations (1.43) and (1.45), are given as:

$$\frac{1}{q_D} = -\frac{\sqrt{\pi}}{4}W_D\left(X_D^2\right)\left(\frac{X_E}{Y_E}\right)^2 t_D^{-0.5} + s_{PB}' + s_{DL}' \tag{11.149}$$

$$\left[t_D*(1/q_D)'\right] = \frac{\sqrt{\pi}}{8}W_D\left(X_D^2\right)\left(\frac{X_E}{Y_E}\right)^2 t_D^{-0.5} \tag{11.150}$$

After the dimensionless parameters are plugged into the former expression, the following equation is obtained:

$$\frac{k^{1.5}Y_E}{b_x^2} = 7705.9213\left(\frac{\mu B}{h\left[t*(1/q)'\right]_{PB}}\right)\left(\frac{\phi\mu c_t}{t_{PB}}\right)^{0.5} \tag{11.151}$$

The geometric parabolic skin factor is obtained by dividing the reciprocal rate equation, Equation (11.149), by its derivative equation, Equation (11.150):

$$S_{PB}' + S_{DL}' = \left(\frac{1/q}{[t*(1/q)']_{PB}} + 2\right)\frac{54.5745 b_x}{Y_E}\sqrt{\frac{\phi\mu c_t}{k\, t_{PB}}} \tag{11.152}$$

where t_{PB} is any convenient time during the parabolic flow regime, and $(1/q)_{PB}$ and $[t*(1/q)']_{PB}$ are the reciprocal rate and its derivative corresponding to t_{PB}, respectively.

The application of conventional analysis requires plugging the dimensionless quantities into Equation (11.149), which leads to:

$$\frac{1}{q} = -\frac{15411.843\, B b_x^2}{h\Delta P}\sqrt{\frac{\phi\mu^3 c_t}{k^3 Y_E^2}}\, t^{-0.5} + \frac{141.2\,\mu B}{kh\Delta P}(S_{PB}' + S_{DL}') \tag{11.153}$$

Therefore, a Cartesian plot of $1/q$ vs. $1/t^{\,0.5}$ will yield a straight line with slope m_{PBF} and intercept b_{PBF}, allowing one to determine that:

$$b_x = -\frac{m_{PBF}^2}{124.1444}\sqrt{\frac{h\Delta P}{B}\frac{k^3 Y_E^2}{\phi\mu^3 c_t}} \tag{11.154}$$

$$S_{PB}' + S_{DL}' = \frac{kh\Delta P\, b_{PBF}}{141.2\,\mu B} \tag{11.155}$$

A unit-slope line is observed on the derivative during a late time. Escobar et al. (2012b) found the following governing equation for that line:

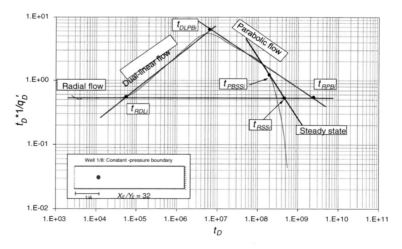

Figure 11.13. Derivative plot showing parabolic flow, steady-state flow period and five intersection points.

$$\left[t_D * (1/q_D)'\right] = \frac{11}{2} \pi\, t_{DA} \tag{11.156}$$

The intercept of this line with the dual-linear, linear and radial lines allows estimations of the reservoir drainage area, such as, respectively:

$$A = \sqrt{\frac{kt_{DLPSSi}Y_E^2}{62.3221\,\varphi\,\mu\,c_t}} \tag{11.157}$$

$$A = \sqrt{\frac{kt_{LPSSi}Y_E^2}{501.2252\,\varphi\,\mu\,c_t}} \tag{11.158}$$

$$A = \frac{kt_{RPSSi}}{109.7355\,\phi\mu c_t} \tag{11.159}$$

The intersection point between the infinite-acting reciprocal rate derivative and the dual-linear and linear lines leads to expressions for estimating reservoir width:

$$Y_E = 0.07195\sqrt{\frac{kt_{RDLi}}{\varphi\,\mu\,c_t}} \tag{11.160}$$

$$Y_E = 0.2040\sqrt{\frac{kt_{RLi}}{\varphi\,\mu\,c_t}} \tag{11.160}$$

The intersections of the parabolic line with dual-linear and linear lines are shown in Figure 11.13. These points are used to find the distance from the well to the closer lateral boundary, b_x.

$$b_x = \frac{1}{38.9470}\sqrt{\left(\frac{kt_{DLPBi}}{\varphi\,\mu\,c_t}\right)} \tag{11.161}$$

$$b_x = \sqrt{\frac{Y_E}{109.2242}\left(\frac{kt_{RPB_i}}{\phi\mu c_t}\right)^{0.5}} \tag{11.162}$$

Escobar et al. (2012a, 2012b) presented more expressions for the *TDS* technique, using more characteristic points among the different flow regimes.

LONG HETEROGENEOUS RESERVOIRS—*TDS* TECHNIQUE

Escobar et al. (2012a, 2012b) introduced the governing dimensionless reciprocal rate and reciprocal rate derivative equations for the dual-linear flow in long heterogeneous systems:

$$1/q_D = \frac{9}{4}\frac{\sqrt{\pi t_D}}{\sqrt{\omega W_D}} + s_{DL}' + s \qquad (11.163)$$

$$[t_D*(1/q_D)'] = \frac{9}{8}\frac{\sqrt{\pi\,t_D}}{W_D\sqrt{\omega}} \qquad (11.164)$$

Substituting Equations (1.35) through (1.39) and (11.17) in Equation (11.164), taking care that permeability now corresponds to the fracture bulk, and solving for the reservoir width results in:

After Escobar et al. (2012b)

$$Y_E = \frac{4.162115B}{h\Delta P[t*(1/q)']_{DL}}\sqrt{\frac{\mu}{k_2\phi c_t\omega}} \qquad (11.165)$$

Again the derivative is read at 1 hr. As for the homogeneous case, the equation for the geometric skin factor is:

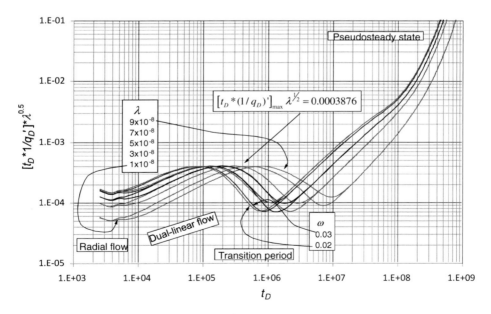

Figure 11.14. Dimensionless reciprocal rate derivative behavior vs. time for a long heterogeneous reservoir with transition period after the dual-linear flow regime.

$$s_{DL} = \left(\frac{(1/q)}{[t*(1/q)']_{DL}} - 2 \right) \frac{1}{54.738732Y_E} \sqrt{\frac{kt_{DL}}{\varphi\mu c_t \omega}} \tag{11.166}$$

Figure 11.14 presents the dimensionless rate derivative behavior against the dimensionless time for several values of λ and two values of ω. Notice in that plot the existence of a characteristic maximum point once the dual-linear flow regime is interrupted by the transition period from heterogeneous to homogeneous behavior. A unique maximum value of the reciprocal rate derivative is obtained when this is multiplied by the square root of the interporosity flow parameter, as shown in Figure 11.14. The governing equation for this maximum point is:

$$\left[t_D *(1/q_D)' \right]_{max} \sqrt{\lambda} = 0.0003876 \tag{11.167}$$

Placing Equation (11.17) into Equation (11.167) and solving for the interporosity flow parameter results in:

$$\lambda = \left[\frac{0.05472912\mu B}{kh\Delta P \left[t*(1/q)' \right]_{max}} \right]^2 \tag{11.168}$$

Escobar et al. (2012b) developed a correlation to determine ω as a function of $1/q_D$, (t_D*1/q_D') and t_D:

$$\omega = \frac{a + b\ln\left(t_D*1/q_D'\right)_{min} + c\left(\ln\left(t_D*1/q_D'\right)_{min}\right)^2 + d\left(\ln\left(t_D*1/q_D'\right)_{min}\right)^3 + e\ln\lambda}{1 + f\ln\left(t_D*1/q_D'\right)_{min} + g\left(\ln\left(t_D*1/q_D'\right)_{min}\right)^2 + h\left(\ln\left(t_D*1/q_D'\right)_{min}\right)^3 + i\ln\lambda} \tag{11.169}$$

where:

$a = -0.15407418 \qquad b = -0.0039050889 \qquad c = 0.0022251207$
$d = 4.2193304\text{x}10^{-4} \quad e = -0.0079273861 \qquad f = 0.2804578$
$g = -0.072289158 \qquad h = 0.023141728 \qquad i = 0.07200547$

Either reservoir width or area can be found from the intersection points given by the pseudosteady-state and dual-linear, radial and pseudosteady-state lines, respectively:

$$A = 0.140998\sqrt{\frac{Y_E^2 \omega kt_{DLPSSi}}{\phi\mu c_t}} \tag{11.170}$$

$$Y_E = 0.0648256\sqrt{\frac{k\,t_{RDLi}}{\phi\mu c_t\omega}} \tag{11.171}$$

$$A = \frac{kt_{RPSSi}}{109.7355\,\phi\mu c_t} \tag{11.159}$$

When the linear flow regime occurs after the transition period, the linear behavior corresponds to a homogeneous reservoir, and the governing expression is Equation (11.135); therefore, expressions derived from that equation also correspond to this analysis. On the contrary, if the linear flow regime occurs before the transition period, the linear flow regime is still affected by the heterogeneous behavior; the reciprocal rate and its derivative were defined by Escobar et al. (2012b) as:

$$\frac{1}{q_D} = \frac{21}{5}\frac{\sqrt{\pi\,t_D}}{W_D\sqrt{\omega}} + s_L{'} + s_{DL}{'} \tag{11.172}$$

$$\left[t_D*(1/q_D)'\right] = \frac{21}{10}\frac{\sqrt{\pi\,t_D}}{W_D\sqrt{\omega}} \tag{11.173}$$

As stated before, Equations (11.160) and (11.161) allow one to obtain:

$$Y_E = \frac{42.480763\,B}{h\Delta P\left[t*(1/q)'\right]_{L1}}\sqrt{\frac{\mu}{k\phi c_t\omega}} \tag{11.174}$$

$$s_L = \left(\frac{1/q}{\left[t*(1/q)'\right]_L} - 2\right)\frac{1}{29.32416\,Y_E}\sqrt{\frac{k\,t_{DL}}{\phi\mu c_t\omega}} \tag{11.175}$$

Again, a direct relationship exists between the square root of the interporosity flow parameter multiplied by the reciprocal rate derivative at the maximum point. This allows the following expression to be obtained:

$$\sqrt{\lambda}\left[t_D*(1/q_D)'\right]_{max} = 0.000746 \tag{11.176}$$

Solving for λ from Equation (11.164) once the respective dimensionless group is replaced yields:

$$\lambda = \left[\frac{0.105335\mu B}{kh\Delta P\left[t*(1/q)'\right]_{max}}\right]^2 \tag{11.177}$$

Readings of the minimum point for different values of the interporosity flow parameter and the dimensionless storativity coefficient were correlated by Escobar et al. (2012b) to obtain the following correlation:

Transient-Rate Analysis

$$\omega = \frac{a + b\ln\lambda + c\ln\lambda^2 + d\left(t_D*1/q_D'\right)_{\min} + e\left(t_D*1/q_D'\right)^2_{\min} + f\left(t_D*1/q_D'\right)^3_{\min}}{1 + g\ln\lambda + h\left(t_D*1/q_D'\right)_{\min} + i\left(t_D*1/q_D'\right)^2_{\min} + j\left(t_D*1/q_D'\right)^3_{\min}} \qquad (11.178)$$

$a = 0.14646803$ $b = 0.013342717$ $c = 0.00030434555$
$d = -0.00026701191$ $e = 2.4497641\times10^{-6}$ $f = -9.2559594\times10^{-9}$
$g = 0.049793248$ $h = 0.0021617165$ $i = -8.3473003\times10^{-6}$
$j = 8.9505062\times10^{-9}$

Finally, the unit-slope line during the transition period was correlated to obey the following governing equation:

$$\lambda = a + b\ln t_{D,usi} + d(\ln t_{D,us})^2 + f\ln t_{D,usi} + g(\ln t_{D,usi})^3 + i\ln t_{D,usi} + j(\ln t_{D,usi})^2\ln y \qquad (11.179)$$

$a = 0.00001727728645$ $b = -3.67144681\times10^{-6}$ $d = 2.2745910723\times10^{-7}$
$f = 3.5320510014\times10^{-7}$ $g = -4.737297387478569\times10^{-9}$ $i = -9.9895945387\times10^{-9}$
$j = -1.135705997\times10^{-8}$

where usi is the intercept of the unit-slope line during the transition period with the radial flow line. λ may also be estimated using Equation (2.13).

LONG HETEROGENEOUS RESERVOIRS—CONVENTIONAL ANALYSIS

The dimensional form of Equation (11.163), a dual-linear flow regime, is:

$$\frac{1}{q} = \frac{9.1442B}{h\Delta P}\sqrt{\frac{\mu t}{\omega k\phi c_t Y_E^2}} + \frac{141.2\mu B}{kh\Delta P}(+s_{DL}'+s) \qquad (11.180)$$

which indicates that a Cartesian plot of $1/q$ vs. $t^{0.5}$ will produce a straight line with slope m_{DLF} and intercept b_{DLF}, resulting in:

$$Y_E = \frac{9.1442B}{h\Delta P m_{DLF}}\sqrt{\frac{\mu}{\omega\phi kc_t}} \qquad (11.181)$$

$$s + s_{DL}' = \frac{kh\Delta P b_{DLF}}{141.2\mu B} \qquad (11.182)$$

If the linear flow regime takes place after the heterogeneous transition is seen, Equations (11.139) and (11.139) apply. Otherwise, the dimensional form of Equation (11.172) is:

$$\frac{1}{q} = \frac{17.0692B}{h\Delta P}\sqrt{\frac{\mu t}{\omega k \phi c_t Y_E^2}} + \frac{141.2\mu B}{kh\Delta P}(s_L' + s_{DL}') \qquad (11.183)$$

As before, a Cartesian plot of $1/q$ vs. $t^{0.5}$ will produce a straight line with slope m_{LF} and intercept b_{LF}, allowing one to obtain:

$$Y_E = \frac{17.0692B}{h\Delta P m_{LF}}\sqrt{\frac{\mu}{\omega \phi k c_t}} \qquad (11.184)$$

$$s_L = \frac{kh\Delta P b_{LF}}{141.2\mu B} \qquad (11.185)$$

In elongated heterogeneous systems, the mass transference between matrix and fractures may take place after the radial flow regime, when the interporosity flow parameter λ is normally lower than 1×10^{-7}. Then either the dual-linear or the linear flow regime may be interrupted by the transition period of the naturally fractured system. However, in any case, dual-linear flow is normally seen after the radial flow regime.

Escobar et al. (2012a) developed a correlation between ω with the intercept at time of 1 hr; the correlation is given as:

$$\omega = \frac{a + bx + cx^2 + dy}{1 + ex + fy + gy^2 + hy^3} \qquad (11.186)$$

$a=$ -0.0293872507 $b=$ 0.0181467652 $c=$ -0.0029463323
$d=$ 0.000216978258 $e=$ 0.00517061559 $f=$ -0.93599701655
$g=$ 0.291208778 $h=$ -0.031024762

Table 11.1. Test data for example 11.3

t, hr	$(1/q)$, D/STB	$t^*(1/q)'$, D/STB	t, hr	$(1/q)$, D/STB	$t^*(1/q)'$, D/STB	t, hr	$(1/q)$, D/STB	$t^*(1/q)'$, D/STB
0.02	1.07E-04	1.49E-05	0.634	1.90E-04	4.41E-05	15.924	2.79E-04	9.86E-06
0.03	1.12E-04	1.30E-05	0.798	2.00E-04	4.60E-05	20.047	2.82E-04	1.22E-05
0.04	1.16E-04	1.43E-05	1.005	2.11E-04	4.68E-05	25.238	2.85E-04	1.54E-05
0.050	1.19E-04	1.47E-05	1.265	2.21E-04	4.62E-05	31.773	2.89E-04	1.94E-05
0.063	1.23E-04	1.61E-05	1.592	2.32E-04	4.39E-05	40.000	2.94E-04	2.46E-05
0.080	1.27E-04	1.77E-05	2.005	2.42E-04	3.99E-05	50.357	3.00E-04	3.11E-05
0.100	1.31E-04	1.99E-05	2.524	2.50E-04	3.46E-05	63.396	3.08E-04	3.94E-05
0.126	1.36E-04	2.24E-05	3.177	2.57E-04	2.83E-05	79.810	3.19E-04	4.98E-05
0.159	1.42E-04	2.52E-05	4.000	2.63E-04	2.20E-05	100.475	3.32E-04	6.31E-05
0.200	1.48E-04	2.83E-05	5.036	2.68E-04	1.64E-05	126.491	3.48E-04	7.98E-05
0.252	1.55E-04	3.16E-05	6.340	2.71E-04	1.20E-05	159.243	3.69E-04	1.01E-04
0.318	1.62E-04	3.49E-05	7.981	2.73E-04	9.31E-06	200.475	3.95E-04	1.27E-04
0.4	1.71E-04	3.83E-05	10.048	2.75E-04	8.19E-06	250.475	4.26E-04	1.58E-04
0.50357	1.80E-04	4.14E-05	12.649	2.77E-04	8.46E-06	300.475	4.58E-04	1.87E-04

where:

$$x = \log \frac{0.0002637k}{\phi \mu c_t r_w^2} \tag{11.187}$$

$$y = \frac{kh\Delta P}{141.2\mu B}\left(\frac{1}{q}\right)_{1\,hr} + 1.0515s \tag{11.188}$$

The mechanical skin factor in Equation (11.188) is estimated by either Equation (11.44) or Equation (11.45). λ may also be estimated by Equation (2.13).

Example 11.3

This example was presented by Escobar et al. (2012a, 2012b). It uses both conventional analysis and *TDS* technique for interpreting a constant-pressure test run in an elongated reservoir in which both lateral boundaries are closed to flow. Other data:

$\Delta P = 2500$ psi $\mu = 2$ cp $\phi = 20\%$ $B = 1.2$ rb/STB
$c_t = 1 \times 10^{-6}$ psi^{-1} $r_w = 0.5$ ft $h = 100$ ft $X_E = 20000$ ft
$Y_E = 1000$ ft $\lambda = 5 \times 10^{-8}$ $\omega = 0.02$

Solution by TDS Technique

In this case the dual-linear flow regime is interrupted by the heterogeneity. Therefore, a pseudosteady state is expected to develop if the test is long enough. The reciprocal rate and its derivative are given in Figure 11.15, from which the below data was obtained:

$(t*1/q')_r = 1.30 \times 10^{-5}$ $t_{DL} = 0.2523$ hr $(t*1/q')_{DL} = 3.16 \times 10^{-5}$

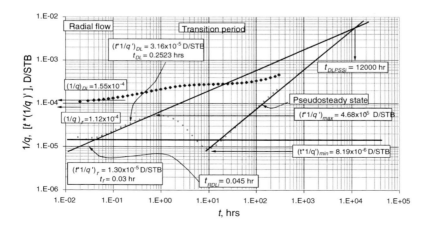

Figure 11.15. Reciprocal rate and its derivative for example 11.3; adapted from Escobar et al. (2012b).

288   Freddy Humberto Escobar

$(1/q')_r = 1.12 \times 10^{-4}$ $(1/q')_{DL} = 1.55 \times 10^{-5}$ $t_{DLPSSi} = 6000$ hr
$t_r = 0.03$ hr $(t*1/q')_{max} = 4.68 \times 10^{-5}$ $t_{RDLi} = 0.035$ hr
$(t*1/q')_{min} = 8.19 \times 10^{-5}$

Find permeability and reservoir width with Equations (11.55) and (11.171), respectively:

$$k_2 = \frac{70.6 \mu B}{h \Delta P [t*(1/q)']_r} = \frac{70.6(2)(1.2)}{100(2500)(1.30 \times 10^{-5})} = 52.14 \text{ md}$$

$$Y_E = 0.0648256 \sqrt{\frac{(52.1353)(0.035)}{(0.2)(2)(1 \times 10^{-6})(0.02)}} = 979.208 \text{ ft}$$

Making use of minimum and maximum points, find the storativity coefficient with the correlation given by Equation (11.178); first, $(t*1/q')_{min}$ has to be converted to a dimensionless form, $(t*1/q_D')_{min}$, with Equation (11.17):

$$[t_D*(1/q_D)']_{min} = \frac{53.117 \times 100 \times 2500}{141.2 \times 2 \times 1.2} * (8.19 \times 10^{-6}) = 0.3209$$

Since λ is known to be 5×10^{-8}, the value of ω is:

$$\omega = \frac{a + b \ln(0.3020) + c(\ln(0.3020))^2 + d(\ln(0.3020))^3 + e \ln 5 \times 10^{-8}}{1 + f \ln(0.3020) + g(\ln(0.3020))^2 + h(\ln(0.3020))^3 + i \ln 5 \times 10^{-8}} = 0.0214$$

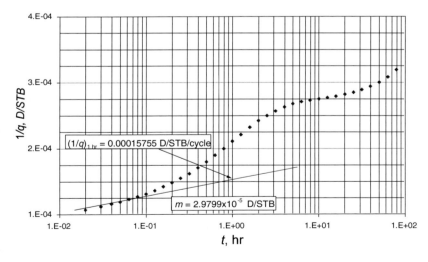

Figure 11.16. Semilog plot of reciprocal rate versus time for example 2.

The reservoir area is found with Equation (11.170):

$$A = \frac{0.140998}{43560} \sqrt{\frac{(1000^2)(0.0198)(52.1353)(6000)}{(0.2)(2)(1\times10^{-6})}} = 402.8 \text{ Ac}$$

The interporosity flow parameter λ is found with Equation (11.168), using the derivative value at the maximum point:

$$\lambda = \left[\frac{(0.05472912)(300)(2)(1.2)}{52.1353(100)(2500)(4.68\times10^{-5})}\right]^2 = 4.64\times10^{-8}$$

The mechanical skin factor is calculated with Equation (11.63):

$$s = 0.5\left(\frac{1.12\times10^{-4}}{1.30\times10^{-5}} - \ln\left(\frac{(52.1353)(0.03)}{(0.2)(2)(1\times10^{-6})(0.0214)(0.5^2)}\right) + 7.43\right) = -2.2$$

Estimate the geometric skin factor with Equation (11.166):

$$s + s_{DL}{}' = \left(\frac{1.55\times10^{-4}}{3.16\times10^{-5}} - 2\right)\frac{1}{54.738732(1000)}\sqrt{\frac{(52.1353)(0.2523)}{(0.2)(2)(1\times10^{-6})(0.0198)}} = 6.738$$

Solution by Conventional Analysis

From the semilog plot in Figure 11.16, slope and intercept values of 3.1×10^{-5} D/STB/cycle and 0.0002 D/STB, respectively, were read, and the mechanical skin factor was estimated from Equation (11.63):

$$s = 1.1513\left[\frac{0.0002}{3.1\times10^{-5}} - \log\left(\frac{50}{(0.2)(0.02)(2)(1\times10^{-6})(0.5^2)}\right) + 3.23\right] = -0.825$$

Parameters x and y from Equations (11.187) and (11.188) are, respectively:

$$x = \log\frac{0.0002637k}{\phi\mu c_t r_w^2} = \log\frac{0.0002637(50)}{(0.2)(2)(1\times10^{-6})(0.5^2)} = 5.120045$$

$$y = \frac{(50)(100)(2500)}{141.2(2)(1.2)}(0.00015755) + 1.0515(-0.825) = 6.5097$$

Applying Equation (11.186) leads to:

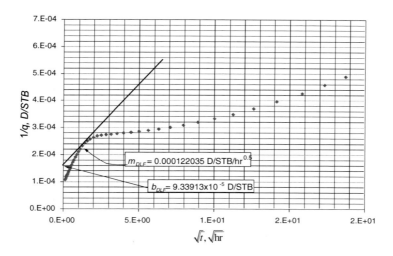

Figure 11.17. Cartesian plot of the reciprocal rate versus the square root of time for example 11.3.

$$\omega = \frac{a + bx + cx^2 + dy}{1 + ex + fy + gy^2 + hy^3} = 0.01$$

A Cartesian plot of the reciprocal rate versus the square root of time is given in Figure 11.17. From this plot, a slope of $m_{DLF} = 0.000128792$ D/STB/hr$^{0.5}$ and an intercept of $b_{DLF} = 9.03618 \times 10^{-5}$ D/STB were read; these are then used to estimate reservoir width and the dual-linear skin factor with Equations (11.181) and (11.182), respectively:

$$Y_E = \frac{9.1442(1.2)}{(100)(2500)(0.000128792)} \sqrt{\frac{2}{(0.02)(0.2)(50)(1 \times 10^{-6})}} = 1077.70 \text{ ft}$$

$$s + s_{DL}' = \frac{(50)(100)(2500)(9.03618 \times 10^{-5})}{141.2(2)(1.2)} = 3.33$$

Finally, the interporosity flow parameter is estimated with Equation (2.13), using a value of $\Delta t_{inf} = 10.047$ hr obtained from the reciprocal derivative plot (not shown here).

$$\lambda = \frac{3792(0.2)(1 \times 10^{-6})(2)(0.5^2)}{(50)(10.047546)} \left[0.02 \ln\left(\frac{1}{0.02}\right) \right] = 5.906 \times 10^{-8}$$

Nomenclature

A	Reservoir drainage area, Ac
B	Oil volume factor, bbl/SCF

b_x	Closer lateral distance from well to boundary, ft
C	Wellbore storage coefficient, bbl/psi
C_A	Dietz shape factor
C_{fD}	Dimensionless fracture conductivity
c_t	Total compressibility, 1/psi
h	Formation thickness, ft
I_0, I_1	Bessel function
K_0, K_1	Bessel function
k	Permeability, md
k_{fb}, k_2	Fracture bulk permeability, md
$k_f w_f$	Fracture conductivity, md-ft
M, m	Slope
$M_{decline}$	Slope of the plot of log q versus t
$m(P)$	Pseudopressure, psi/cp
P	Pressure, psi
P_D	Dimensionless pressure
P_{wf}	Well-flowing pressure, psi
q	Flow rate, STB/Day
$1/q$	Reciprocal rate, Day/STB
r	Radius, ft
r_e	Reservoir external radius, psi
r_w	Wellbore radius, ft
r_{weff}	Effective wellbore radius, $r_w e^{-s}$, ft
s	Mechanical skin factor
s'	Pseudomechanical skin factor for gas wells
s_{DL}'	Geometric skin factor due to conversion from radial to dual-linear
s_L'	Geometric skin factor due to conversion from dual-linear to linear
s_{PB}'	Geometric skin factor due to conversion from dual-linear to parabolic
t	Time, hr
u	Laplace parameter
T	Temperature, °R
t_D	Dimensionless time based on wellbore radius
t_{DA}	Dimensionless time based on area
t_{Dxf}	Dimensionless time based on half-fracture length
$t_D * P_D'$	Dimensionless pressure derivative
$t * (1/q)'$	Reciprocal rate derivative, Day/STB
$t_D * (1/q_D)'$	Dimensionless reciprocal rate derivative
$t_a(P)$	Pseudotime, hr-psi/cp
t_{pss}	Starting time of pseudosteady state, hr
W_D	Dimensionless reservoir width
w_f	Fracture width, ft
X_D	Dimensionless reservoir length
x_e	Reservoir external side, ft
x_f	Half-fracture length, ft
Y_E	Reservoir width, ft

Greek

Δ	Change
ϕ	Porosity, fraction
λ	Interporosity flow parameter
μ	Viscosity, cp
ω	Dimensionless storativity ratio

Suffixes

$1hr$	One hour
app	Apparent
$b2$	Start of second radial flow
BL	Bilinear
$'BLLi$	Intercept of bilinear and linear on the derivative curve
$BL1$	Bilinear, read at 1 hr
$BLBRi$	Intercept of bilinear and biradial
BLF	Bilinear flow
$BLPSSi$	Intercept of bilinear and pseudosteady state
BR	Biradial
$BR1$	Biradial, read at 1 hr
$BRPSSi$	Intercept of biradial and pseudosteady state
D	Dimensionless
$DLPSSi$	Intercept of dual-linear and pseudosteady state
$e1$	End of first radial flow
f	Fracture
i	Initial
inf	Inflection point
$inter$	Intercept
$int1$	Intercept of horizontal semilog line during transition period and first radial flow
$int2$	Intercept of horizontal semilog line during transition period and second radial flow
L	Linear
$L1$	Linear, read at 1 hr
LF	Linear flow
$LPSSi$	Intercept of linear and pseudosteady state
m	Matrix
$m+f$	Matrix plus fracture
max	Maximum point on the derivative curve
min	Minimum point on the derivative curve
N	Point on the early unit-slope line
PB	Parabolic
PBF	Parabolic flow
pss	Pseudosteady state
$pss1$	Pseudosteady state, read at 1 hr

r	Radial flow
$r1$	First radial flow regime
$r2$	Second radial flow regime
rdi	Intercept between the reciprocal rate and reciprocal rate derivative curves
$RBLi$	Intercept of radial and bilinear
$RBRi$	Intercept of radial and biradial
$RDLi$	Intercept of radial and dual-linear
RLi	Intercept of radial and linear
$RPBi$	Intercept of radial and parabolic
$RPPSi$	Intercept of radial and pseudosteady state
rpi	Intercept of radial and pseudosteady state
sp	Start of pseudosteady state
sr	Start of radial flow
t	Total
us	Point on the unit-slope line
us,i	Intercept of radial and unit-slope during the transition period
w	Well
x	Maximum during wellbore storage

REFERENCES

Agarwal, R. G. (1979, January 1). "Real Gas Pseudo-Time - A New Function For Pressure Buildup Analysis Of MHF Gas Wells." *Society of Petroleum Engineers.* doi:10.2118/8279-MS.

Arab, N. (2003). "*Application of Tiab's Direct Synthesis Technique to Constant Bottom Hole Pressure Test.*" The University of Oklahoma. M.Sc. Thesis.

Cinco-Ley, H., Samaniego, F. & Dominguez, N. (1978, August 1). "Transient Pressure Behavior for a Well with a Finity-Conductivity Vertical Fracture." *Society of Petroleum Engineers.* doi:10.2118/6014-PA.

DaPrat, G., Cinco-Ley., H. & Ramey, H. J, Jr. (1981). "Decline Curve Analysis Using Type Curves for Two-Porosity Systems." *Society of Petroleum Engineers Journal.* Jun. 354-62.

Engler, T. & Tiab, D. (1996). "Analysis of Pressure and Pressure Derivative without Type Curve Matching, 4. Naturally Fractured Reservoirs." *Journal of Petroleum Science and Engineering, 15.* 127-138.

Escobar, F. H., Lopez, A. M. & Cantillo, J. H. (2007). "Effect of the Pseudotime Function on Gas Reservoir Drainage Area Determination." *CT&F – Ciencia, Tecnología y Futuro.* Vol. *3*, No. 3. 113-124. Dec.

Escobar, F. H., Sanchez, J. A. & Cantillo, J. H. (2008). "Rate Transient Analysis for Homogeneous and Heterogeneous Gas Reservoirs using The TDS Technique." *CT&F – Ciencia, Tecnología y Futuro.* Vol. *4*, No. 4. 45-59. Dec.

Escobar, F. H., Rojas, M. M. & Cantillo, J. H. (2012a). "Straight-Line Conventional Transient Rate Analysis for Long Homogeneous and Heterogeneous Reservoirs." *Dyna.* Year 79, Nro. 172, 153-163. April.

Escobar, F. H., Rojas, M. M. & Bonilla, L. F. (2012b). "Transient-Rate Analysis for Long Homogeneous and Naturally Fractured Reservoir by The TDS Technique." *Journal of Engineering and Applied Sciences.*, Vol. *7*. Nro. 3. P. 353-370. Mar.

Escobar, F. H., Castro, J. R. & Mosquera, J. S. (2014). "Rate-Transient Analysis for Hydraulically Fractured Vertical Oil and Gas Wells." *Journal of Engineering and Applied Sciences.*, Vol. *9*. Nro. 5. P. 739-749. May.

Fetkovich, M. J. (1980). "Decline Curve Analysis Using Type Curves." *Journal of Petroleum Technology.* 1065-77. June.

Kazemi, H. (1969, December 1). "Pressure Transient Analysis of Naturally Fractured Reservoirs with Uniform Fracture Distribution." *Society of Petroleum Engineers.* doi:10.2118/2156-A.

Mavor, M. J. & Cinco-Ley, H. (1979, January 1). "Transient Pressure Behavior Of Naturally Fractured Reservoirs." *Society of Petroleum Engineers.* doi:10.2118/7977-MS.

Samaniego-V. F. & Cinco-Ley, H. (2009). "Transient Well Testing." *Society of Petroleum Engineers.* SPE Monograph Vol. *23*. Chapter 10. Edited by M. M. Kamal.

Tiab, D. & Escobar, F. H. (2003) "Determinación del Parámetro de Flujo Interporoso a Partir de un Gráfico Semilogarítmico." *X Congreso Colombiano del petróleo* (Colombian Petroleum Symposium). Oct. 14-17, 2003. ISBN 958-33-8394-5. Bogotá (Colombia).

Tiab, D. (1993, January 1). "Analysis of Pressure and Pressure Derivatives Without Type-Curve Matching: I-Skin and Wellbore Storage." *Society of Petroleum Engineers.* This paper was prepared for presentation at the Production Operations Symposium held in Oklahoma City, OK, U.S.A., March 21-23. doi:10.2118/25426-MS.

Tiab, D. (1994). "Analysis of Pressure Derivative without Type-Curve Matching: Vertically Fractured Wells in Closed Systems." *Journal of Petroleum Science and Engineering*, 11 (1994) 323-333.

Tiab, D., Azzougen, A., Escobar, F. H. & Berumen, S. (1999, January 1). "Analysis of Pressure Derivative Data of Finite-Conductivity Fractures by the 'Direct Synthesis' Technique." *Society of Petroleum Engineers.* doi:10.2118/52201-MS.

Van Everdingen, A. F. & Hurst. W., 1949. "The Application of the Laplace Transformation to Flow Problems in Reservoirs," Trans., *AIME*, 186 (1949) 305-324.

Warren, J. E. & Root, P. J. (1963). "The Behavior of Naturally Fractured Reservoir" *Society of Petroleum Engineers Journal.* Sep.

Chapter 12

CONDUCTIVITY FAULTS

BACKGROUND

Many hydrocarbon-bearing formations are faulted, and often, little information is available about the actual physical characteristics of such faults. Some faults are known to be sealing to the migration of hydrocarbons while others are not. While sealing faults block fluid and pressure communication with other regions of the reservoir, infinite-conductivity faults act as pressure support sources and allow fluid transfer across and along the fault planes. Finite-conductivity faults fall between these two limiting cases of sealing and totally non-sealing faults, and they are believed to be included in the majority of faulted systems. A sealing fault is often generated when the throw of the fault plane is such that a permeable stratum on one side of the fault plane is completely juxtaposed against an impermeable stratum on the other side. On the contrary, a non-sealing fault usually has an insufficient throw to cause a complete separation of productive strata on opposite sides of the fault plane. Depending on the permeability of the fault, fluid flow may occur along the fault within the fault plane or just across it, laterally from one stratum to another. In general, a finite-conductivity fault exhibits a combined behavior of flow along and across its plane.

While seismic analysis can detect a fault distance to a well with a margin of error near two kilometers, transient-pressure analysis is the best tool to detect the well-fault distance with a margin of error in feet. However, conventional transient-pressure analysis methods have been used only for detection of fault-well distance without taking into account such variables as conductivity, damage and fault length. Before the works by Escobar, Martinez and Montealegre-Madero (2013a, 2013b), it was only possible to estimate the fault conductivity using straight-line conventional analysis and an equation proposed by Trocchio (1990).

Rahman, Miller and Mattar (2003) presented an analytical solution to the transient flow problem of a well located near a finite-conductivity fault in a two-zone composite reservoir. This solution considered flow within the fault and allowed understanding the behavior of a conductive fault. When the pressure disturbance reaches an undamaged conductivity fault, the pressure derivative goes down, forming a negative unit-slope line because the steady-state flow regime takes place and the fault acts as a bridge with an underlying aquifer. However, if the fault is damaged, and the test runs under radial flow, the pressure derivative rises up more than a log cycle (depending on the degree of damage), and once the damaged zone is passed, the pressure derivative goes down, eventually forming the negative unit-slope line. Afterwards, either a half-slope or a quarter-slope is seen if the conductivity in the fault is either infinite or finite This solution was used by Escobar et al. (2013a, 2013b) to find the characteristic signature observed on the pressure and pressure derivative plot with the purpose of developing appropriate expressions to characterize the typical parameters involved in finite-conductivity faults following the *TDS* technique philosophy of Tiab (1993), so fault damage, fault conductivity and fault length can be easily estimated using data read from the pressure and pressure derivative plot. Furthermore, the well-known straight-line conventional methodology was augmented so that the above-named parameters (fault damage, fault conductivity and fault length) can be easily estimated using the slope and intercept of a Cartesian plot.

PRESSURE BEHAVIOR OF CONDUCTIVITY FAULTS WITHOUT MOBILITY CONTRAST

In the finite-conductivity fault model used by Rahman et al. (2003), the fault permeability is larger than the reservoir permeability. Fluid flow is allowed to occur both across and along the fault plane, and the fault enhances the drainage capacity of the reservoir. it is assumed that the reservoir properties are the same on both sides of the fault.

The typical influence of a semi-permeable fault is shown in Figure 12.1, in which we can observe that the response starts following the usual infinite-acting regime at an early time. Once the finite-conductivity fault is felt by the pressure disturbance, the pressure derivative drops along a straight line of slope -1 because a radial stabilization phenomenon takes place. Later, as the pressure drops in the fault, a flow is established in the thickness of the fault plane that results in a bilinear flow regime, as depicted in Figure 12.2. One linear flow takes place in the reservoir when the fluid enters and exits the fault; the second linear flow describes the flux inside the fault thickness. As seen in Figure 12.1, once the negative unit-slope line disappears as time progresses, the quarter-slope line develops. Finally, the pressure derivative response becomes flat again, describing the infinite-acting radial regime when the fault no longer has an effect on the pressure response.

Figure 12.3 shows the pressure and pressure derivative behavior when the skin factor across the fault plane is equal to zero and the reservoir properties on both sides of the fault are the same (no mobility contrast). Wellbore storage and wellbore skin effects are not included. Several pressure derivative curves as a function of conductivity of the fault plane, ranging from 0.1 to 10^7, are shown on this plot. At an early time, the pressure derivative is flat, representing infinite-acting radial flow in the left side of the reservoir. At a dimensionless time, t_{DF}, of 0.25, the pressure derivative curves deviate from the radial flow when the transient-pressure reaches the fault plane. The deviation degree depends upon the conductivity of the fault plane. For fault conductivities less than 0.1, the pressure derivative essentially remains on the radial flow regime, indicating that there is no flow along the fault plane and that fluid transfer occurs only across the fault. This is due to the fact that very low fault conductivities create a large flow resistance occurring along the fault plane while a zero skin factor creates no resistance to flow across the fault. Therefore, fluid flow comes from the right to the left side of the reservoir, across the fault, as if the fault plane did not exist.

For high-conductivity cases, the fault plane initially acts as a linear constant-pressure boundary, and the pressure derivative becomes a straight line with a slope of minus unity. As time progresses, pressure in the fault plane decreases, and fluid enters the fault linearly from the reservoir, moves linearly along it and exits from the fault plane toward the producing well. This flow characteristic is seen as a quarter-slope straight-line bilinear flow regime on the pressure derivative curves. At later times, when the disturbance has practically passed the fault system, the behavior reflects the entire reservoir response, and the derivative curves asymptotically reach the radial flow regime again. It is interesting to note that the transient-pressure behavior for intermediate values of fault conductivity is similar to that of naturally fractured reservoirs. Thus, in a pressure test, a single conductive fault can give the appearance of a naturally fractured reservoir.

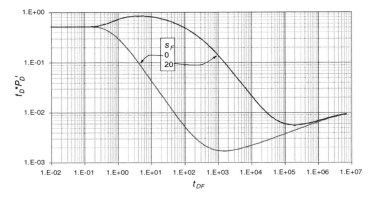

Figure 12.1. Dimensionless pressure derivative for a well near a finite-conductivity fault for $s_F = 0$ and 20; adapted from Escobar et al. (2013a).

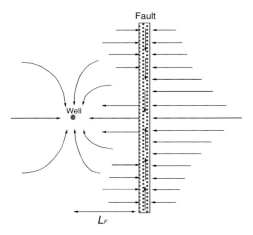

Figure 12.2. Schematic of a typical fault system and flow lines; adapted from Abbaszadeh and Cinco-Ley (1995).

Figure 12.4 shows pressure derivative behaviors for finite-conductivity fault systems under fault skin factor conditions. The reservoir properties are the same everywhere. As expected, the skin creates additional resistance to flow within the fault plane for some period of time, resembling a sealing fault for all conductivity values. Pressure derivatives after the onset of the fault effects tend to approach the well-known behavior of the doubling of the semilog straight-line slope (dimensionless pressure derivative equals 1) for $s_F > 100$. At larger times, when pressure on the left side of the fault becomes low enough to allow for appreciable flow to cross the fault plane, the pressure waves propagate through the right side of the reservoir, and the behavior becomes similar to the undamaged fault case, $s_F = 0$. The negative unitslope line of the constant-pressure linear boundary and the quarter-slope line of the bilinear flow regime characteristics are developed for high conductivity values, and eventually the dimensionless pressure derivative curves approach the value of 0.5 (combined reservoir behavior).

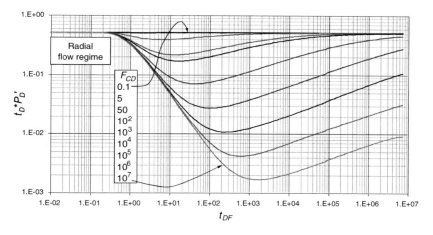

Figure 12.3. Effect of fault conductivity on dimensionless pressure derivative, $s_F = 0.$; adapted from Escobar et al. (2013a).

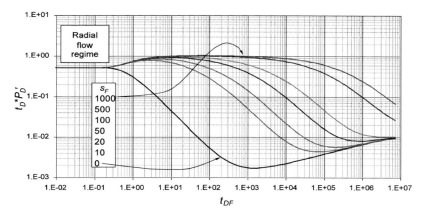

Figure 12.4. Effect of fault skin factor on dimensionless pressure derivative, $h_D = 1$. After Escobar et al. (2013a).

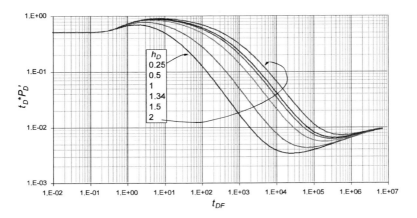

Figure 12.5. Effect of h_D on dimensionless pressure derivative, $s_F = 20$; adapted from Escobar et al. (2013a).

Figure 12.5 shows dimensionless pressure derivative curves at several dimensionless pay thicknesses, h_D. At low h_D, the negative unit-slope line is more visible than it is at higher values.

TDS TECHNIQUE FOR CONDUCTIVITY FAULTS WITHOUT MOBILITY CONTRAST

The application of the *TDS* technique to finite-conductivity faulted systems in Escobar et al. (2013a) was based upon the mathematical model presented by Rahman et al. (2003). The dimensionless quantities used are defined as:

$$P_D = \frac{kh}{141.2q\mu B}\Delta P \tag{1.3}$$

$$t_D * P_D' = \frac{kh(t*\Delta P')}{141.2q\mu B} \tag{12.1}$$

$$t_{DF} = \frac{0.0002637kt}{\phi\mu c_t L_F^2} \tag{12.2}$$

$$h_D = \frac{h}{L_F} \tag{12.3}$$

$$F_{CD} = \frac{k_f w_f}{kL_F} \tag{12.4}$$

The formulation of the equations follows the philosophy of the *TDS* technique outlined by Tiab (1993). Several specific regions and "fingerprints" found on the pressure and pressure derivative behavior are dealt with. The permeability and skin factors are found using the following equations from Tiab (1993):

$$k = \frac{70.6q\mu B}{h\left(t*\Delta P'\right)_r} \tag{1.18}$$

$$s = \frac{1}{2}\left(\frac{\Delta P_r}{\left(t*\Delta P'\right)_r} - \ln\left(\frac{kt_r}{\phi\mu c_t r_w^{2}}\right) + 7.43\right) \tag{1.21}$$

Two observed radial flow regimes can be observed if the test has run long enough: one before the fault and one after the fault. The early radial flow ends at:

$$t_{DFer} = 0.25 \tag{12.5}$$

Placing Equation (12.2) into the above expression and solving for the distance from the well to the fault yields:

$$L_F = 0.0325\sqrt{\frac{kt_{er}}{\phi\mu c_t}} \tag{12.6}$$

Escobar et al. (2013a) found that the dimensionless pressure derivative for the steady-state flow caused by the fault is governed by:

$$\left(t_D * P_D'\right)_{ss} = \frac{1}{2}\left(1 + s_F h_D\right)^2 \frac{1}{t_{DF}} \tag{12.7}$$

Placing the dimensionless quantities given by Equations (12.1) through (12.3) into Equation (12.7) and solving for the fault skin factor results in:

$$s_F = \frac{L_F}{h}\left[\sqrt{\left(\frac{3.7351\times10^{-6}k^2 ht_{ss}\left(t*\Delta P'\right)_{ss}}{q\mu^2 B\phi c_t L_F^2}\right)} - 1\right] \tag{12.8}$$

The dimensionless pressure and pressure derivative expressions for the bilinear flow regime are:

$$P_D = \frac{2.45}{\sqrt{F_{CD}}} t_{DF}^{0.25} + s_{BL} \tag{12.9}$$

$$t_D * P_D ' = \frac{0.6125}{\sqrt{F_{CD}}} t_{DF}^{0.25} \tag{12.10}$$

Placing the dimensionless quantities given by Equations (12.1), (12.2) and (12.4) into Equation (12.10) will result in an expression to estimate fault conductivity using any arbitrary point on the pressure derivative during the bilinear-flow regime:

$$k_f w_f = 121.461 \left(\frac{q\mu B}{h(t * \Delta P')_{BL}} \right)^2 \left(\frac{t_{BL}}{k\phi\mu c_t} \right)^{0.5} \tag{12.11}$$

Using the minimum pressure derivative coordinate, Escobar et al. (2013a) obtained another expression for fault conductivity:

$$k_f w_f = \frac{a + c(t_D * P_D')_{min} + e(t_D * P_D')_{min}^2 + g(t_D * P_D')_{min}^3}{1 + b(t_D * P_D')_{min} + d(t_D * P_D')_{min}^2 + f(t_D * P_D')_{min}^3 + h(t_D * P_D')_{min}^4} (kL_F + S_F kh) \tag{12.13}$$

where the constants are $a = 11198700$, $b = -1235,2895$, $c = 256626000$, $d = 712041,381$, $e= -491990000$, $f = 64974400$, $g = -154650000$ and $h = 116739000$.

The dimensionless pressure derivative lines obtained from the radial flow and the steady-state flow regimes intercept at:

$$0.5 = \frac{1}{2}(1 + s_F h_D)^2 \frac{1}{t_{DF}} \tag{12.14}$$

$$t_{DF\,rssi} = (1 + s_F h_D)^2 \tag{12.15}$$

Replacing the dimensionless time in Equation (12.15) and solving for the well-fault distance will result in:

$$L_F = \sqrt{\frac{0.0002637 k t_{rssi}}{\phi\mu c_t}} - s_F h \tag{12.16}$$

The unit-slope line going through the steady-state period and the bilinear flow regime line on the dimensionless pressure derivative curve intersect at:

$$\frac{0.6125}{\sqrt{F_{CD}}} t_{DF}^{0.25} = \frac{1}{2}(1 + s_F h_D)^2 \frac{1}{t_{DF}} \tag{12.17}$$

$$t_{DF\,ssBli} = \left[\frac{\left(1+s_F h_D\right)^2 \sqrt{F_{CD}}}{1.225}\right]^{0.8} \tag{12.18}$$

Placing the dimensionless time defined by Equation (12.2) into Equation (12.18) and solving for the conductivity fault will result in:

$$k_f w_f = 1.694 \times 10^{-9} k L_F \left(\frac{k t_{ssBLi}}{\phi \mu c_t L_F^2}\right)^{2.5} 1/\left(1+s_F \frac{h}{L_F}\right)^4 \tag{12.19}$$

If the dimensionless fault conductivity is bigger than 2.5×10^8, the bilinear flow disappears, and the linear flow appears, exhibiting a half-slope straight line on the pressure derivative curve. In this case we have an infinite-conductivity fault. The dimensionless pressure derivative expression for the above-mentioned linear flow regime is:

$$t_D * P_D' = 2.8 \times 10^{-6} \sqrt{t_{DF}} \tag{12.20}$$

Placing the dimensionless quantities given by Equations (12.1) and (12.2) into Equation (12.20) will result in another expression that is useful for estimating the distance from the well to the fault:

$$L_F = 6.42 \times 10^{-6} \frac{qB}{h(t*\Delta P')_L} \sqrt{\frac{\mu t_L}{k \phi c_t}} \tag{12.21}$$

New expressions for the straight-line conventional analysis developed in this work, along with some expressions reported in the literature, are reported in Appendix A.

CONVENTIONAL METHODOLOGY FOR CONDUCTIVITY FAULTS WITHOUT MOBILITY CONTRAST

Placing the dimensionless quantities given by Equations (1.3), (12.2) and (12.4) into Equation (12.9) will result in:

$$\Delta P = \frac{44.1 q \mu B t^{1/4}}{h\sqrt{k_f w_f} \sqrt[4]{\phi \mu c_t k}} + \frac{141.2 q \mu B}{kh} s_{BL} \tag{12.22}$$

Notice that Equation (12.22) suggests that a Cartesian plot of ΔP_{wf} vs. $t^{0.25}$ gives a linear trend with slope m_{BL}, allowing for the estimation of fault conductivity, such as:

$$k_f w_f = \left[\frac{44.1 q \mu B}{m_{BL} h^4 \sqrt{\phi \mu c_t k}}\right]^2 \tag{12.23}$$

The above equation was also found by Trocchio (1990). The dimensionless pressure for the linear flow is:

$$P_D = 5.8 \times 10^{-6} \sqrt{t_{DF}} + s_L \tag{12.24}$$

Placing the dimensionless quantities given by Equations (1.3) and (12.2) into Equation (12.24) will result in:

$$\Delta P = 1.33 \times 10^{-5} \frac{q \mu B t^{1/2}}{h L_F \left(\phi \mu c_t k\right)^{1/2}} + \frac{141.2 q \mu B}{kh} s_L \tag{12.25}$$

Equation (12.25) suggests that a Cartesian plot of ΔP_{wf} vs. $t^{0.5}$ gives a linear trend with slope m_L, allowing for the estimation of the distance from the well to the fault:

$$L_F = 1.33 \times 10^{-5} \frac{q \mu B}{m_L h \left(\phi \mu c_t k\right)^{1/2}} \tag{12.26}$$

The governing dimensionless pressure for the steady-state period caused by the fault is:

$$P_D = -\frac{1}{2}\left(1 + s_F h_D\right)^2 \frac{1}{t_{DF}} + \frac{1}{2}\ln\left(\frac{4L_F^2}{r_w^2} + 8 \times 10^5 \left(s_F h_D\right)^2\right) \tag{12.27}$$

Placing the dimensionless quantities given by Equations (1.3), (12.2) and (12.3) into Equation (12.27) results in:

$$\Delta P = -\frac{q \mu^2 B \phi c_t L_F^2}{3.7351 \times 10^{-6} k^2 h}\left(1 + s_F \frac{h}{L_F}\right)^2 \frac{1}{t} + \frac{70.6 q \mu B}{kh}\ln\left(\frac{4L_F^2}{r_w^2} + 8 \times 10^5 \left(s_F \frac{h}{L_F}\right)^2\right) \tag{12.28}$$

A Cartesian plot of ΔP_{wf} vs. $1/t$ gives a linear trend whose slope allows for the estimation of the fault skin factor:

$$s_F = \frac{L_F}{h}\left[\sqrt{\left(\frac{3.7351 \times 10^{-6} k^2 h m_{ss}}{q \mu^2 B \phi c_t L_F^2}\right)} - 1\right] \tag{12.29}$$

Trocchio (1990) presented the following expression to find the dimensionless end time of the bilinear flow regime and the minimum fracture length:

$$t_{eBLD} = \frac{0.0002637 k t_{eBL}}{\phi \mu c_t L_F^2} = \left(\frac{4.55}{\sqrt{F_{CD}}} - 2.5\right)^{-4} \quad (12.30)$$

Solving for the minimum fault length results in:

$$x_{f_{min}} = \left(\frac{2.5}{4.55\sqrt{\dfrac{k}{k_f w_f}} \pm \sqrt[4]{\dfrac{\phi \mu c_t}{0.0002637 k t_{eBL}}}}\right)^2 \quad (12.31)$$

Example 12.1

Abbaszadeh and Cinco-Ley (1995) presented well test data from a heavily faulted carbonate reservoir. Data for the reservoir, fluids and well are given below. Pressure and pressure derivative data are reported in both Figure 12.6 and Table 12.1. Estimates of permeability, skin factor formation, distance to fault, fault conductivity and fault skin factor are required.

$q = 2151$ bbl/D $\qquad B = 2.112$ rb/STB $\qquad \mu = 0.147$ cp
$h = 134.5$ ft $\qquad r_w = 0.208$ ft $\qquad \phi = 3.2\%$
$c_t = 3.175 \times 10^{-5}$ 1/psi $\qquad k = 8$ md $\qquad L_F = 50$ ft
$F_{CD} = 1 \times 10^7$ $\qquad s_F = 2$

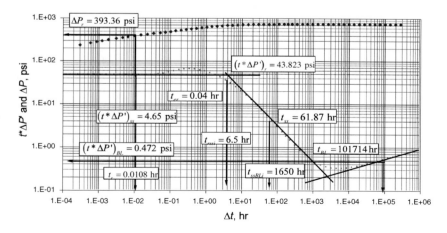

Figure 12.6. Pressure and pressure derivative for example 12.1.

Conductivity Faults

Table 12.1. Pressure and pressure derivative versus time data for example 12.1

t, hr	ΔP, psi	$t*\Delta P'$, psi	t, hr	ΔP, psi	$t*\Delta P'$, psi	t, hr	ΔP, psi	$t*\Delta P'$, psi
0.00030	236.70	43.76	0.7523	608.19	56.50	495.25	732.14	0.77
0.00059	265.81	43.76	1.049	626.45	53.05	690.81	732.36	0.60
0.00089	284.02	43.78	1.464	643.40	48.68	963.60	732.54	0.49
0.00124	298.60	43.79	2.042	658.79	43.69	1344.10	732.69	0.41
0.00188	316.81	43.80	2.848	672.45	38.36	1874.85	732.82	0.35
0.00310	338.68	43.81	3.973	684.31	32.96	2615.17	732.93	0.32
0.00470	356.91	43.81	5.541	694.40	27.73	3647.84	733.03	0.30
0.00656	371.49	43.81	7.729	702.81	22.86	5088.27	733.13	0.29
0.00994	389.71	43.82	10.78	709.67	18.47	7097.49	733.23	0.29
0.0151	407.94	43.82	15.04	715.16	14.65	9900.10	733.33	0.30
0.0228	426.17	43.80	20.98	719.48	11.43	13809.39	733.43	0.31
0.0346	444.42	44.02	29.26	722.83	8.78	19262.35	733.54	0.33
0.0525	462.92	45.20	40.81	725.38	6.66	26868.54	733.65	0.35
0.0732	478.28	47.29	56.93	727.31	5.01	37478.21	733.77	0.38
0.1021	494.49	50.30	79.41	728.75	3.73	52277.36	733.90	0.41
0.1425	511.80	53.70	110.77	729.83	2.78	72920.30	734.04	0.44
0.1987	530.21	56.78	154.51	730.63	2.06	101714.58	734.20	0.47
0.2772	549.49	58.86	198.32	731.09	1.65	141878.96	734.36	0.51
0.3867	569.24	59.56	254.54	731.46	1.33	197903.18	734.53	0.55
0.5393	588.97	58.74	355.05	731.84	1.00	276049.88	734.72	0.59

Solution by TDS Technique

The log-log plot of pressure and pressure derivative against production time is given in Figure 12.6. From that plot, the following information can be read:

$t_r = 0.0108$ hr $\Delta P_r = 393.36$ psi $(t*\Delta P')_r = 43.823$ psi

$t_{ss} = 61.87$ hr $(t*\Delta P')_{ss} = 4.65$ psi $t_{BL} = 101714$ hr

$(t*\Delta P')_{BL} = 0.472$ psi $t_{er} = 0.04$ hr $t_{rssi} = 6.5$ hr

$t_{ssBLi} = 1650$ hr $(t*\Delta P')_{min} = 0.3$ psi

First, the formation permeability is evaluated with Equation (1.18), and the skin factor is evaluated with Equation (1.21):

$$k = \frac{70.6(2151)(0.147)(2112)}{134.5(43.823)} = 7.99 \text{ md}$$

$$s = \frac{1}{2}\left(\frac{393.36}{43.823} - \ln\left(\frac{7.99(0.0108)}{0.032(0.147)(3.175\times10^{-5})(0.208^2)} \right) + 7.43 \right) = -0.00127$$

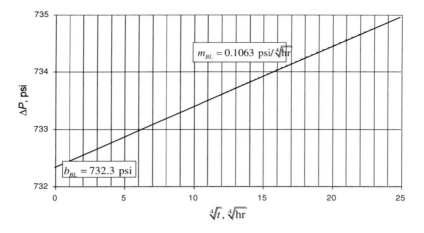

Figure 12.7. Cartesian plot of pressure drop vs. the fourth-root of time for example 12.1.

The distance to the fault is evaluated with Equation (12.6), and the fault skin factor is estimated with Equation (12.8):

$$L_F = 0.0325\sqrt{\frac{7.99(0.04)}{0.032(0.147)(3.175\times 10^{-5})}} = 47.54 \text{ ft}$$

$$S_F = \frac{47.54}{134.5}\sqrt{\left(\frac{3.7351\times 10^{-6}(7.99^2)(134.5)(61.87)(4.65)}{2151(0.147^2)(2112)(0.032)(3.175\times 10^{-5})(47.54^2)}-1\right)} = 2.23$$

The distance to the fault is re-estimated with Equation (12.16):

$$L_F = \sqrt{\frac{0.0002637(7.99)(6.5)}{0.032(0.147)(3.175\times 10^{-5})} - [2.23(134.5)]^2} = 41.67 \text{ ft}$$

The fault conductivity is evaluated with Equation (12.11) and re-estimated with Equations (12.13) and (12.19):

$$k_f w_f = 121.461\left(\frac{2151(0.147)(2112)}{134.5(0.472)}\right)^2$$

$$\left(\frac{101714}{7.99(0.032)(0.147)(3.175\times 10^{-5})}\right)^{0.5} = 3.92\times 10^9 \text{ md-ft}$$

The minimum-point pressure derivative is converted to a dimensionless form using equation (12.1):

$$\left(t_D * P_D{}'\right)_{min} = \frac{7.99(134.5)(0.3)}{141.2(2151)(0.147)(2112)} = 0.00342$$

$$k_f w_f = \frac{a + c*0.00342 + e*0.00342^2 + g*0.00342^3}{1 + b*0.00342 + d*0.00342^2 + f*0.00342^3 + h*0.00342^4}$$
$$\left(7.99*47.54 + 2.23(7.99)(134.5)\right) = 4.34\times10^9 \text{ md-ft}$$

$$k_f w_f = 1.694\times10^{-9}(7.99)(47.54)\left(\frac{7.99(1650)}{0.032(0.147)(3.175\times10^{-5})(47.54^2)}\right)^{2.5}$$

$$\left(\frac{1}{1+\left[2.23\dfrac{134.5}{47.54}\right]^2}\right)^2 = 3.68\times10^9 \text{ md-ft}$$

Finally, the dimensionless fracture conductivity is calculated with Equation (12.4)

$$F_{CD} = \frac{3.92\times10^9}{7.99(47.54)} = 1.032\times10^7$$

Solution by Conventional Analysis

The late-time data of pressure drop versus the fourth-root of time are plotted in Cartesian coordinates in Figure 12.7, where the bilinear flow regime takes place. This plot provides a slope of 0.1063 psi/hr$^{0.25}$, allowing for the estimation of the fault's conductivity value of 3.882×10^9 md-ft with Equation (12.23):

$$k_f w_f = \left[\frac{44.1(2151)(0.147)(2112)}{(0.1063)(134.5)\sqrt[4]{(0.032)(0.147)(3.175\times10^{-5})(7.99)}}\right]^2 = 3.882\times10^9 \text{md} - \text{ft}$$

PRESSURE BEHAVIOR OF CONDUCTIVITY FAULTS WITH MOBILITY CONTRAST

When properties of reservoirs on two sides of the fault plane are not the same, complexities other than fault conductivity and skin effects are introduced. Figures 12.8 and 12.9, generated for finite-conductivity faults, show dimensionless pressure-derivative curves at several dimensionless fault conductivity and mobility ratios. At mobility ratios greater than one, the minimum point increases with the mobility ratio, but the distance between the curves for the same conductivity decreases considerably. At mobility ratios less than 1, the minimum point decreases with the mobility ratio, but the distance between the curves for the same conductivity increases.

Figure 12.10 shows pressure derivative behaviors for finite-conductivity fault systems under fault skin factor conditions and mobility ratios. As expected, the skin creates additional resistance to flow within the fault plane for some period of time, resembling a sealing fault for all conductivity values. The mobility ratio only affects the minimum point in the pressure derivative.

If the dimensionless fault conductivity is assumed to be zero, and the hydraulic diffusivity is also set to zero, there is essentially no fault, and the solution degenerates to that of a single composite system, as presented by Bixel, Larkin and Van Poollen (1963). Additionally, in these conditions, if the mobility ratio is assumed to be infinite, reservoir zone 2 provides a strong pressure support to reservoir zone 1. Then the behavior looks like the case of a well near a constant-pressure boundary. Raman et al. (2003) provided other limitations of the model.

TDS TECHNIQUE FOR CONDUCTIVITY FAULTS WITH MOBILITY CONTRAST

The dimensionless quantities used in this work are defined as:

$$P_D = \frac{k_1 h}{141.2 q \mu B} \Delta P \tag{12.32}$$

$$t_D * P_D = \frac{k_1 h (t * \Delta P')}{141.2 q \mu B} \tag{12.33}$$

$$t_{DF} = \frac{0.0002637 k_1 t}{\phi \mu c_t L_F^2} \tag{12.34}$$

$$h_D = \frac{h}{L_F} \tag{12.3}$$

$$F_{CD} = \frac{k_f w_f}{k_1 L_F} \tag{12.35}$$

$$M = \frac{k_2}{k_1} \tag{12.36}$$

Permeability and skin factors are found using the following equations from Tiab (1993):

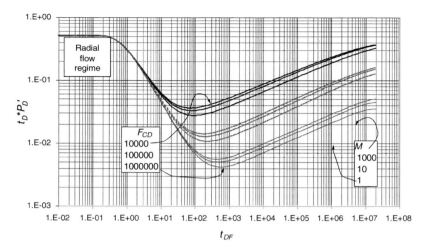

Figure 12.8. Effect of mobility ratio on dimensionless pressure derivative. $s_F = 0$ and $M > 1$; adapted from Escobar et al. (2013b).

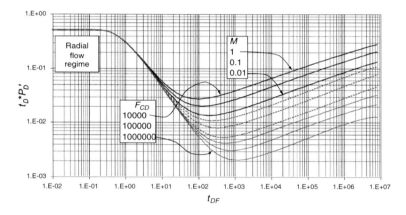

Figure 12.9. Effect of mobility ratio on dimensionless pressure derivative, $s_F = 0$ and $M < 1$; adapted from Escobar et al. (2013b).

$$k_1 = \frac{70.6q\mu B}{h(t*\Delta P')_r} \tag{12.37}$$

$$s = \frac{1}{2}\left(\frac{\Delta P_r}{(t*\Delta P')_r} - \ln\left(\frac{k_1 t_r}{\phi \mu c_t r_w^2}\right) + 7.43\right) \tag{12.38}$$

According to Figures (12.8) through (12.10), the early radial flow ends at:

$$t_{DFer} = 0.25 \tag{12.39}$$

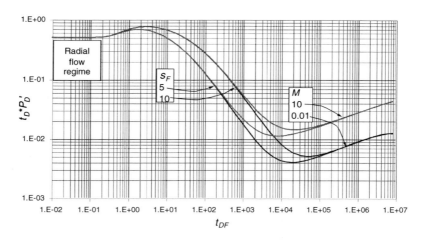

Figure 12.10. Effect of fault skin factor and mobility ratio on dimensionless pressure derivative, $h_D = 1$; adapted from Escobar et al. (2013b).

Placing Equation (12.34) into the above expression and solving for the distance from the well to the fault leads to:

$$L_F = 0.0325 \sqrt{\frac{k_1 t_{er}}{\phi \mu c_t}} \tag{12.40}$$

The governing dimensionless pressure derivative for the steady-state flow caused by the fault is:

$$(t_D * P_D')_{ss} = \frac{1}{2}(1 + s_F h_D)^2 \frac{1}{t_{DF}} \tag{12.41}$$

Equation (12.41) is a corrected form (by Escobar et al. [2013b]) of an expression introduced by Abbaszadeh and Cinco-Ley (1995), and it considers the dimensionless pay thickness. Placing the dimensionless quantities given by Equations (12.3), (12.34) and (12.35) into Equation (12.41) and solving for the fault skin factor will result in:

$$s_F = \frac{L_F}{h}\left[\sqrt{\left(\frac{3.7351 \times 10^{-6} k_1^2 h t_{ss}(t*\Delta P')_{ss}}{q \mu^2 B \phi c_t L_F^2}\right)} - 1\right] \tag{12.42}$$

The pressure and pressure derivative dimensionless expressions for the bilinear flow regime given by Abbaszadeh and Cinco-Ley (1995) and corrected by Escobar et al. (2013b) are:

Conductivity Faults

$$P_D = \frac{2.6084}{\sqrt{F_{CD}\dfrac{\sqrt{M}+1}{2\sqrt{M}}}}\sqrt[4]{t_{DF}} + s_{BL} \tag{12.43}$$

$$t_D * P_D' = \frac{0.6521}{\sqrt{F_{CD}\dfrac{\sqrt{M}+1}{2\sqrt{M}}}}\sqrt[4]{t_{DF}} \tag{12.44}$$

Placing the dimensionless quantities given by Equations (12.34) and (12.35) into Equation (12.44) will result in an expression to estimate effective fault conductivity using any arbitrary point on the pressure derivative during the bilinear-flow regime:

$$\left(k_f w_f\right)_{eff} = k_f w_f \frac{\sqrt{M}+1}{2\sqrt{M}} = 137.67\left(\frac{q\mu B}{h\left(t*\Delta P'\right)_{BL}}\right)^2 \sqrt{\frac{t_{BL}}{\phi\mu c_t k_1}} \tag{12.45}$$

Using the minimum pressure derivative coordinate, correlating with the fault skin factor and the dimensionless pay thickness, Escobar et al. (2013b) obtained a correlation for the estimation of the effective fault conductivity:

$$\sqrt{F_{CDeff}} = \sqrt{F_{CD}\frac{\sqrt{M}+1}{2\sqrt{M}}} = \frac{a+c\left(t_D*P_D'\right)_{min}}{1+b\left(t_D*P_D'\right)_{min}+d\left(t_D*P_D'\right)_{min}^2}\sqrt{\left(1+s_F h_D\right)} \tag{12.46}$$

Replacing the dimensionless quantities leads to:

$$\left(k_f w_f\right)_{eff} = k_f w_f \frac{\sqrt{M}+1}{2\sqrt{M}} = k_1 L_F\left(\frac{a+c\left(t_D*P_D'\right)_{min}}{1+b\left(t_D*P_D'\right)_{min}+d\left(t_D*P_D'\right)_{min}^2}\right)^2\left(1+s_F\frac{h}{L_F}\right) \tag{12.47}$$

where the constants are $a = $ -14048.04, $b = $ -3044.648, $c = $ -513947.31 and $d= $ -279062.51

Having the effective fault conductivity and using an iterative procedure, assuming a value of dimensionless conductivity, the mobility ratio can be found using:

$$F_{CD} = \frac{a+c*\ln(X)+e*\ln(Y)+g*\left(\ln(X)\right)^2+i*\left(\ln(Y)\right)^2+k*\ln(X)*\ln(Y)}{1+b*\ln(X)+d*\ln(Y)+f*\left(\ln(X)\right)^2+h*\left(\ln(Y)\right)^2+j*\ln(X)*\ln(Y)} \tag{12.48}$$

Where:

$$X = \frac{\sqrt{M}+1}{2\left(1+0.84 S_F h_D\right)\sqrt{M}} = \frac{F_{CDeff}}{F_{CD}\left(1+0.84 S_F h_D\right)} \tag{12.49}$$

$$Y = F_{CD} * (t_D * P_D')_{\min} \qquad (12.50)$$

where constants are: $a = -149853.12$, $b = -0.09009733$, $c = 19693.58889$, $d = -0.22773859$, $e = 47430.36913$, $f = 0.001902798$, $g = -701.266695$, $h = 0.012815243$, $i = -3646.98008$, $j = -0.010110749$ and $k = -3024.39415$.

The dimensionless pressure derivative lines obtained from the early radial flow $(t_D*P_D'=0.5)$ and the steady-state flow regimes from Equation (12.41) intersect at:

$$0.5 = \frac{1}{2}(1 + s_F h_D)^2 \frac{1}{t_{DF}} \qquad (12.51)$$

$$t_{DFrssi} = (1 + s_F h_D)^2 \qquad (12.52)$$

Placing the dimensionless time into Equation (12.52) and solving for the fault-well distance will result in:

$$L_F = \sqrt{\frac{0.0002637 k t_{rssi}}{\phi \mu c_t}} - s_F h \qquad (12.53)$$

The lines corresponding to the steady state and the bilinear flow line of the dimensionless pressure derivative intersect; Equations (12.41) and (12.44) lead to:

$$\frac{0.6125}{\sqrt{F_{CD} \dfrac{\sqrt{M}+1}{2\sqrt{M}}}} t_{DF}^{0.25} = \frac{1}{2}(1 + s_F h_D)^2 \frac{1}{t_{DF}} \qquad (12.54)$$

$$t_{DFssBli} = \left[\frac{(1 + s_F h_D)^2 \sqrt{F_{CD} \dfrac{\sqrt{M}+1}{2\sqrt{M}}}}{1.3042} \right]^{0.8} \qquad (12.55)$$

Placing the dimensionless time defined by Equation (12.34) into Equation (12.55) and solving for the effective conductivity fault will result in:

$$k_f w_f \frac{\sqrt{M}+1}{2\sqrt{M}} = 1.9207 \times 10^{-9} k L_F \left(\frac{k t_{ssBLi}}{\phi \mu c_t L_F^2} \right)^{2.5} 1/\left(1 + s_F \frac{h}{L_F}\right)^4 \qquad (12.56)$$

If the dimensionless effective fault conductivity is larger than 2.5×10^8, the bilinear flow disappears and the linear flow appears, exhibiting a half-slope straight line on the pressure

Conductivity Faults

derivative curve. In this case, we have an infinite-conductivity fault. The dimensionless pressure derivative expression for the above-mentioned linear flow regime is:

$$t_D * P_D' = \frac{5.6 \times 10^{-6} \sqrt{M}}{\sqrt{M}+1} \sqrt{t_{DF}} \tag{12.57}$$

Placing the dimensionless quantities given by Equations (12.33) and (12.34) into Equation (12.57) will result in another expression that helps to estimate the distance from the well to the fault:

$$L_F = 1.284 \times 10^{-5} \frac{\sqrt{M}}{\sqrt{M}+1} \frac{qB}{h(t*\Delta P')_L} \sqrt{\frac{\mu t_L}{k_1 \phi c_t}} \tag{12.58}$$

Example 12.2

Escobar et al. (2013b) presented a synthetic pressure test for a well inside an infinite reservoir with the information given below. Pressure and pressure derivative data are reported in Figure 12.11. Estimates of permeability, skin factor formation, distance to fault, fault conductivity and mobility ratio are required.

q = 300 bbl/D	B = 1.553 rb/STB	μ = 0.147 cp
h = 100 ft	r_w = 0.5 ft	ϕ = 15%
c_t = 1.4576x10^{-5} 1/psi	k_1 = 100 md	L_F = 100 ft
F_{CD} = 1x10^7	s_F = 0	k_2 = 1200 md

Solution

The log-log plot of pressure and pressure derivative against production time is given in Figure 12.11, from which the following information can be read:

t_r = 0.00538 hr	ΔP_r = 16.88 psi	$(t*\Delta P')_r$ = 2.546 psi
t_{er} = 0.153 hr	t_{ss} = 4.02 hr	$(t*\Delta P')_{ss}$ = 0.386 psi
t_{BL} = 11791.5hr	$(t*\Delta P')_{BL}$ = 0.0483 psi	t_{rssi} =0.62 hr
t_{ssBLi} = 105 hr	$(t*\Delta P')_{min}$ = 0.02499 psi	

First, the formation permeability is evaluated with Equation (12.37), and skin factor is evaluated with Equation (12.38):

$$k_1 = \frac{70.6(300)(0.7747)(1.553)}{100(2.546)} = 100.086 \text{ md}$$

$$s = \frac{1}{2}\left(\frac{16.88}{2.546} - \ln\left(\frac{100.086(0.00538)}{0.15(0.7747)(1.4576\times10^{-5})(0.5^2)}\right) + 7.43\right) = 0.00211$$

The distance to the fault is evaluated with Equation (12.40), and the fault skin factor is evaluated with Equation (12.42):

$$L_F = 0.0325\sqrt{\frac{100.086(0.153)}{0.15(0.7747)(1.4576\times10^{-5})}} = 97.72 \text{ ft}$$

$$s_F = \frac{97.72}{100}\left[\sqrt{\left(\frac{3.7351\times10^{-6}(100.086^2)(100)(4.03)(0.388)}{300(0.7747^2)(1.553)(0.15)(1.4576\times10^{-5})(97.72^2)}\right)} - 1\right] = 0.00105$$

The distance to fault is re-estimated with Equation (12.53):

$$L_F = \sqrt{\frac{0.0002637(100.086)(0.62)}{0.15(0.7747)(1.4576\times10^{-5})}} - 0.00105(100) = 98.18 \text{ ft}$$

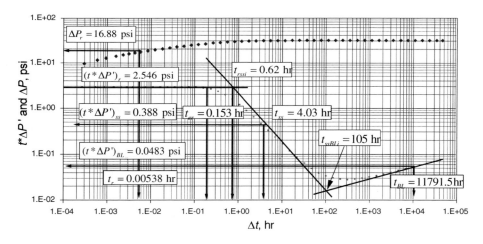

Figure 12.11. Pressure and pressure derivative for example 12.2.

The effective fault conductivity is evaluated with Equation (12.45) and re-estimated with Equations (12.47) and (12.56):

$$k_f w_f \frac{\sqrt{M}+1}{2\sqrt{M}} = 137.67\left(\frac{300(0.7747)(1.553)}{100(0.0483)}\right)^2$$

$$\left(\frac{11791.5}{100.086(0.15)(0.7747)(1.4576\times10^{-5})}\right)^{0.5} = 6.411\times10^9 \text{ md-ft}$$

Conductivity Faults

Table 12.1. Pressure and pressure derivative versus time data for example 12.2

t, hr	ΔP, psi	$t*\Delta P'$, psi	t, hr	ΔP, psi	$t*\Delta P'$, psi	t, hr	ΔP, psi	$t*\Delta P'$, psi
0.00030	9.55	2.53	0.2604	26.69	2.33	166.874	30.59	0.03
0.00059	11.28	2.53	0.3808	27.53	2.08	244.058	30.60	0.03
0.00094	12.43	2.54	0.5569	28.26	1.75	356.943	30.61	0.02
0.00160	13.79	2.54	0.8145	28.86	1.40	522.041	30.62	0.03
0.00234	14.75	2.54	1.1913	29.33	1.07	763.502	30.63	0.03
0.00316	15.53	2.54	1.7423	29.69	0.80	1116.65	30.64	0.03
0.00429	16.30	2.55	2.5482	29.95	0.58	1633.13	30.65	0.03
0.00581	17.07	2.55	3.7268	30.13	0.41	2388.51	30.66	0.03
0.00850	18.04	2.55	5.4505	30.27	0.29	3493.27	30.68	0.04
0.0124	19.01	2.55	7.9716	30.36	0.21	5109.02	30.69	0.04
0.0182	19.98	2.55	11.659	30.43	0.15	7472.10	30.71	0.04
0.0266	20.95	2.55	17.051	30.47	0.11	10928.19	30.72	0.05
0.0389	21.92	2.55	24.938	30.51	0.08	15982.83	30.74	0.05
0.0569	22.88	2.55	36.473	30.53	0.06	23375.40	30.76	0.06
0.0832	23.85	2.55	53.342	30.55	0.04	34187.28	30.79	0.06
0.1217	24.82	2.54	78.015	30.57	0.04	46339.24	30.80	0.07
0.1780	25.78	2.48	114.099	30.58	0.03			

$$(t_D * P_D')_{min} = \frac{100.086(100)(0.02499)}{141.2(300)(0.7747)(1.553)} = 0.004908$$

$$k_f w_f \frac{\sqrt{M}+1}{2\sqrt{M}} = 100.086*97.72 \left(\frac{a+c(0.004908)}{1+b(0.004908)+d(0.004908)^2} \right)^2$$

$$\left(1+0.00105\frac{100}{97.72}\right) = 6.295\times10^9 \text{ md-ft}$$

Averaging the above values, the effective dimensionless fault conductivity is:

$$F_{CD} \frac{\sqrt{M}+1}{2\sqrt{M}} = \frac{6.357\times10^9}{100.086(97.72)} = 649973.16$$

Applying Equation (12.48) by iterative procedure, the dimensionless fault conductivity and the mobility ratio are:

$$F_{CD} = 999671.676$$

$$M = 11.08$$

Finally, the fault conductivity is:

$$k_f w_f = 9.777 \times 10^9 \text{ md-ft}$$

Nomenclature

B	Oil formation factor, rb/STB
c_t	Total system compressibility, 1/psi
F_{CD}	Dimensionless fault conductivity
h	Formation thickness, ft
h_D	Dimensionless pay thickness
k	Reservoir permeability, md
k_1	Permeability in the well zone, md-ft
k_2	Reservoir permeability at the other side of the fault, md
$k_f w_f$	Fault conductivity, md-ft
L_F	Distance from the well to the fault, ft
M	Mobility ratio
m	Slope
q	Flow rate, STB/D
r	Radius, ft
s	Mechanical skin factor
s_F	Fault skin factor
s_{BL}	Bilinear flow skin factor
s_L	Linear flow skin factor
t	Time, hr
$t*\Delta P'$	Pressure derivative, psi

Greek

Δ	Change, drop
ϕ	Porosity, fraction
μ	Viscosity, cp

Suffixes

BL	Bilinear flow
D	Dimensionless
DF	Dimensionless with respect to the fault
eBL	End of bilinear flow
$eBLD$	End of bilinear flow, dimensionless
er	End of radial flow
F	Fault
i	Intersection

L	Linear
min	Minimum
r	Radial
rssi	Radial and steady-state intersection
ss	Steady state
ssBLi	Steady-state and bilinear intersection
w	Wellbore

REFERENCES

Abbaszadeh, M. D.& Cinco-Ley, H. (1995, March 1). "Pressure-Transient Behavior in a Reservoir With a Finite-Conductivity Fault." *Society of Petroleum Engineers.* doi:10.2118/ 24704-PA.

Bixel, H. C., Larkin, B. K. & Van Poollen, H. K. (1963, August 1). "Effect of Linear Discontinuities on Pressure Build-Up and Drawdown Behavior." *Society of Petroleum Engineers.* doi:10.2118/611-PA.

Escobar, F. H., Martinez, J. A. & Montealegre-Madero, M. (2013a). "Pressure Transient Analysis for a Reservoir with a Finite-Conductivity Fault." *CT&F – Ciencia, Tecnología y Futuro.*, Vol. 5. Num. 2. P. 5-18. Jun.

Escobar, F. H., Martinez, J. A. & Montealegre-Madero, M. (2013b). *"Pressure and Pressure Derivative Analysis in a Reservoir with a Finite-Conductivity Fault and Contrast of Mobilities." Fuentes Journal.*, Vol. 11. Nro. 2. p. 17-25. ISSN 1657-6527. Jul/Dic.

Rahman, N. M. A., Miller, M. D. & Mattar, L. (2003, January 1). *"Analytical Solution to the Transient-Flow Problems for a Well Located near a Finite-Conductivity Fault in Composite Reservoirs." Society of Petroleum Engineers.* doi:10.2118/84295-MS

Tiab, D. (1993, January 1). *"Analysis of Pressure and Pressure Derivatives Without Type-Curve Matching: I-Skin and Wellbore Storage." Society of Petroleum Engineers.* This paper was prepared for presentation at the Production Operations Symposium held in Oklahoma City, OK, U.S.A., March 21-23. doi:10.2118/25426-MS.

Trocchio, J. T. (1990). *"Investigation on fateh mishrif fluid-conductivity faults." Journal of Petroleum Technology.*, *42*(8). P. 1038-1045.

Chapter 13

Variable Temperature

Background

Heat transfer has to take place when temperature differences exist inside a medium. When cold water is injected into a hot reservoir, the formation surrounding the well cools to the level of the injected water temperature. The heat exchange in the reservoir mainly occurs by conduction and by convection between injected fluid and matrix. Platenkamp (1985) showed the importance of heat transfer when cold water is injected in a reservoir containing hot oil. He concluded that, during an injection period, heat transfer by conduction is negligible compared to that by convection as long as the test duration is not long and the injection rate is sufficiently high.

The analysis of injection tests under nonisothermal conditions is important for the accurate estimation of reservoir permeability and the well's skin factor; an isothermal system was previously assumed, without taking into account either a moving temperature front that expands with time or the consequent changes in viscosity and mobility between the reservoir's cold and hot zones. This leads to unreliable estimation of the reservoir and well parameters.

For interpretation purposes, a technique based upon the unique features of the pressure and pressure derivative curves and without type-curve matching (the *TDS* technique) was proposed by Escobar, Martinez and Montealegre (2008). The formulation was verified by its application to field and synthetic examples. As expected, increasing reservoir temperature causes a decrement in the mobility ratio, and the estimation of reservoir permeability is somewhat less accurate from the second radial flow, especially as the mobility ratio increases.

Modeling

To construct the interpretation solution, Escobar et al. (2008) used an analytical approach presented by Boughrara et al. (2007a). That solution was initially introduced for the calculation of the injection pressure in an isothermal system. It was later modified by Boughrara and Reynolds (2007b) to consider a system with variable temperature in vertical wells. In this work, the pressure response was obtained by a numerical solution of the anisothermal model, using the Gauss Quadrature method to solve the integrals and assuming that both injection and reservoir temperatures were kept constant during the injection process and that the water saturation was uniform throughout the reservoir.

The dimensionless variables are defined in terms of oil properties at the irreducible water saturation, S_{wi}. The dimensionless variables were defined by Boughrara et al. (2007a):

$$P_D(r_D,t_D) = \frac{k\hat{\lambda}_{oh}h(P(r,t)-P_i)}{141.2*q_{inj}} \tag{13.1}$$

where:

$$\hat{\lambda}_{oh} = \frac{k_{ro}(S_{wi})}{\mu_o(T_{oi})} \tag{13.2}$$

$$r_D = \frac{r}{r_w} \tag{1.4}$$

$$t_D = \frac{0.0002637k\hat{\lambda}_{oh}t}{\phi c_{to}r_w^2} \tag{13.3}$$

where:

$$c_{to} = c_w S_{wi} + c_o(1-S_{wi}) + c_r \tag{13.4}$$

Boughrara and Reynolds (2007) provided a solution that considered the constant rate of injection, q_{inj}, of cold water at temperature T_{wi} in a vertical well drilled in a homogeneous reservoir with an initial temperature T_{oi}. It was also assumed that the initial saturation distribution was constant and equal to the irreducible water saturation, S_{wi}, and that the fluids' viscosities were only a function of temperature. The temperature distribution can be approximated to:

$$T(r,t) = \begin{cases} T_{wi} & r \leq r_T(t) \\ T_{oi} & r \geq r_T(t) \end{cases} \tag{13.5}$$

where $r_T(t)$ is the radial position of the temperature front, which is calculated by an expression provided by Benson and Bodvarsson (1986), who studied the anisothermal effects and built a numerical model for the estimation of permeability and skin factor in injection tests:

$$r_T = 2.37\sqrt{\frac{\rho_w c_w I}{\rho_R c_R \pi h}} \tag{13.6}$$

Variable Temperature

Platenkamp (1985) derived an equation to estimate the position of the moving front:

$$r_f = r_T \sqrt{\frac{\phi\left(\rho_w c_w S_w + \rho_o c_o \left(1 - S_w\right)\right) + \left(1 - \phi\right)\rho_R c_R}{\phi \rho_w c_w S_w}} \qquad (13.7)$$

Using the assumption that both the water front and the temperature front fall inside a "steady-state region" so that total rate is equal to the injection rate q_{inj}, and following a procedure similar to the one presented by Boughrara and Reynolds (2007), Escobar et al. (2008) presented a modified analytical solution for cold-water injection in a hot-oil reservoir drained by a vertical well:

$$\Delta P = P_{wf}\left(t\right) - P_i = \frac{141.2}{h \hat{\lambda}_{oh}} \int_{r_w}^{r_e} q_t\left(r,t\right) \frac{dr}{r\overline{k}\left(r\right)} + \frac{141.2 q_{inj}}{hG} \int_{r_w}^{r_f(t)} \left(\frac{\hat{\lambda}_{wc}}{\lambda_{th}\left(r,t\right)} - 1\right)\frac{dr}{r\overline{k}\left(r\right)} +$$

$$\frac{141.2 q_{inj}}{h} \int_{r_w}^{r_T(t)} \left(\frac{1}{\lambda_{tc}\left(r,t\right)} - \frac{1}{\lambda_{th}\left(r,t\right)}\right)\frac{dr}{r\overline{k}\left(r\right)} \qquad (13.8)$$

where:

$$q_t = q_{inj}\left[\exp\left(-\frac{\phi c_t r^2}{0.0010548 k \lambda_{th} t}\right)\right] \qquad (13.9)$$

$$c_t = c_w S_w + c_o \left(1 - S_w\right) + c_r \qquad (13.10)$$

$$\overline{k}\left(r_n\right) = \begin{cases} \overline{k}_s = \sqrt[3]{k_{xs}k_{ys}k_{zs}} \text{ for } r_w < r_n < r_s, \\ \overline{k} = \sqrt[3]{k_x k_y k_z} \text{ for } r_n < r_s \end{cases} \qquad (13.11)$$

Since the system is subjected to two different temperatures, λ_{tc} represents the total mobility estimated at T_{wi}, which is valid for $r < r_T$; λ_{th} represents the total mobility calculated at T_{oi}, which is valid for $r > r_T$. These are defined by:

$$\lambda_{tc} = \frac{k_{ro}\left(S_w\right)}{\mu_o\left(T_{wi}\right)} + \frac{k_{rw}\left(S_w\right)}{\mu_w\left(T_{wi}\right)} \qquad (13.12)$$

$$\lambda_{th} = \frac{k_{ro}\left(S_w\right)}{\mu_o\left(T_{oi}\right)} + \frac{k_{rw}\left(S_w\right)}{\mu_w\left(T_{oi}\right)} \qquad (13.13)$$

The end-point oil mobility evaluated at the water injection temperature is defined as:

$$\hat{\lambda}_{oc} = \frac{k_{ro}(S_{wi})}{\mu_o(T_{wi})} \tag{13.14}$$

Similarly, the terminal water mobility evaluated at T_{oi} and T_{wi} is given by:

$$\hat{\lambda}_{wh} = \frac{k_{rw}(1-S_{or})}{\mu_w(T_{wi})} \tag{13.15}$$

$$\hat{\lambda}_{wc} = \frac{k_{rw}(1-S_{or})}{\mu_w(T_{oi})} \tag{13.16}$$

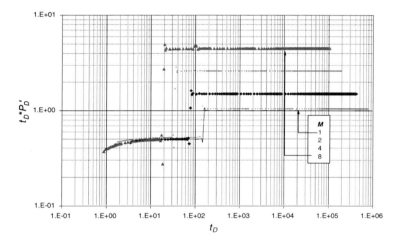

Figure 13.1. Dimensionless pressure derivative for $M \geq 1$; adapted from Escobar et al. (2008).

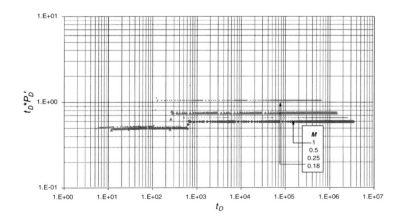

Figure 13.2. Dimensionless pressure derivative for $M \leq 1$; adapted from Escobar et al. (2008).

Variable Temperature

The variable G introduced in this work depends upon the end-point mobility ratio. This term has been introduced in the second integral of the transient-pressure model, Equation (13.8), to appropriately account for the temperature variations. It is defined as follows:

$$G = \begin{cases} \hat{\lambda}_{wc} & M > 1 \\ \hat{\lambda}_{oh} & M \leq 1 \end{cases} \tag{13.17}$$

$$M = \frac{\hat{\lambda}_{wc}}{\hat{\lambda}_{oh}} \tag{13.18}$$

The pressure behavior obtained from the application of Equation (13.8) is reported in Figures (13.1) and (1.3.2). In both cases, as the mobility ratio increases due to temperature effects, the pressure derivative in the second plateau also increases, and the moving front position gets closer. Notice in Figures (13.1) and (1.3.2) that if M is much larger than one, the first radial flow is shorter.

TDS TECHNIQUE FOR VARIABLE TEMPERATURE INJECTION

Escobar et al. (2008) formulated an interpretation methodology, following the philosophy of the *TDS* technique, as introduced by Tiab (1993). As observed in Figures (13.1) and (13.2), the pressure derivatives during the radial flow regimes are given by:

$$\left(t_D * P_D{}'\right)_{r1} = 0.5 \tag{13.19}$$

$$\left(t_D * P_D{}'\right)_{r2} = 0.5(M+1) \tag{13.20}$$

Replacing the dimensionless quantities in the above expressions yields:

$$k = \frac{70.6 q_{inj}}{h\hat{\lambda}_{oh}\left(t * \Delta P'\right)_{r1}} \tag{13.21}$$

$$k = \frac{70.6 q_{inj}(M+1)}{h\hat{\lambda}_{wc}\left(t * \Delta P'\right)_{r2}} \tag{13.22}$$

By analogy with the procedure presented by Tiab (1995), the skin factor formulae are:

$$s = \frac{1}{2}\left[\left(\frac{\Delta P}{t * \Delta P'}\right)_{r1} - \ln\left(\frac{k\hat{\lambda}_{oh} t_{r1}}{\phi c_{to} r_w^2}\right) + 7.43\right] \tag{13.23}$$

$$s = \frac{1}{2}\left[\left(\frac{\Delta P}{t * \Delta P'} \right)_{r2} - \ln\left(\frac{k\hat{\lambda}_{oh}t_{r2}}{\phi c_{to}r_w^2} \right) + 7.43 \right] \qquad (13.24)$$

Example 13.1

Estimates of permeability and skin factor, from a field example provided by Boughrara and Reynolds (2007b), are required. Table 13.1 and Figure 12.3 present the pressure and pressure derivative data for an injection test in a hot reservoir at a temperature of 180 °F when $\mu_o(T_{oi}) = 1.553$ cp. The following are some other relevant data:

Table 13.1. Pressure and pressure derivative versus time data for example 13.1

t, hr	ΔP, psi	$t*\Delta P'$, psi	t, hr	ΔP, psi	$t*\Delta P'$, psi	t, hr	ΔP, psi	$t*\Delta P'$, psi
0.0002	134.62	41.22	0.018	318.42	55.03	0.9	723.55	104.07
0.00024	142.37	42.47	0.02	326.98	99.46	1	734.52	104.05
0.00028	148.94	42.71	0.022	336.92	105.61	1.2	753.48	103.96
0.00032	154.65	42.81	0.028	362.04	104.16	1.5	776.66	103.82
0.00038	162.02	42.92	0.034	382.27	104.18	1.8	795.58	103.67
0.00046	170.23	43.02	0.04	399.21	104.20	2.1	811.55	103.48
0.0006	181.67	43.13	0.046	413.77	104.20	2.4	825.35	103.29
0.0007	188.33	43.18	0.06	441.45	104.16	2.8	841.26	103.03
0.0009	199.19	43.25	0.07	457.51	104.18	3.3	858.15	102.67
0.001	203.75	43.29	0.08	471.42	104.18	3.8	872.61	102.26
0.0012	211.64	43.31	0.09	483.69	104.18	4.2	882.83	99.81
0.0016	224.11	43.36	0.1	494.67	104.21	4.4	887.88	111.71
0.002	233.79	43.39	0.12	513.66	104.16	4.6	892.51	104.77
0.0024	241.71	43.41	0.14	529.72	104.18	5.2	905.29	104.15
0.003	251.39	43.43	0.16	543.64	104.18	5.9	918.44	104.18
0.0036	259.31	43.44	0.2	566.88	104.18	6.6	930.12	104.22
0.0042	266.01	43.87	0.24	585.88	104.19	7.4	942.04	104.24
0.0044	268.00	39.38	0.28	601.94	104.19	8.5	956.48	104.15
0.0046	269.61	35.34	0.32	615.85	104.17	10.8	981.43	104.24
0.006	279.19	35.99	0.38	633.76	104.17	13.5	1004.68	104.14
0.008	289.51	35.79	0.44	649.03	104.18	17.2	1029.92	104.13
0.01	297.49	35.73	0.5	662.34	104.20	21.4	1052.68	104.08
0.012	303.99	35.38	0.6	681.33	104.12	24	1064.62	101.01
0.014	309.48	36.68	0.7	697.39	104.13			
0.016	314.23	30.93	0.8	711.29	104.10			

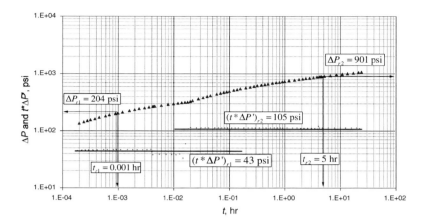

Figure 13.3. Pressure derivative for the field example using $T_{wi}=95$ °F; adapted from Escobar et al. (2008).

$P_i = 3922$ psi
$r_w = 0.35$ ft
$S_w = 0.5$
$\mu_o(T_{wi}) = 8.261$ cp
$k_{rw}(S_{or}) = 0.18$

$q_{inj} = 3000$ BPD
$r_e = 2000$ ft
$S_{wi} = 0.25$
$\mu_w(T_{wi}) = 0.749$ cp
$t_p = 24$ hr

$h = 50$ ft
$\phi = 32\%$
$k = 270$ md
$k_{ro}(S_{wi}) = 0.56$
$T_{wi} = 95$ °F

Solution

The following information is read from Figure 12.3:

$(t*\Delta P')_{r1} = 43$ psi $t_{r1} = 0.001$ hr $(\Delta P)_{r1} = 204$ psi
$(t*\Delta P')_{r2} = 105$ psi $t_{r2} = 5$ hr $(\Delta P)_{r2} = 901$ psi

For the first radial flow, Equation (13.14) and (13.21) are used to find:

$$\hat{\lambda}_{oc} = \frac{k_{ro}(S_{wi})}{\mu_o(T_{wi})} = \frac{0.56}{1.553} = 0.361 \text{ cp}^{-1}$$

$$k = \frac{70.6 q_{inj}}{h\hat{\lambda}_{oh}(t*\Delta P')_{r1}} = \frac{70.6(3000)}{50(0.361)(43)} = 272.88 \text{ md}$$

Skin factor is obtained from Equation (12.23):

$$s = \frac{1}{2}\left[\left(\frac{204}{43}\right)_{r1} - \ln\left(\frac{272.88(0.361)(0.001)}{0.32(1.234\times10^{-5})0.35^2}\right) + 7.43\right] = -0.025$$

For the second radial flow, Equations (13.15) and (13.22) are used to obtain:

$$\hat{\lambda}_{wc} = \frac{k_{rw}\left(1-S_{or}\right)}{\mu_w\left(T_{oi}\right)} = \frac{0.18}{0.749} = 0.24 \text{ cp}^{-1}$$

and $M = 0.666$; then:

$$k = \frac{70.6 Q_{inj}\left(M+1\right)}{h\hat{\lambda}_{wc}\left(t*\Delta P'\right)_{r2}} = \frac{70.6(3000)}{50(0.24)105}\left(1+0.666\right) = 280.05 \text{ md}$$

Skin factor is obtained from Equation (13.24):

$$s = \frac{1}{2}\left[\left(\frac{901}{105}\right)_{r2} - \ln\left(\frac{280.05(0.361)(5)}{0.32(1.234\times10^{-5})(0.35^2)}\right)+7.43\right] = -2.38$$

The permeability results agree with the value reported by Boughrara and Reynolds (2007). Simulation runs were also performed for four other temperatures.

TWO-REGION COMPOSITE RESERVOIR

In recent years, constantly increasing oil prices and declining reserves of conventional crude oils have caused a reevaluation of those light oil deposits that were previously considered economically unattractive. These deposits can now be exploited as an alternative way to maintain the world oil supply volume. Such is the case with heavy oil deposits, which are mainly characterized by having a high resistance to flow (high viscosity) that makes then difficult to produce. Since oil viscosity is reduced by increasing the temperature, thermal recovery techniques, such as steam injection or in-situ combustion, have become the main tool for tertiary recovery of these oils.

Composite reservoirs can occur naturally or may be artificially created. Changes in reservoir width, facies or the type of fluid (hydraulic contact) forming two different regions can create naturally occurring two-zone composite reservoirs. On the other hand, such enhanced oil-recovery projects as water flooding, polymer floods, gas injection, in-situ combustion, steam drive and CO_2 miscible artificially create conditions in which the reservoir can be considered as a composite system. A reservoir undergoing a thermal recovery process is typically idealized as a two-zone composite reservoir in which the inner region represents the swept region surrounding the injection well and the outer region represents the larger portion of the reservoir. Additionally, the mobilities of the two zones are quite different, and the storativity ratio is not 1.

As far as well test interpretation is concerned, a few researchers have worked on thermal recovery processes. Satman, Eggenschwiler and Ramey (1980) presented an analytical solution for a two-zone, infinitely large composite reservoir undergoing a thermal recovery process. They specified a constant rate as the inner boundary condition and neglected wellbore storage effects. They used the conventional straight-line method as the interpretation technique. A year later, Walsch, Ramey and Brigham (1981) conducted an analysis of pressure fall-off testing using a simplistic model. They found a long transition zone between two semilog straight lines for the swept and upswept regions, which obeyed a pseudosteady-state behavior. They calculated the swept zone volume using mean values of temperature and pressure by applying the conventional straight-line method. Barau and Horne (1987) applied the automated type-curve matching to reservoirs subjected to thermal recovery. They reported that their solution improved upon the conventional technique for low mobility ratios. Ambastha and Ramey (1989) first used Satman et al.'s (1980) model, to report the application of the pressure derivative for the systems under discussion. They used the conventional method and type-curve matching to estimate the mobility ratio and the distance to the front.

The aforementioned models consider the fluids behind the combustion front to be slightly compressible, which is incorrect since they are mainly inert gases, according to Soliman, Brigham and Raghavan (1981). These models also assume that the gas flow is restricted in the region ahead of the combustion front, which is also incorrect, as the gases (mostly nitrogen) go through the front and reach the production wells very quickly, according to Islam, Chakma and Farouq Ali (1989). In this sense, assuming a large mobility contrast between regions behind and ahead of the combustion front is not accurate.

Escobar, Martinez and Bonilla (2011) used Satman et al.'s (1980) model to generate the pressure and pressure derivative behavior for different mobility and diffusivity ratios so that the *TDS* was extended to analyze well test data under thermal recovery conditions.

MODELING OF A TWO-REGION COMPOSITE RESERVOIR

According to Stannislav and Kabir (1990), thermal recovery is defined "as a process in which a heat is generated in-situ or injected into the reservoir with the purpose of recovering more oil." In any of these processes, a region around the injection well is mostly completed and is referred to as swept volume. Thermal systems exhibit a significant difference in mobility and diffusivity between the two major areas. The length of the transition period is directly dependent on the magnitude of this contrast in mobility. Both types of thermal recovery processes are characterized by several zones of different properties. For in-situ combustion, there are areas of coke, hot water and light hydrocarbons ahead of the combustion front, followed by an oil bank. The in-situ combustion and steam stimulation technique differs from water injection processes in that it exhibits the following characteristics, as stated by Stannislav and Kabir (1990): (a) The contrast between the mobilities of the two regions is very pronounced, (b) the hydraulic diffusivities are significantly different in the two regions, (c) the two processes are accompanied by heat transfer and (d) the in-situ combustion is characterized by a non-isothermal temperature distribution.

Satman et al. (1980) developed a pressure transient model for a composite reservoir, which represents a thermal oil-recovery process. Figure 13.5 is an idealization of the type of system under consideration. There is a swept region from the injection sandface to the displacement front. Region I is the area dominated by the injected fluid, steam, air or, in the case of a forward in-situ combustion recovery process, any suitable oxidizing gas, with R

being the distance from the injection well to front. Region II is the zone ahead of the front. There are several implicit assumptions in the development of the model:

- The formation is horizontal, homogeneous and of uniform thickness.
- The front has an infinitesimal thickness in the radial direction.
- The region behind the front contains only gas, but the gas flow is restricted in the region ahead of the front. However, the mobility of the gas is much greater than that of the liquid phases, and only gas flow needs to be considered.
- The flow is radial, and the effects of gravity and capillarity are negligible.
- The front can be considered stationary for the few hours of the testing period.

Assuming that the fluid is slightly compressible, the diffusivity equation for two different regions can be written in the following manner:

For Region I:

$$\frac{1}{r}\frac{\partial P_1}{\partial r}\left[r\frac{\partial P_1}{\partial r}\right] = \left(\frac{\phi\mu c_t}{k}\right)_1 \frac{\partial P_1}{\partial t} \qquad r_w \leq r \leq R \qquad (13.25)$$

For Region II:

$$\frac{1}{r}\frac{\partial P_2}{\partial r}\left[r\frac{\partial P_2}{\partial r}\right] = \left(\frac{\phi\mu c_t}{k}\right)_2 \frac{\partial P_2}{\partial t} \qquad R \leq r \leq r_e \qquad (13.26)$$

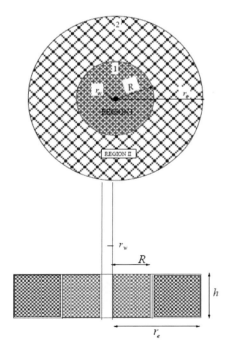

Figure 13.5. Radial composite reservoir; adapted from Gates and Ramey (1978).

where R is the distance to the radial discontinuity. The pressure at the radial discontinuity has the same value, so the continuity condition indicates that:

$$P_1 = P_2 \ @ \ r = R \tag{13.27}$$

$$\frac{\partial P_2}{\partial r} = M \frac{\partial P_1}{\partial r} \tag{13.28}$$

Using the Laplace transformation, Satman et al. (1980) provided an analytical solution to the given problem. The dimensionless quantities used in this work are defined as:

$$P_D = \left(\frac{k_1 h}{141.2 q \mu_1 B} \right) \Delta P \tag{13.29}$$

$$t_D = \left(\frac{0.0002637 k_1}{\phi_1 \mu_1 c_{t1} r_w^2} \right) t \tag{13.30}$$

$$t_{RD} = \left(\frac{0.0002637 k_1}{\phi_1 \mu_1 c_{t1} R^2} \right) t \tag{13.31}$$

$$t_{DA} = \left(\frac{0.0002637 k_1}{\phi_1 \mu_1 c_{t1} A} \right) t \tag{13.32}$$

$$M = \frac{\lambda_1}{\lambda_2} = \frac{(k/\mu)_1}{(k/\mu)_2} \tag{13.33}$$

$$F_s = \left(\frac{\phi_1 c_{t1}}{\phi_2 c_{t2}} \right) \tag{13.34}$$

EFFECT OF MOBILITY AND STORATIVITY RATIO FOR A TWO-REGION COMPOSITE RESERVOIR

Figure 13.6 shows the effect of mobility ratio on the pressure derivative behavior for a fixed storativity of 100. After the end of first radial flow regime, the pressure derivative rises for $M \geq 1$. During the pseudosteady-state period, if mobility ratio, storativity ratio, or both are greater than unity, the pressure derivative reaches a maximum value above the second radial flow regime, corresponding to the outer-region mobility. For large mobility and storativity ratios, the inner region may behave like a closed system for some time during the pseudosteady-state period after the end of the first radial flow. Figure 13.6 shows the following characteristics:

- The first radial flow ends at $t_{RD} = 0.18$ for any value of mobility ratio studied.
- There is a long transition period between the end of the first radial flow and the beginning of the second radial flow for large mobility ratios.
- The transition period is longer for large mobility ratios. This translates into a longer time for the beginning of the second radial flow regime.
- The time of the maximum pressure derivative and the magnitude of the maximum pressure derivative are affected by the mobility ratio.

Figure 13.7 shows the effect of storativity ratio on pressure derivative behavior for a mobility ratio of 10. For storativity ratios greater than unity, the pressure derivative rises above $0.5M$ during the pseudosteady-state period and passes through a maximum value. Thus, a hump takes place in the pressure derivative behavior for mobility and storativity ratios larger than unity. In Figure 13.7, the following characteristics are exhibited:

- Storativity ratio does not affect the time to the end of the first radial flow regime corresponding to the inner-region mobility, and it only mildly affects the beginning of the second radial flow corresponding to the outer-region mobility. The transition time between the two radial flows takes approximately three cycles.
- Storativity ratio affects the pressure derivative behavior at intermediate times. The storativity ratio mildly affects the time and magnitude of the maximum pressure derivative.

TDS TECHNIQUE RATIO FOR A TWO-REGION COMPOSITE RESERVOIR

Escobar et al. (2011) presented the *TDS* technique for two-region composite reservoirs. The permeability and skin factor of the inner zone or swept region, respectively, are found using the following equations from Tiab (1993):

$$k_1 = \frac{70.6 q_a \mu_1 B}{h (t * \Delta P')_{r1}} \qquad (13.35)$$

$$s = \frac{1}{2} \left(\frac{\Delta P_{r1}}{(t * \Delta P')_{r1}} - \ln \left(\frac{k_1 t_{r1}}{(\phi \mu c_t)_1 r_w^2} \right) + 7.43 \right) \qquad (13.36)$$

The pressure derivative during the pseudosteady-state between the first radial flow and the second radial flow is:

$$t_D * P'_D = 2 t_{RD} \qquad (13.37)$$

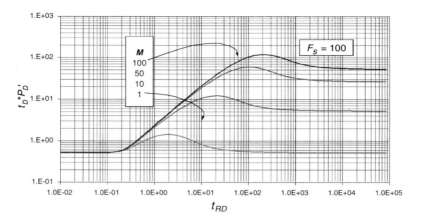

Figure 13.6. Effect of mobility ratio on pressure derivative; adapted from Escobar et al. (2011).

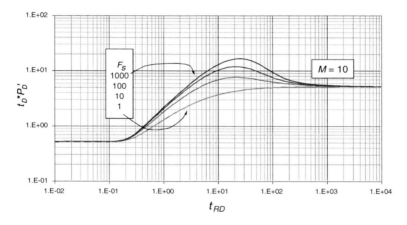

Figure 13.7. Effect of storativity ratio on pressure derivative; adapted from Escobar et al. (2011).

Replacing the dimensionless terms, the distance to the discontinuity or thermal front using the pressure derivative of the pseudosteady-state line, extrapolated to a time of 1 hr, is:

$$R = \sqrt{\frac{0.0745 q_a B}{h(\phi c_t)_1 (t*\Delta P')_{ps1,\,r1}}} \qquad (13.38)$$

The outer-region effect can be seen in Figure 13.8. This plot shows a graph of the pressure derivative versus t_{RD} for several mobilities and storativity ratios in composite reservoirs. The features are:

- The pressure derivative has a maximum in the transition. The developed equations of dimensionless time at this point are:

$$(t_{RD})_{max} = M(2.76 - 0.276\log F_s), \text{ for } 1 < F_s < 45 \qquad (13.39)$$

$$(t_{RD})_{max} = M(1.68 + 0.38\log F_s), \text{ for } F_s \geq 45 \qquad (13.40)$$

Replacing the dimensionless terms in Equations (13.39) and (13.40), the mobility in the outer region, according to the value of F_s, is given by:

$$\left(\frac{k}{\mu}\right)_2 = \frac{(\phi_1 c_{t1})R^2}{0.0002637 t_{max}}\left(2.76 - 0.276\log\left[\frac{\phi_1 c_{t1}}{\phi_2 c_{t2}}\right]\right) \qquad (13.41)$$

$$\left(\frac{k}{\mu}\right)_2 = \frac{(\phi_1 c_{t1})R^2}{0.0002637 t_{max}}\left(1.68 + 0.38\log\left[\frac{\phi_1 c_{t1}}{\phi_2 c_{t2}}\right]\right) \qquad (13.42)$$

- The equation of the dimensionless pressure derivative at the maximum point (peak), corrected by Ambastha and Ramey (1989), is:

$$(t_D * P_D')_{max} = M(0.304 + 0.4343\log F_s), \text{ for } M > 1 \qquad (13.43)$$

Replacing the dimensionless terms in Equation (14.43) yields:

$$\left(\frac{k}{\mu}\right)_2 = \frac{141.2 q_a B}{h(t*\Delta P')_{max}}\left[0.304 + 0.4343\log\left(\frac{\phi_1 c_{t1}}{\phi_2 c_{t2}}\right)\right] \qquad (13.44)$$

Setting Equations (13.41) and (13.44) equal to each other and solving for F_s ($1 < F_s < 45$) yields:

$$\log F_s = \frac{0.01132 q_a B t_{max} - 2.76(\phi c_t)_1(t*\Delta P')_{max} R^2 h}{-0.276(\phi c_t)_1 R^2 h(t*\Delta P')_{max} - 0.01617 q_a B t_{max}} \qquad (13.45)$$

Setting Equations (13.42) and (13.44) equal to each other (for $F_s \geq 45$) results in:

$$\log F_s = \frac{0.01132 q_a B t_{max} - 1.68(\phi c_t)_1(t*\Delta P')_{max} R^2 h}{0.38(\phi c_t)_1 R^2 h(t*\Delta P')_{max} - 0.01617 q_a B t_{max}} \qquad (13.46)$$

The calculated and appropriate value of F_s must be in the specified range.
- During the second radial flow regime, the pressure derivative is:

$$t_D * P'_D = 0.5M \qquad (13.47)$$

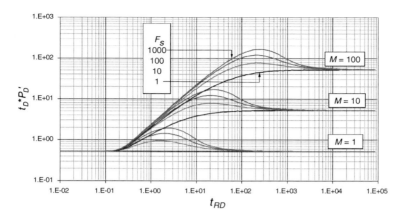

Figure 13.8. Pressure derivative type-curve for composite reservoirs; adapted from Escobar et al. (2011).

From this, the following expression is obtained:

$$k_2 = \frac{70.6 q_a \mu_2 B}{h(t*\Delta P')_{r2}} \quad (13.48)$$

Setting Equations (13.37) and (13.47) equal to each other results in an equation for calculating the distance to the discontinuity or thermal front, with the intersection time between the second radial flow and the pseudosteady-state in the transition:

$$R = \sqrt{\frac{1.0548 \times 10^{-3} t_{ir2ps1}}{(\phi c_t)_1} \left(\frac{k}{\mu}\right)_2} \quad (13.49)$$

For long producing times, the pressure derivative function yields a straight line of unit slope. This line, which corresponds to the pseudosteady-state flow regime, starts at a t_{DA} equal to $0.2M/F_s$. The equation of this straight line is:

$$t_{DA} * P'_D = 2\pi F_s t_{DA} \quad (13.50)$$

After replacing the dimensionless quantities, an expression to estimate reservoir drainage area is found:

$$A = \frac{q_a B}{4.27 \phi_2 c_{t2} h(t*\Delta P')_{ps,r2}} t_{ps,r2} \quad (13.51)$$

Setting Equations (13.50) and (13.47) equal to each other results in:

334 Freddy Humberto Escobar

Table 13.2. Pressure and pressure derivative data for example 13.2

t, hr	ΔP, psi	$t*\Delta P'$, psi	t, hr	ΔP, psi	$t*\Delta P'$, psi	t, hr	ΔP, psi	$t*\Delta P'$, psi
0.000327	2.851	0.453	0.274	6.357	1.353	140.57	308.329	157.154
0.000631	3.148	0.453	0.381	6.879	1.864	186.44	353.229	159.745
0.001033	3.372	0.454	0.529	7.597	2.565	271.09	412.515	155.629
0.001435	3.521	0.455	0.735	8.584	3.523	443.63	486.285	142.849
0.001992	3.670	0.455	1.02	9.939	4.830	616.06	531.582	133.173
0.002767	3.820	0.455	1.42	11.794	6.604	855.51	573.838	124.610
0.003842	3.969	0.456	1.97	14.327	8.999	1188.04	613.563	117.796
0.005335	4.119	0.456	2.73	17.771	12.212	1649.82	651.339	112.679
0.007409	4.269	0.456	3.80	22.433	16.485	2291.09	687.670	108.902
0.01029	4.418	0.456	5.27	28.708	22.107	3181.61	722.941	106.147
0.01429	4.568	0.456	7.32	37.089	29.401	5206.60	774.454	103.308
0.01984	4.718	0.456	10.16	48.180	38.698	8520.44	824.852	101.505
0.02755	4.868	0.456	14.11	62.690	50.279	13943.43	874.538	100.352
0.03826	5.018	0.457	19.60	81.400	64.282	22817.99	923.770	99.603
0.05313	5.168	0.462	27.22	105.095	80.570	37340.91	972.700	99.136
0.07379	5.323	0.489	37.80	134.455	98.564	61107.21	1021.444	98.799
0.102	5.493	0.567	52.49	169.873	117.112	84858.92	1053.869	98.703
0.142	5.702	0.726	72.89	211.268	134.458	100000	1070.068	0
0.198	5.979	0.984	101.23	257.900	148.469			

$$A = \frac{k_2 t_{ir2ps2}}{301.77(\phi \mu c_t)_2} \tag{13.52}$$

The intersection point of the first radial flow and the pseudosteady-state line, from Equations (13.37) and (13.50), allows another expression for the estimation of the area to be developed:

$$A = \frac{k_1 t_{ir1ps2}}{301.77 \mu_1 (\phi c_t)_2} \tag{13.53}$$

If the peak is not correctly determined, the effect of the outer boundary in Equations (13.39) through (13.53) may not be applicable.

Example 13.2

Satman et al. (1980) presented a well test field example during an in-situ combustion operation. Data for the reservoir, fluids and well are given below. Pressure and pressure derivative data are reported in both Figure 13.9 and Table 13.2. The mobility in each zone,

Variable Temperature   335

the skin factor in the inner region and the distance to the combustion front all must be estimated.

$P_R = 134.5$ psi $\qquad q_a = 623333$ BPD $\qquad h = 30$ ft

$r_w = 0.3$ ft $\qquad r_e = 2000$ ft $\qquad \phi_1 = 37\%$

$\phi_2 = 34\%$ $\qquad S_{wi} = 0.25$ $\qquad \lambda_1 = 290909.91$ md/cp

$c_{t1} = 0.003333$ 1/psi $\qquad c_{t2} = 0.000206$ 1/psi $\qquad \lambda_2 = 1351.351$ md/cp

$R = 150$ ft

Solution

The log-log plot of pressure and pressure derivative against injection time is given in Figure 13.9. From that plot, the following information can be read:

$t_{r1} = 0.03247$ hr $\qquad \Delta P_{r1} = 4.943$ psi $\qquad (t^*\Delta P')_{r1} = 0.456$ psi

$(t^*\Delta P')_{r2} = 99.277$ psi $\qquad t_{max} = 186.44$ hr $\qquad (t^*\Delta P')_{max} = 159.745$ psi

$(t^*\Delta P')_{ps1,r1} = 4.9$ psi $\qquad t_{r2pi1} = 20$ hr

First, the mobility of the inner region is evaluated with Equation (13.35), and the skin factor is evaluated with Equation (13.36),

$$\left(\frac{k}{\mu}\right)_1 = \frac{70.6 q_a B}{h(t^*\Delta P')_{r1}} = \frac{70.6(623333)(0.090528)}{30(0.456)} = 291220.3 \text{ md/cp}$$

$$s = \frac{1}{2}\left(\frac{4.943}{0.456} - \ln\left(\frac{291220.3(0.03247)}{0.37(0.003333)(0.3^2)}\right) + 7.43\right) = 0.0047$$

The mobility of the outer region is evaluated with Equation (13.48):

$$\left(\frac{k}{\mu}\right)_2 = \frac{70.6 q_a B}{h(t^*\Delta P')_{r2}} = \frac{70.6(623333)(0.090528)}{30(99.277)} = 1337.63 \text{ md/cp}$$

The distance to the discontinuity or combustion front is found with Equation (13.49) and re-estimated with Equation (13.38):

$$R = \sqrt{\frac{0.0745 q_a B}{h(\phi c_t)_1 (t^*\Delta P')_{ps1,r1}}} = \sqrt{\frac{0.0745(623333)(0.090528)}{30(0.37)(0.003333)(4.9)}} = 152.28 \text{ ft}$$

$$R = \sqrt{\frac{1.0548 \times 10^{-3} t_{r2pi1}}{(\phi c_t)_1}\left(\frac{k}{\mu}\right)_2} = \sqrt{\frac{1.0548 \times 10^{-3}(20)(1337.67)}{0.37(0.003333)}} = 151.26 \text{ ft}$$

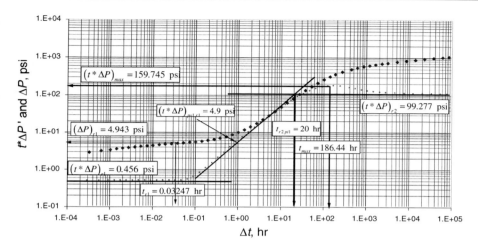

Figure 13.9. Pressure and pressure derivative for example 13.2; adapted from Escobar et al. (2011).

Equation (13.33) is used to calculate the storativity ratio, and Equations (13.41) and (13.44) are used to re-estimate the mobility of the outer region:

$$F_s = \frac{\phi_1 c_{t1}}{\phi_2 c_{t2}} = \frac{0.37(0.003333)}{0.34(0.000206)} = 17.607$$

$$\left(\frac{k}{\mu}\right)_2 = \frac{0.37(0.003333)(152.28^2)}{0.0002637(186.44)}(2.76 - 0.276\log[17.607]) = 1405.41 \text{ md/cp}$$

$$\left(\frac{k}{\mu}\right)_2 = \frac{141.2(623333)(0.090528)}{30(159.745)}\left[0.304 + 0.4343\log(17.607)\right] = 1404.9 \text{ md/cp}$$

THREE-REGION COMPOSITE RESERVOIR

In recent years, the declining reserves of conventional crude have changed views on the exploitation of deposits that were previously considered economically unattractive, as these deposits can be used to maintain the world's oil supply. Heavy oil deposits are mainly characterized by having high resistance to flow (high viscosity), which makes then difficult to produce. Since oil viscosity is reduced by increasing the temperature, thermal recovery techniques, such as steam injection or in-situ combustion, have become the main tool for tertiary recovery of heavy oil. Usually, well tests from enhanced oil-recovery projects, such as steam injection, in-situ combustion, and CO_2 flooding projects, are analyzed using a radial, two-region composite reservoir model. However, a three-region model may be more appropriate in many cases since a transition zone may be developed.

The determination of the swept volume in a thermal oil-recovery process is of primary concern. Estimation of the swept volume at intermediate stages of the operation, for either in-

Variable Temperature337

situ combustion or steam injection, makes early economic evaluation of the field operations possible.

Two-region composite reservoirs models have been used to analyze pressure transient data from enhanced oil-recovery projects. Three-region composite reservoir models have been used less frequently to analyze well tests from enhanced oil-recovery projects. Ambastha and Ramey (1989) presented a review of methodologies used to interpret well test data from enhanced recovery projects and developed several design and interpretation relationships from an analysis of a well test response for a well located in a two-region composite reservoir. An analytical solution in the Laplace space for the transient-pressure behavior of a well in a three-region composite reservoir was presented by Onyekonwu (1985) and by Barua and Horner (1987). To study the effects of an intermediate region on the deviation time method and the pseudosteady-state method, the analytical solution for a three-region reservoir presented by Onyekonwu (1985) is useful. Escobar, Martinez and Bonilla (2012) used the model proposed by Onyekonwu to generate the pressure and pressure derivative behavior for different mobility and diffusivity ratios in order to present a methodology without using type-curve matching to analyze well tests under thermal recovery conditions. This is the first analytical methodology available. The model does not take into account compressibility effects due to the possible presence of the gas phase from the combustion process.

MODELING A THREE-REGION COMPOSITE RESERVOIR

Figure 13.10 shows an idealized three-zone model based on Onyekonwu (1985). Region 1 is the swept volume. Region 2 is the transition zone, which is the region of rapidly changing mobility. Region 3 contains low mobility fluid. Assumptions implicit in the development of the model include:

1. The formation is homogeneous, horizontal, and of uniform thickness.
2. Flow is radial, and gravity and capillary effects are negligible.
3. In the three regions, the fluid is considered to be of slight constant compressibility, but the fluid mobility and compressibility may be different.
4. The pressure gradient in the reservoir is considered to be small.
5. Other assumptions are inherent to Darcy's law.

The diffusivity equation in dimensionless form for the three regions can be written as follows:

For Region 1:

$$\frac{\partial^2 P_{D1}}{\partial r_D^2} + \frac{1}{r_D}\frac{\partial P_{D1}}{\partial r_D} = \frac{\partial P_{D1}}{\partial t_D}, \text{ for } 1 \leq r_D \leq R_{D1} \tag{13.54}$$

For Region 2:

$$\frac{\partial^2 P_{D2}}{\partial r_D^2} + \frac{1}{r_D}\frac{\partial P_{D2}}{\partial r_D} = \eta_{12}\frac{\partial P_{D2}}{\partial t_D}, \text{ for } R_{D1} \le r_D \le R_{D2} \qquad (13.55)$$

where:

$$\eta_{12} = \left(\frac{k}{\phi\mu c_t}\right)_1 \Big/ \left(\frac{k}{\phi\mu c_t}\right)_2 \qquad (13.56)$$

For Region 3:

$$\frac{\partial^2 P_{D3}}{\partial r_D^2} + \frac{1}{r_D}\frac{\partial P_{D3}}{\partial r_D} = \eta_{13}\frac{\partial P_{D3}}{\partial t_D}, \text{ for } R_{D2} \le r_D \le \infty \qquad (13.57)$$

where:

$$\eta_{13} = \left(\frac{k}{\phi\mu c_t}\right)_1 \Big/ \left(\frac{k}{\phi\mu c_t}\right)_3 \qquad (13.58)$$

The inner and outer boundary conditions can be found in Onyekonwu (1985). The dimensionless quantities used by Onyekonwu (1985) are defined as:

$$P_D = \left(\frac{k_1 h}{141.2q\mu_1 B}\right)\Delta P \qquad (13.29)$$

$$t_{DR1} = \left(\frac{0.0002637k_1}{\phi_1\mu_1 c_{t1}R_1^2}\right)t \qquad (13.59)$$

$$t_{DR2} = \left(\frac{0.0002637k_1}{\phi_1\mu_1 c_{t1}R_2^2}\right)t \qquad (13.60)$$

$$M_{12} = \frac{\lambda_1}{\lambda_2} = \frac{(k/\mu)_1}{(k/\mu)_2} \qquad (13.61)$$

$$M_{13} = \frac{\lambda_1}{\lambda_3} = \frac{(k/\mu)_1}{(k/\mu)_3} \qquad (13.62)$$

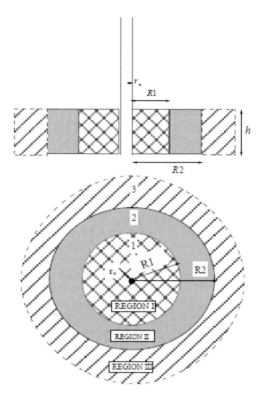

Figure 13.10. Three-Zone Model of the porous system; adapted from Onyekonwu (1985).

$$F_{S12} = \left(\frac{\phi_1 c_{t1}}{\phi_2 c_{t2}}\right) \qquad (13.34)$$

$$F_{S13} = \left(\frac{\phi_1 c_{t1}}{\phi_3 c_{t3}}\right) \qquad (13.63)$$

EFFECT OF MOBILITY AND STORATIVITY RATIO ON A THREE-REGION COMPOSITE RESERVOIR

Figure 13.11 shows the effect of M_{12} on pressure on pressure derivative behavior for fixed values of F_{S12}, F_{S13}, M_{13} and R_2. As mobility between regions 1 and 2 increases, the pressure derivative goes through a first maximum value, while the latter remains almost constant until the relationship of the mobility ratios is equal to unity. The second radial flow is not visible due to the contrast in the properties between regions 2 and 3. Figure 12.12 shows the effect of M_{13} on the pressure derivative behavior for fixed values of F_{S12}, F_{S13}, M_{12} and R_2. As mobility between regions 1 and 3 increases, the pressure derivative goes through a first maximum value which remains unaltered, while the latter increases its value. Figure

12.13 shows the effect of F_{S12} on the pressure derivative behavior for fixed values of F_{S13}, M_{12}, M_{13} and R_2. As the storativity ratio between regions 1 and 2 increases, the value of the maximum second derivative also increases and moves to the right-hand side. Figure 12.14 shows the effect of F_{S13} on pressure derivative behavior for fixed values of F_{S12}, M_{12}, M_{13} and R_2. As the storativity ratio between regions 1 and 3 increases, the value of the maximum second derivative also increases, but the time of the second peak remains unaltered.

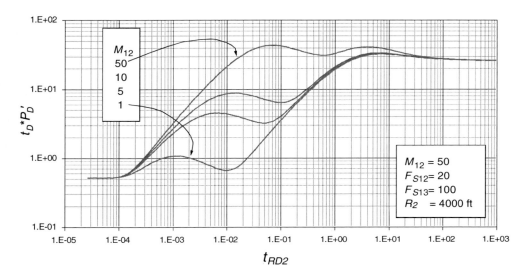

Figure 13.11. Effect of mobility ratio between regions 1 and 2; adapted from Escobar et al. (2012).

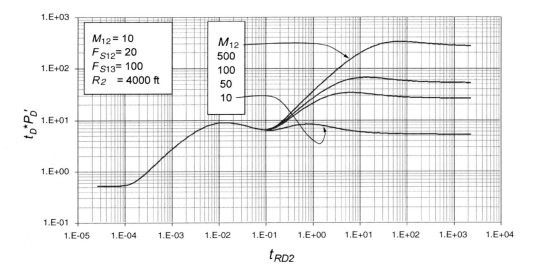

Figure 13.12. Effect of mobility ratio between regions 1 and 3; adapted from Escobar et al. (2012).

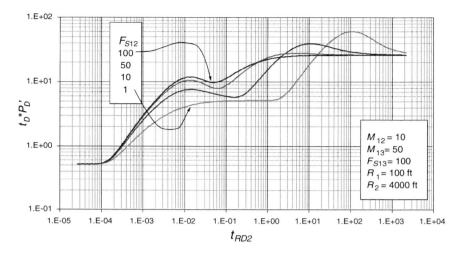

Figure 13.13. Effect of storativity ratio between regions 1 and 2; adapted from Escobar et al. (2012).

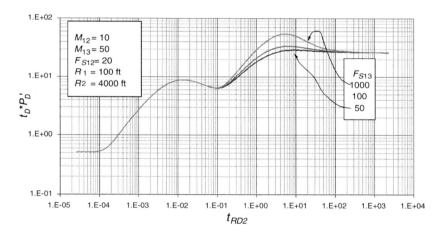

Figure 13.14. Effect of storativity ratio between regions 1 and 3; adapted from Escobar et al. (2012).

The permeability and the skin factor of the inner zone or swept region are found using the following equation from Tiab (1993):

$$k_1 = \frac{70.6 q_a \mu_1 B}{h(t*\Delta P')_{r1}} \tag{13.64}$$

$$s = \frac{1}{2}\left(\frac{\Delta P_{r1}}{(t*\Delta P')_{r1}} - \ln\left(\frac{k_1 t_{r1}}{(\phi \mu c_t)_1 r_w^2}\right) + 7.43 \right) \tag{13.65}$$

The radius of swept region is estimated with the end of first radial flow at $t_{RD1} = 0.18$. See Figure (13.15).

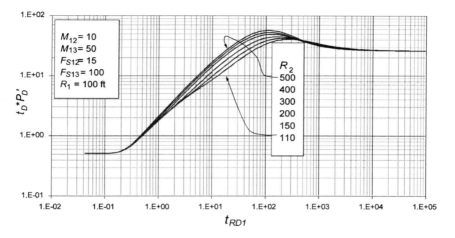

Figure 13.15. Effect of R_2 on pressure derivative with t_{DR1}; adapted from Escobar et al. (2012).

$$R_1 = \sqrt{\frac{0.001465 k_1 t_{er1}}{\mu_1 (\phi c_t)_1}} \qquad (13.66)$$

The pressure derivative has a first maximum in the transition between the first and second region. The equations of dimensionless time developed at this point are:

$$(t_{RD})_{max1} = M_{12}(2.76 - 0.276 \log F_{S12}), \text{ for } 1 < F_{S12} < 45 \text{ and } R_2/R_1 > 15 \qquad (13.67)$$

$$(t_{RD})_{max1} = M_{12}(1.68 + 0.38 \log F_{S12}), \text{ for } F_{S12} \geq 45 \text{ and } R_2/R_1 > 15 \qquad (13.68)$$

Replacing the dimensionless terms in Equations (13.67) and (13.68), the mobility in the outer region according to the value of F_{s12} is given by:

$$\left(\frac{k}{\mu}\right)_2 = \frac{(\phi_1 c_{t1}) R_1^2}{0.0002637 t_{max1}} \left(2.76 - 0.276 \log\left[\frac{\phi_1 c_{t1}}{\phi_2 c_{t2}}\right]\right) \qquad (13.69)$$

$$\left(\frac{k}{\mu}\right)_2 = \frac{(\phi_1 c_{t1}) R_1^2}{0.0002637 t_{max1}} \left(1.68 + 0.38 \log\left[\frac{\phi_1 c_{t1}}{\phi_2 c_{t2}}\right]\right) \qquad (13.70)$$

The equation of the dimensionless pressure derivative at the first maximum point (peak), corrected by Ambastha and Ramey (1989), is:

$$(t_D * P_D')_{max1} = M_{12}(0.304 + 0.4343 \log F_{S12}), \text{ for } M_{12} > 1 \text{ and } R_2/R_1 > 15 \qquad (13.71)$$

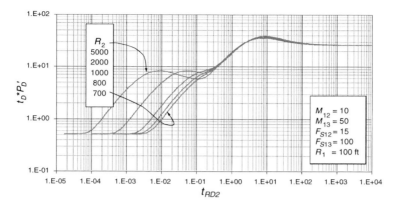

Figure 13.16. Effect of R_2 on pressure derivative; adapted from Escobar et al. (2012).

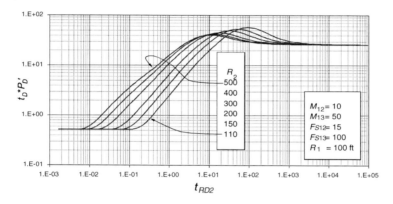

Figure 13.17. Effect of R_2 on pressure derivative; adapted from Escobar et al. (2012).

Replacing the dimensionless terms in Equation (13.71) yields:

$$\left(\frac{k}{\mu}\right)_2 = \frac{141.2 q_a B}{h(t*\Delta P')_{max1}} \left[0.304 + 0.4343 \log\left(\frac{\phi_1 c_{t1}}{\phi_2 c_{t2}}\right)\right] \qquad (13.72)$$

Figure 13.16 shows when then ratio of the radius of the second region to that of the swept region is greater than or equal to 7. Two maxima are observed in the pressure derivative, and based on the value of t_{Dr2}, the second maximum remains constant. For $R_2/R_1 \geq 7$, the dimensionless pressure derivative of the second peak is estimated with:

$$(t_D * P_D)_{max2} = M_{13}\left(-0.16077835 \ln\left(\frac{F_{S12}}{F_{S13}}\right) + 0.36060599\right) + M_{12}\left(-0.034756599 \ln\left(\frac{F_{S12}}{F_{S13}}\right) + 0.10347416\right) \qquad (13.73)$$

Replacing the dimensionless terms leads to:

$$\lambda_2 = \lambda_1 \frac{\left(-0.034756599\ln\left(\dfrac{F_{S12}}{F_{S13}}\right)+0.10347416\right)}{\left(t_D * P_D\right)_{\max 2} - \dfrac{\lambda_1}{\lambda_3}\left(-0.16077835\ln\left(\dfrac{F_{S12}}{F_{S13}}\right)+0.36060599\right)} \tag{13.74}$$

This is valid for $F_{S12}/F_{S13} \le 0.4$. The dimensionless time of the second peak is estimated with:

$$\left(t_{DR2}\right)_{\max 2} = \frac{M_{13}}{F_{S12}}\left[7.9020822\left(\frac{F_{S12}}{F_{S13}}\right)+1.5808639\right]+\frac{M_{12}}{F_{S12}}\left[-10.9839962\left(\frac{F_{S12}}{F_{S13}}\right)+0.51691818\right] \tag{13.75}$$

This is valid for $0.1 \le F_{S12}/F_{S13} \le 0.4$ when $M_{13}/M_{12} > 1$. Alternately, the same value can be estimated with:

$$\left(t_{DR2}\right)_{\max 2} = \frac{M_{13}}{F_{S12}}\left[-1.09372076\left(\frac{F_{S12}}{F_{S13}}\right)+2.4179022\right]-\frac{M_{12}}{F_{S12}}\left[7.4853828\left(\frac{F_{S12}}{F_{S13}}\right)+0.085786786\right] \tag{13.76}$$

This is valid for $0.01 \le F_{S12}/F_{S13} < 0.1$ and $0.1 \le F_{S12}/F_{S13} \le 0.4$ when $M_{13}/M_{12} = 1$. Replacing the dimensionless terms in Equations (13.75) and (13.76), the radius of the second region according to the value of F_{S12}/F_{S13} is given by:

$$R_2 = \sqrt{\frac{\left(\dfrac{0.0002637\lambda_1 t_{\max 2}}{\phi_1 c_{t1}}\right)}{\dfrac{M_{13}}{F_{S12}}\left[7.9020822\left(\dfrac{F_{S12}}{F_{S13}}\right)+1.5808639\right]+\dfrac{M_{12}}{F_{S12}}\left[-10.9839962\left(\dfrac{F_{S12}}{F_{S13}}\right)+0.51691818\right]}} \tag{13.77}$$

$$R_2 = \sqrt{\frac{\left(\dfrac{0.0002637\lambda_1 t_{\max 2}}{\phi_1 c_{t1}}\right)}{\dfrac{M_{13}}{F_{S12}}\left[-1.09372076\left(\dfrac{F_{S12}}{F_{S13}}\right)+2.4179022\right]-\dfrac{M_{12}}{F_{S12}}\left[7.4853828\left(\dfrac{F_{S12}}{F_{S13}}\right)+0.085786786\right]}} \tag{13.78}$$

Figure 13.17 describes when then the ratio of the radius of the second region to that of the first region is less than 7. The pressure derivative behaves as a reservoir composed of two zones, and only the second maximum, which is not constant, is observed. If $R_2/R_1 < 7$, it is not possible to determine the mobility of the second region because it depends on all the variables studied, which are not kept constant. For $1.1 \le R_2/R_1 < 3.5$ and $0.01 < F_{S12}/F_{S13} < 0.1$:

Variable Temperature

Table 13.3. Constants values for Equations (13.79) through (13.82)

Coefficient	Equation		
	(13.79)	**(13.80)**	**(13.81)**
a	1.148056075	-1.65348167	-123.171636754
b	-0.054300723	0.359685904	-96.912384629
c	-0.1111916097	3.633603633	518.685599496
d	-29.28054451	0.7867843589	-1098.051770102
e	-31.94854034	-0.53215988	1004.175610598
f	0.0085109453	0.011789055	2761.218331423
g	-0.024785904	0.024704553	-7677.7631303581
h	194.799992696	0.140924797	7924.945179820
i	208.154701539	0.18132819	-2877.195951804
j	0.15346215629	0.202918526	-
k	0.84388292168	1.930425704	-

$$\frac{R_2}{R_1} = \frac{a + c\ln\left(\frac{t_{DR1max}}{M13}\right) + e\left(\frac{F_{S12}}{F_{S13}}\right) + g(\ln\left(\frac{t_{DR1max}}{M_{13}}\right))^2 + i\left(\frac{F_{S12}}{F_{S13}}\right)^2 + k\left(\frac{F_{S12}}{F_{S13}}\right)\ln\left(\frac{t_{DR1max}}{M_{13}}\right)}{1 + b\ln\left(\frac{t_{DR1max}}{M13}\right) + d\left(\frac{F_{S12}}{F_{S13}}\right) + f(\ln\left(\frac{t_{DR1max}}{M_{13}}\right))^2 + h\left(\frac{F_{S12}}{F_{S13}}\right)^2 + j\left(\frac{F_{S12}}{F_{S13}}\right)\ln\left(\frac{t_{DR1max}}{M_{13}}\right)} \quad (13.79)$$

For $0.1 < F_{S12}/F_{S13} < 0.3$:

$$\frac{R_2}{R_1} = \frac{a + c\ln\left(\frac{t_{DR1max}}{M13}\right) + e\ln\left(\frac{F_{S12}}{F_{S13}}\right) + g\left(\ln\left(\frac{t_{DR1max}}{M13}\right)\right)^2 + i\left(\ln\left(\frac{F_{S12}}{F_{S13}}\right)\right)^2 + k\ln\left(\frac{t_{DR1max}}{M13}\right)\ln\left(\frac{F_{S12}}{F_{S13}}\right)}{1 + b\ln\left(\frac{t_{DR1max}}{M13}\right) + d\ln\left(\frac{F_{S12}}{F_{S13}}\right) + f\left(\ln\left(\frac{t_{DR1max}}{M13}\right)\right)^2 + h\left(\ln\left(\frac{F_{S12}}{F_{S13}}\right)\right)^2 + j\ln\left(\frac{t_{DR1max}}{M13}\right)\ln\left(\frac{F_{S12}}{F_{S13}}\right)} \quad (13.80)$$

For $0.3 < F_{S12}/F_{S13} < 0.6$:

$$\ln\left(\frac{R_2}{R_1}\right) = a + \frac{b}{\ln\left(\frac{t_{DR1max}}{M13}\right)} + \frac{c}{\left(\frac{t_{DR1max}}{M13}\right)} + \frac{d}{\left(\frac{t_{DR1max}}{M13}\right)^{1.5}} + \frac{e}{\left(\frac{t_{DR1max}}{M13}\right)^2}$$
$$+ f\left(\frac{F_{S12}}{F_{S13}}\right) + g\left(\frac{F_{S12}}{F_{S13}}\right)^{1.5} + h\left(\frac{F_{S12}}{F_{S13}}\right)^2 + i\left(\frac{F_{S12}}{F_{S13}}\right)^{2.5} \quad (13.81)$$

Constants for Equations (13.79) through (13.81) are given in Table (13.3).

For $3.5 \leq R_2/R_1 < 7$:

$$R_2 = R_1\left(\frac{2.6766673 - 0.22589317(t_{DR})_{max} + 0.17183862(M_{13}) - 0.0014327696(M_{13})^2 + 5.3310075\times10^{-6}(M_{13})^3}{1 - 0.014790427(t_{DR})_{max} - 0.1460338(M_{13}) - 0.0004564009(M_{13})^2 + 1.523638\times10^{-6}(M_{13})^3}\right)$$
$$(13.82)$$

Table 13.4. Pressure and pressure derivative data for example 13.3

t, hr	ΔP, psi	$t*\Delta P'$, psi	t, hr	ΔP, psi	$t*\Delta P'$, psi	t, hr	ΔP, psi	$t*\Delta P'$, psi
0.00100	24.21	4.15	0.3718	72.65	34.56	138.22	7872.92	5420.27
0.00134	25.44	4.17	0.4998	84.48	46.03	185.81	9576.80	6075.57
0.00181	26.67	4.18	0.6719	100.25	61.40	249.80	11455.32	6587.55
0.00243	27.91	4.19	0.9033	121.29	81.92	335.82	13456.50	6895.45
0.00327	29.16	4.20	1.2143	149.36	109.23	451.46	15514.51	6970.18
0.00439	30.40	4.21	1.6325	186.76	145.47	606.92	17561.40	6827.71
0.00590	31.65	4.22	2.1947	236.54	193.44	815.92	19540.52	6525.78
0.00794	32.90	4.23	2.9504	302.65	256.70	1096.88	21416.15	6143.52
0.01067	34.15	4.23	3.9664	390.29	339.82	1474.60	23175.92	5754.66
0.01434	35.40	4.22	5.3322	506.13	448.50	1982.39	24826.00	5408.70
0.01928	36.65	4.23	7.1684	658.74	589.80	2665.03	26382.95	5126.79
0.02592	37.91	4.36	9.6369	858.96	772.07	3582.75	27866.16	4909.06
0.03485	39.25	4.76	12.9554	1120.33	1004.96	4816.49	29293.26	4745.27
0.04685	40.77	5.60	17.4166	1459.39	1299.22	6972.26	31019.39	4597.46
0.06298	42.62	6.97	23.41	1895.92	1665.60	10092.90	32699.63	4494.07
0.08467	44.95	8.95	31.48	2452.70	2113.46	14610.28	34347.60	4420.94
0.1138	47.97	11.60	42.32	3154.80	2648.32	21149.55	35972.65	4368.77
0.1530	51.90	15.10	56.89	4027.97	3268.34	30615.67	37581.30	4331.37
0.2057	57.02	19.78	76.48	5095.78	3959.93	44318.63	39178.18	4304.50
0.2765	63.75	26.07	102.81	6375.51	4693.39	64154.76	40766.60	4285.19
						92869.15	42348.94	4271.30

The permeability of the outer region is found using the following equation:

$$k_3 = \frac{70.6 q_a \mu_3 B}{h(t*\Delta P')_{r3}} \tag{13.83}$$

Example 13.3

Relevant data for reservoir and fluid properties for an example presented by Onyekonwu and Ramey (1986) are given below. Pressure and pressure derivative data are reported in Figure 13.18 and Table 13.4. These values characterize the reservoir and estimate the radii of the inner and intermediate regions.

$q_a = 100$ BPD

$(\phi c_t)_1 = 4\times10^{-7}$ 1/psi

$\lambda_2 = 2$ md/cp

$c_{t2} = 0.000206$ 1/psi

$R_2 = 105$ ft

$h = 50$ ft

$(\phi c_t)_2 = 8\times10^{-8}$ 1/psi

$\lambda_3 = 0.05$ md/cp

$(\phi c_t)_3 = 1.16\times10^{-8}$ 1/psi

$B = 1.5$ rb/STB

$r_w = 0.5$ ft

$\lambda_1 = 50$ md/cp

$c_{t1} = 0.003333$ 1/psi

$R_1 = 68.5$ ft

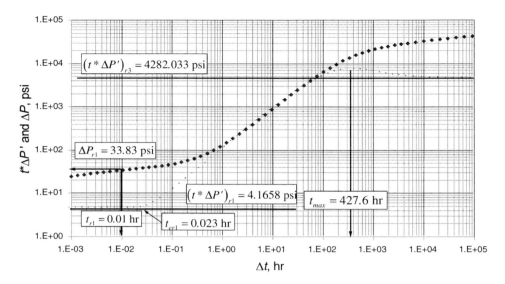

Figure 13.18. Pressure and pressure derivative for example 3.3; adapted from Escobar et al. (2012).

Solution

The log-log plot of pressure and pressure derivative against injection time is given in Figure 13.18. From that plot, the following information can be read:

$t_{r1} = 0.01$ hr $\qquad \Delta P_{r1} = 33.83$ psi $\qquad (t*\Delta P')_{r1} = 4.1658$ psi

$t_{max} = 427.6$ hr $\qquad (t*\Delta P')_{r3} = 4282.033$ psi $\qquad t_{er1} = 0.023$ hr

First, the mobility of the inner region is evaluated with Equation (13.64), and the skin factor is evaluated with Equation (13.65),

$$\left(\frac{k}{\mu}\right)_1 = \lambda_1 = \frac{70.6 q_a B}{h(t*\Delta P')_{r1}} = \frac{70.6(100)(1.5)}{50(4.1658)} = 50.84 \text{ md/cp}$$

$$s = \frac{1}{2}\left(\frac{33.83}{4.1658} - \ln\left(\frac{50.84(0.01)}{4\times10^{-7}(0.5^2)}\right) + 7.43\right) = 0.055$$

Next, the radius of inner region is estimated with Equation (13.66):

$$R_1 = \sqrt{\frac{0.001465 \lambda_1 t_{er1}}{(\phi c_t)_1}} = \sqrt{\frac{0.001465(50.84)(0.023)}{4\times10^{-7}}} = 65.44 \text{ ft}$$

The mobility of the outer region is evaluated with Equation (13.83):

$$\left(\frac{k}{\mu}\right)_3 = \lambda_3 = \frac{70.6 q_a B}{h(t*\Delta P')_{r3}} = \frac{70.6(100)(1.5)}{50(4282.033)} = 0.049 \text{ md/cp}$$

The storativity ratios between regions 1 and 2 and regions 1 and 3, and the mobility ratio between regions 1 and 3, are then calculated:

$$F_{S12} = \frac{(\phi c_t)_1}{(\phi c_t)_2} = \frac{4\times10^{-7}}{8\times10^{-8}} = 5$$

$$F_{S13} = \frac{(\phi c_t)_1}{(\phi c_t)_3} = \frac{4\times10^{-7}}{1.16\times10^{-8}} = 34,483$$

$$M13 = \frac{50.84}{0.049} = 1037.55$$

The distance to the discontinuity of the intermediate region is found with Equation (38):

$$t_{DR1\max} = \frac{0.0002637\lambda_1 t_{\max}}{(\phi c_t)_1 R_1^2} = \frac{0.0002637(50.84)(427.6)}{4\times10^{-7}(65.44^2)} = 3347.89$$

$$\frac{R_2}{R_1} = \frac{a + c\ln(3.227) + e\ln(0.145) + g(\ln(3.227))^2 + i(\ln(0.145))^2 + k\ln(3.227)\ln(0.145)}{1 + b\ln(3.227) + d\ln(0.145) + f(\ln(3.227))^2 + h(\ln(0.145)^2 + j\ln(3.227)\ln(0.145)} = 1.702$$

$R_2 = 111.38$ ft

Nomenclature

c_o	Oil compressibility, psi^{-1}
c_w	Water compressibility, psi^{-1}
c_r	Rock compressibility, psi^{-1}
c_t	Total compressibility, psi^{-1}
c_{to}	Total compressibility at S_{wi}, psi^{-1}
F_s	Storativity ratio
h	Formation thickness, ft
I	Cumulative injection, STB
k	Formation permeability, md
k_{ro}	Relative oil permeability, md
k_{rw}	Relative water permeability, md
M	Mobility ratio
q_a	Flow/injection rate, STB/D
q_{inj}	Injection rate, BPD
R	Discontinuity radius, ft

r	Radius, ft
r_e	Reservoir radius, ft
r_f	Distance to the flooding front, ft
r_T	Distance to the thermal front, ft
S_{wi}	Irreducible water saturation
t	Time, hr
$t*\Delta P'$	Logarithmic pressure derivative, psi
t_{DR}	Dimensionless time based on R
t_D*P_D'	Dimensionless logarithmic pressure derivative
T_{oi}	Initial temperature, °F
T_{wi}	Injection temperature, °F
$\rho_o c_o$	Oil heat capacity, BTU/ft^3/°F
$\rho_R c_R$	Rock heat capacity, BTU/ft^3/°F
$\rho_w c_w$	Water heat capacity, BTU/ft^3/°F

Greek

Δ	Change, drop
ϕ	Porosity, fraction
λ	Mobility, cp^{-1}
$\hat{\lambda}_{oh}$	Oil end-point mobility at irreducible water saturation and initial temperature
μ	Viscosity, cp

Suffixes

1	Inner region
2	Intermediate region for a three-region reservoir
2	Outer region for a two-region reservoir
3	Outer region for a three-region reservoir
D	Dimensionless
e	External
i	Intersection
max	Maximum dimensionless pressure derivative or time
o	Oil
ps	Pseudosteady state
t	Total
w	Water, wellbore
wf	Well flowing

REFERENCES

Ambastha, A. K. & Ramey, H. J. (1989, June 1). "Thermal Recovery Well Test Design and Interpretation." *Society of Petroleum Engineers*. doi:10.2118/16746-PA.

Barua, J. & Horne, R. N. (1987, December 1). "Computerized Analysis of Thermal Recovery Well Test Data." *Society of Petroleum Engineers*. doi:10.2118/12745-PA.

Benson, S. M. & Bodvarsson, G. S. (1986, February 1). "Nonisothermal Effects During Injection and Falloff Tests." *Society of Petroleum Engineers*. doi:10.2118/11137-PA.

Boughrara, A. A., Peres, A. M., Chen, S., Machado, A. A. V. & Reynolds, A. C. (2007a, March 1). "Approximate Analytical Solutions for the Pressure Response at a Water-Injection Well." *Society of Petroleum Engineers*. doi:10.2118/90079-PA.

Boughrara, A. A. & Reynolds, A. C. (2007b, January 1). "Analysis of Injection/Falloff Data From Horizontal Wells." *Society of Petroleum Engineers*. doi:10.2118/109799-MS.

Escobar, F. H., Martinez, J. A. & Montealegre-M., M. (2008). "Pressure and Pressure Derivative Analysis for Injection Tests with Variable Temperature without Type-Curve Matching." *CT&F – Ciencia, Tecnología y Futuro.*, Vol. *4*, No. 4. p. 83-91. Dec.

Escobar, F. H. Martinez, J. A. & Bonilla, L. F. (2011). Pressure and Pressure Derivative Analysis Without Type-Curve Matching For Thermal Recovery Processes.*" CT&F – Ciencia, Tecnología y Futuro.*, Vol. *4*. No. 4. p. 23-35. ISSN 0122-5383. Dec.

Escobar, F. H., Martinez, J. A. & Bonilla, L.F. (2012). "Pressure and Pressure Derivative Analysis for a Three-Region Composite Reservoir." *Journal of Engineering and Applied Sciences.*, Vol. *7*. Nro. 10. Oct. 2012.

Gates, C. F. & Ramey, H. J. (1980, February 1). *"*A Method for Engineering In-Situ Combustion Oil Recovery Projects." *Society of Petroleum Engineers*. doi:10.2118/7149-PA.

Onyekonwu, M. O. (1985). *"Interpretation of In-Situ Combustion Thermal Recovery Falloff Tests."* Ph.D. Dissertation. Stanford University, California.

Platenkamp, R. J. (1985, January 1). "Temperature Distribution Around Water Injectors: Effects on Injection Performance." *Society of Petroleum Engineers*. doi:10.2118/13746-MS.

Satman, A., Eggenschwiler, M. & Ramey, H. J. (1980, January 1). "Interpretation of Injection Well Pressure Transient Data in Thermal Oil Recovery." *Society of Petroleum Engineers*. doi:10.2118/8908-MS.

Soliman, M. Y., Brigham, W. E. & Raghavan, R. (1981, January 1). *"*Numerical Simulation Of Thermal Recovery Processes." *Society of Petroleum Engineers*. doi:10.2118/9942-MS.

Stannislav, J. F. & Kabir, C. S. (1990). *"Pressure Transient Analysis."* Prentice Hall. New Jersey. 320p.

Tiab, D. (1993, January 1). "Analysis of Pressure and Pressure Derivatives Without Type-Curve Matching: I-Skin and Wellbore Storage." *Society of Petroleum Engineers*. This paper was prepared for presentation at the Production Operations Symposium held in Oklahoma City, OK, U.S.A., March 21-23. doi:10.2118/25426-MS.

Walsh, J. W., Ramey, H. J. & Brigham, W. E. (1981, January 1). "Thermal Injection Well Falloff Testing." *Society of Petroleum Engineers*. doi:10.2118/10227-MS.

Chapter 14

BOTTOM AQUIFERS

BACKGROUND

There are three typical cases in which a constant-pressure boundary is combined with some other transient period, causing the development of new flow regimes. Such cases are radial stabilization, linear stabilization and spherical stabilization. The first occurs when a radial flow regime finds a constant-pressure boundary, causing the late pressure derivative to display a straight line with a negative unit-slope. Once all the boundaries have been felt by the transient wave, the pressure derivative will exhibit the classic cascade behavior. The characterization of such a regime is given in Tables 1.3 and 1.4. The second case takes place in an elongated system; when the well is near a lateral pressure-constant boundary, a transient period is expected along the other side of the reservoir. This, combined with the effect of the constant–pressure boundary, leads to the formation of the linear stabilization or parabolic flow regime, according to Escobar et al. (2005a, 2005b). The characterization of this case is given in Equation (1.43). The third case corresponds to a limited-entry well completed near a constant-pressure boundary (top gas gap or bottom aquifer). In that case, a -3/2 slope is seen in the pressure derivative plot; no characterization of this had been presented in the literature before 2015. Thus, the first part of this chapter will deal with the characterization of such a flow regime. Escobar, Ghisays-Ruiz and Staristav (2015) developed a governing equation for this regime using both conventional analysis and the *TDS* technique. Both vertical and horizontal permeabilities can be estimated.

The second part of this chapter is focused on determining the aquifer leakage factor from well tests in coalbed methane reservoirs (CBMs). Either top gas caps or aquifers cause the pressure derivative to abruptly decline because of the constant-pressure boundary, without the development (or masking) of the radial flow regime. Escobar, Staristav and Wu (2015) obtained an expression to find the radial permeability from the application of the second derivative concept. This equation may be also used in conventional reservoirs.

SPHERICAL STABILIZATION

With the advent of newer technologies like wireline formation testing (WFT) and tools for formation pressure testing, it is now possible to gain critical information about a reservoir just after drilling a well. It is now possible in many cases, as shown by Frimann-Dahl et al. (1998), to replace the expensive drill-stem tests, which may take days or months. The new technique involves short formation pressure tests also called Mini-DSTs. Although these tests provide a smaller radius of investigation than normal well tests, they can be nevertheless ne analyzed using the same principles, as illustrated by Daungkaew et al. (2004). In addition, Daungkaew et al. (2004) showed that WFT provides information about localized wellbore phenomena that may be masked by convectional well tests. While several probe configurations exist for conducting the test, almost all WFT relies on inserting a probe in a virgin reservoir and pumping reservoir fluids from the probe to create drawdown, and accurate quartz gauges allow measurement of sand face pressure with time. All WFT pressure data is expected to have spherical flow; Stewart and Wittmann (1979) introduced analytical solutions for the analysis of a spherical flow regime in WFT. If we want to extract maximum information from these tests, it becomes critically important to properly classify spherical flow and its transition to other regimes. Tippie and Abbot (1978) showed that analysis of pressure transient data in a bottom water drive with partial completion displays flow geometry that changes from spherical to hemispherical to linear flow, depending on the distance of perforations or the position of the probe (in the case of WFT with vertical reservoir boundaries). For this type of reservoir setting, existing flow regimes do not adequately describe the flow system.

The radial stabilization flow regime has been taken into account by Escobar, Hernandez and Tiab (2010a), using the intercept formed by the negative unit-slope line and radial flow regime to estimate reservoir area. Much information has been produced on linear stabilization, which was characterized by Escobar et al. (2004), Escobar et al. (2005a) and Escobar et al. (2005b). These studies plotted isobaric lines and found that they took the shape of a parabola. Thus, this phenomenon was named parabolic flow. Previously, Escobar et al. (2004) had called this a pseudo-hemispherical flow regime. Later, Escobar, Hernandez and Hernandez (2007b) estimated the reservoir length, skin factor, well position and reservoir width with the help of that flow regime. An even deeper characterization of skin factors in elongated systems was presented by Escobar and Montealegre (2006). A characterization of such regimes in elongated systems with area anisotropy was conducted by Escobar, Tiab and Tovar (2007a), and a comprehensive study of elongated systems, including the parabolic fluid, was presented by Escobar (2008). Escobar, Hernandez and Saavedra (2010b) included the study of naturally fractured elongated reservoirs. As far as transient rate analysis is concerned, the research of Escobar, Rojas and Bonilla (2012a) and Escobar, Rojas and Cantillo (2012b) included the characterization of the parabolic flow regime.

The only work on the characterization of the spherical stabilization flow regime that takes place in a limited-entry well near a constant-pressure boundary was presented by Escobar et al. (2015b). A governing model for limited-entry wells with a bottom constant-pressure boundary was presented by Ichara (1981), and a conventional analysis was introduced by Tippie and Abbot (1978), although they did not elaborate on the -3/2 slope. Then, Escobar et al. (2015b) presented the recognition of a new flow regime for the case of spherical flow under partial penetration, where the main hydrocarbon column is supported by a bottom water drive or a top gas cap drive. This spherical partial penetration or spherical stabilization flow regime has a pressure derivative characteristic slope of -3/2 (however, since it is near the boundary, hemispherical stabilization would be a better name), which contrasts with the -1/2 slope that is observed during normal spherical flow. This increase in slope can be envisioned as support provided by a constant pressure boundary (gas cap or water drive), and thus, development of this flow regime is dependent upon distance from the perforation to the boundaries and upon permeability anisotropy. These studies also demonstrate the significance of newly found spherical partial penetration flow regime for estimation of permeability anisotropy in a reservoir system. This can provide us with an additional tool for estimation of reservoir permeability anisotropy when well-defined spherical flow is not present in the pressure derivative diagnostic curves. Both the *TDS* technique and conventional analysis were implemented for the characterization of such a regime. These techniques were tested with synthetic examples.

MATHEMATICAL FORMULATION FOR SPHERICAL STABILIZATION

Ichara (1981) presented the solution for pressure behavior of a limited-entry well with a constant-pressure bottom boundary. The dimensionless time, pressure and pressure derivative used by Escobar et al. (2015b) are given as, respectively:

$$t_D = \frac{0.0002637 kt}{\phi \mu c_t r_w^2} \tag{1.2}$$

$$P_D = \frac{kh \Delta P}{141.2 q \mu B} \tag{1.3}$$

$$t_D * P_D' = \frac{kh(t * \Delta P')}{141.2 q \mu B} \tag{1.40}$$

The vertical anisotropy or permeability ratio, I_v, is here defined as:

$$I_v = \frac{k_z}{k} \tag{14.1}$$

Finally, the penetration ratio, b, is defined as:

$$b = \frac{h_p}{h} \tag{14.2}$$

Spherical stabilization takes place when a partially completed well is perforated near a constant-pressure boundary, meaning that either a gas cap or a bottom aquifer is overlying or underlying, respectively, the oil reservoir. This flow regime has a characteristic slope of -3/2 on the pressure derivative versus time log-log plot, as seen in Figure 14.1. This flow regime can be seen if the formation is thick enough to provide its development. As seen in Figure 14.1, for reservoir thicknesses less than 50 ft, the spherical stabilization flow regime is not seen even though the vertical/horizontal permeability ratio is small. Notice that in the case of a smaller permeability ratio, as in last curve in the right, the spherical stabilization is seen for a reservoir thickness of 200 ft. Penetration ratios higher than 40% avoid developing this flow regime. Notice that for low permeability contrast, the radial flow is almost seen, as the effect of the constant-pressure boundary is retarded. The complete steady-state period is fully developed once the transient wave has reached the no-flow pressure boundary; as the reservoir becomes thicker, the maximum point is seen later. This maximum corresponds to the presence of both the no-flow boundary and the penetration ratio.

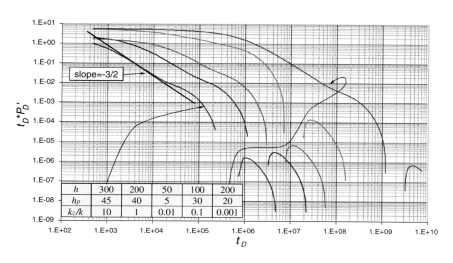

Figure 14.1. Dimensionless pressure derivative versus time log-log plot for several values of reservoir thickness, perforated thickness and vertical/horizontal permeability ratio; adapted from Escobar et al. (2015b).

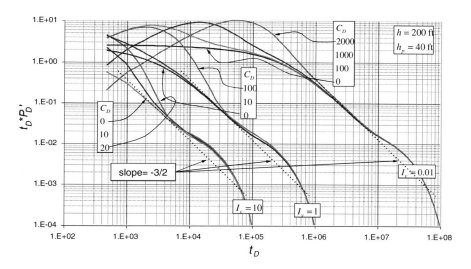

Figure 14.2. Effect of wellbore storage on the spherical stabilization flow regime; adapted from Escobar et al. (2015b).

Figure 14.2 is a dimensionless pressure derivative versus dimensionless time log-log plot for several values of dimensionless wellbore storage and anisotropy ratio. The spherical stabilization flow regime is affected more by wellbore storage for reservoirs with higher anisotropy ratios. For instances in which the anisotropy ratio is 10, meaning that the vertical permeability is 10 times higher than the horizontal permeability, dimensionless wellbore storage greater than 20 will mask the spherical stabilization flow regime. However, for the isotropic case, onset is at dimensionless wellbore storage of 100. Finally, when the permeability ratio is 0.01, dimensionless wellbore storage values up to 2000 allow the spherical stabilization flow to be seen. Generally, for reservoirs containing hydrocarbons due to geological sedimentary deposition, horizontal permeability is greater than vertical permeability; hence, the anisotropy ratio will be less than 1, and we will expect to see a spherical stabilization flow regime even for higher wellbore storage Figure 14.3 presents a unified pressure derivative curve for different values of reservoir thickness, thickness penetration ratio and vertical/horizontal permeability ratio. All the curves fall into a single one if the dimensionless time is multiplied by the permeability ratio and the pressure derivative is multiplied by the penetration ratio and raised to the power -3/2. This allows the following mathematical model by regression analysis to be developed:

$$t_D * P_D' = \frac{\pi}{11}\sqrt{\frac{r_w}{h}}\left(\frac{1}{h_p}\frac{k_z}{k}\frac{r_w^2}{h}t_D\right)^{-3/2} \tag{14.3}$$

Integrating the above expression leads to:

$$P_D = -\frac{2\pi}{33}\sqrt{\frac{r_w}{h}}\left(\frac{1}{h_p}\frac{k_z}{k}\frac{r_w^2}{h}t_D\right)^{-3/2} + s_{sps} \tag{14.4}$$

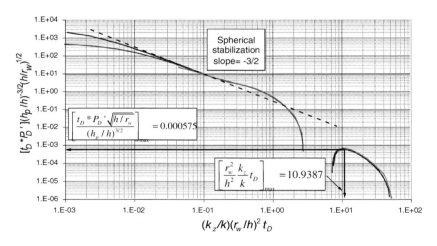

Figure 14.3. Unified behavior of the dimensionless pressure derivative versus time log-log plot; adapted from Escobar et al. (2015b).

The spherical stabilization skin factor, s_{sps}, is assumed to be a combination of the mechanical skin factor and the vertical flow.

TDS TECHNIQUE FOR SPHERICAL STABILIZATION

Escobar et al. (2015b) extended the *TDS* technique of Tiab (1993) based upon characteristic features and points in the pressure and pressure derivative log-log plot. Placing the dimensionless parameters given by Equations (1.2) and (1.40) into Equation (14.3) and solving for the radial permeability results in:

$$k = \frac{9417306.154 q \mu^{5/2} B \sqrt{r_w}}{(t*\Delta P')_{sps}} \left(\frac{h_p \phi c_t}{k_z t_{sps}} \right)^{3/2}$$

(14.5)

The pressure derivative $(t*\Delta P')_{sps}$ is read at any convenient time, t_{sps}, during the spherical stabilization flow regime. Equation (14.5) assumes that the value of the vertical permeability is known; this may be obtained from a repeat WFT. When the reservoir anisotropy, as defined here, is very low (see Figure 14.1), it is possible to see the radial flow regime. In such cases, a horizontal line is drawn along the radial flow regime, and the pressure derivative corresponding to such a line is read; horizontal permeability can be obtained from the following expression developed by Tiab (1993):

$$k = \frac{70.6 q \mu B}{h(t*\Delta P')_r}$$

(1.18)

If this is the case, then it is better to solve for the vertical permeability from Equation (14.5):

$$k_z = 44594.83 \left(\frac{q \mu^{5/3} B}{k(t*\Delta P')_{sps}} \right)^{2/3} \left(\frac{h_p \phi c_t}{t_{sps}} \right)$$

(14.6)

The spherical stabilization skin factor is obtained by dividing Equation (14.4) by Equation (14.3):

$$s = 66694.8 h \sqrt{r_w} \left(\frac{h_p \phi \mu c_t}{k_z t_{sps}} \right) \left(\frac{\Delta P_{sps}}{(t*\Delta P')_{sps}} + \frac{2}{3} \right)$$

(14.7)

Once horizontal permeability is calculated from Equation (14.5), the radial pressure derivative can be estimated from Equation (1.18):

$$(t * \Delta P')_r = \frac{70.6 q \mu B}{hk} \tag{14.8}$$

This value may be drawn as a horizontal line on the pressure derivative plot. This horizontal line takes the value of 0.5 in dimensionless form. Therefore, the intersection of the horizontal line with the spherical stabilization line allows one to obtain the vertical permeability. The below equation is derived by setting the left-hand side of Equation (14.3) equal to 0.5 and solving for k_z:

$$k_z = \frac{2610.65 h_p \phi \mu c_t r_w^{1/3} h^{2/3}}{t_{spsri}} \tag{14.9}$$

The normal case is that both vertical and horizontal permeabilities are unknown. In such cases, the *TDS* technique shows its power, capability and practicality. As observed in Figure 14.3, the coordinates of the maximum point are:

$$\left[\frac{t_D * P_D' \sqrt{h/r_w}}{(h_p/h)^{3/2}} \right]_{max} = 0.000575 \tag{14.10}$$

$$\left[\frac{r_w^2}{h^2} \frac{k_z}{k} t_D \right]_{max} = 10.9387 \tag{14.11}$$

When the dimensionless pressure derivative given by Equation (1.40) is placed into Equation (14.10), and the dimensionless time given by Equation (1.2) is plugged into Equation (14.11), it is possible to obtain expressions for estimating both horizontal and vertical permeabilities, respectively:

$$k = \frac{q \mu B h_p^{3/2} r_w^{1/2}}{12.3168 h^3 (t * \Delta P')_{max}} \tag{14.12}$$

$$k_z = \frac{41481.608 \phi \mu c_t h^2}{t_{max}} \tag{14.13}$$

Notice that the second derivative is not used since the pressure derivative tendency is negative, which makes the second derivative negative; this is impossible to plot on a log-log scale.

CONVENTIONAL ANALYSIS FOR SPHERICAL STABILIZATION

Replacing the dimensionless quantities in Equation (14.4) yields:

$$\Delta P = \frac{6278204.1 q \mu^{5/2} B \sqrt{r_w}}{k} \left(\frac{h_p \phi c_t}{k_z} \right)^{3/2} t_{sps}^{-3/2} + \frac{141.2 q \mu B}{kh} s_{sps} \qquad (14.14)$$

Equation (14.14) suggests that a Cartesian plot of P_{wf} versus $t^{-3/2}$ for drawdown or of P_{wf} versus $(t_p + \Delta t)^{-3/2} + \Delta t^{-3/2}$ will provide a straight line with slope m_{sps} and intercept, b_{sps}, allowing one to obtain horizontal permeability (if k_z is known) and spherical stabilization skin:

$$k = \frac{6278204.1 q \mu^{5/2} B \sqrt{r_w}}{|m_{sps}|} \left(\frac{h_p \phi c_t}{k_z} \right)^{3/2} \qquad (14.15)$$

$$s_{sps} = \frac{khb_{sps}}{141.2 q \mu B} \qquad (14.16)$$

As observed here, conventional analysis has a strong limitation if the vertical is unknown. In such cases, the best result from conventional analysis is to find the product of k and $k_z^{3/2}$:

$$kk_z^{3/2} = \frac{6278204.1 q \mu^{5/2} B \sqrt{r_w} (h_p \phi c_t)^{3/2}}{|m_{sps}|} \qquad (14.17)$$

Example 14.1

Escobar et al. (2015b) presented synthetic pressure test data generated with the information given below. Pressure drop and pressure derivative data are given in Figures 14.4 and 14.5 and in Table 14.1, respectively.

$B = 1.0$ bbl/STB	$q = 200$ STB/D	$h = 200$ ft
$\mu = 1$ cp	$r_w = 0.3$ ft	$c_t = 3 \times 10^{-6}$ psi^{-1}
$P_i = 5000$ psi	$\phi = 10\%$	$k = 50$ md
$h_p = 40$ ft	$C_D = 0$	$s = 0$
$k_z = 5$ md		

One can estimate horizontal and vertical permeabilities using *TDS* and conventional techniques.

Solution by TDS technique

The following information can be read from Figure 14.4:

Bottom Aquifers

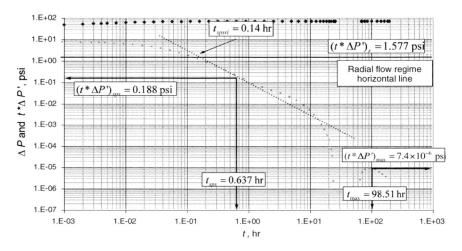

Figure 14.4. Pressure and pressure derivative versus time log-log plot for example 14.1.

Table 14.1. Pressure and pressure derivative versus time data for example 14.1

t, hr	ΔP, psi	$t*\Delta P'$, psi	t, hr	ΔP, psi	$t*\Delta P'$, psi	t, hr	ΔP, psi	$t*\Delta P'$, psi
0.001	48.03	6.42	0.2261	70.50	7.09E-01	16.51	71.05	8.99E-04
0.002	52.38	6.14	0.3193	70.70	4.66E-01	18.51	71.05	5.06E-04
0.003	54.84	5.95	0.4511	70.83	2.99E-01	20.51	71.05	2.71E-04
0.004	56.53	5.79	0.6372	70.91	1.87E-01	22.01	71.05	1.61E-04
0.005	57.80	5.65	0.9	70.96	1.16E-01	24.01	71.05	6.02E-05
0.007	59.66	5.39	1.271	70.99	7.34E-02	25.01	71.05	2.82E-05
0.0101	61.58	5.06	1.796	71.01	4.92E-02	26.01	71.05	1.89E-06
0.0143	63.27	4.69	2.537	71.03	3.53E-02	68.51	71.05	9.09E-07
0.0201	64.82	4.25	3.583	71.04	2.56E-02	70.51	71.05	2.18E-06
0.0285	66.20	3.74	5.011	71.04	1.75E-02	76.51	71.05	4.74E-06
0.0402	67.39	3.17	6.511	71.05	1.19E-02	84.51	71.05	6.59E-06
0.0568	68.38	2.57	8.511	71.05	7.07E-03	101.01	71.05	7.49E-06
0.0802	69.17	1.99	10.51	71.05	4.26E-03	123.51	71.05	6.36E-06
0.1133	69.77	1.47	12.51	71.05	2.55E-03	143.01	71.05	5.06E-06
0.1600	70.20	1.04	14.51	71.05	1.52E-03	161.51	71.05	4.01E-06
						182.51	71.05	3.02E-06

t_{max} = 98.51 hr $\qquad$ $(t*\Delta P')_{max}$ = 7.4x10^{-6} psi
t_{sps} = 0.637 hr $\qquad$ $(t*\Delta P')_{sps}$ = 0.188 psi

The horizontal permeability is found with Equation (14.12), using the maximum point pressure derivative:

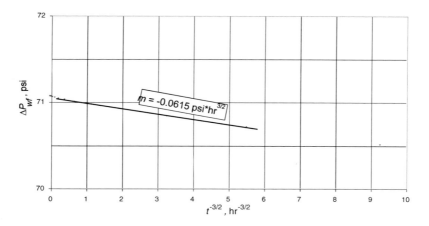

Figure 14.5. Pressure versus time to the power -3/2 for example 14.1.

$$k = \frac{200(1)(1)(40)^{3/2}\sqrt{0.3}}{12.3168(200)^3(7.4\times10^{-6})} = 38 \text{ md}$$

The vertical permeability is estimated with Equation (14.13), using the time at which the maximum derivative occurs:

$$k_z = \frac{41481.608(0.1)(1)(3\times10^{-6})(200)^2}{98.51} = 5.05 \text{ md}$$

Any arbitrary point on the spherical stabilization straight line is used to find permeability with Equation (14.5):

$$k = \frac{9417306.154(200)(1)^{5/2}(1)\sqrt{0.3}}{0.188}\left(\frac{(40)(0.1)(3\times10^{-6})}{(5.08)(0.637)}\right)^{3/2} = 39.5 \text{ md}$$

Once permeability is known, Equation (14.8) is used to find the value of the pressure derivative during the radial flow regime, masked by the effect of a constant-pressure boundary:

$$(t*\Delta P')_r = \frac{70.6(200)(1)(1)}{(200)(39.5)} = 1.7874 \text{ psi}$$

A horizontal line is drawn through 1.7874 psi (see Figure 14.4), and the intersection of that line with the spherical stabilization straight line is extrapolated if needed:

$$t_{spsrsi} = 0.14 \text{ hr}$$

The vertical permeability is again estimated using Equation (14.13):

$$k_z = \frac{5.0627\pi^{15}(40)(0.1)(3\times10^{-6})(0.3)^{1/3}(200)^{2/3}}{0.14} = 5.12 \text{ psi}$$

Solution by Conventional Analysis

A slope value of -0.0615 psi*hr$^{3/2}$ can be read from Figure 14.5. Assuming vertical permeability is known, horizontal permeability can be estimated by means of Equation (14.17):

$$k = \frac{6278204.1(200)(1)^{5/2}(1)\sqrt{0.3}}{0.0615}\left(\frac{(40)(0.1)(3\times10^{-6})}{5}\right)^{3/2} = 40.94 \text{ md}$$

DETERMINING AQUIFER LEAKAGE FACTOR IN CBM RESERVOIRS

Water influx is an important factor that needs to be quantified during early stages of reservoir development to justify project economics. For a CBM reservoir, it is much more important to quantify degree of connection between coal seams and aquifers (if any), as the reservoir's production mechanism is based on an efficient dewatering process. However, it is difficult to quantify connection factor values early in field life. Many traditional models, ranging from the simple steady-state model of Schilthuis and Fetkovich (year), which utilizes material balance, to the unsteady-state solution of the diffusivity equation crated by Van Everdingen and Hurst (year), exist for finding water influx. Nearly all of them, however, have inherited assumptions related to influx rate or aquifer/reservoir boundary pressure, and they require accurate historical production data, which is often unavailable, to estimate the correct influx. However, during the appraisal/exploratory stage, we do have accurate measurements of wellbore pressure and production/injection rate when we conduct transient-pressure testing. Well test analysis plays an important role in reservoir characterization and can aid in correct water influx calculations. Currently, pressure falloff test responses to quantify water influx in the reservoir with wells that exist near constant pressure boundary can be analyzed either by type-curve matching or by non-linear regression analysis. The former is basically a trial-and-error procedure, and the latter can lead to incorrect or impractical results. Thus, we need a more robust and accurate method for water influx calculation. Escobar et al. (2015a) presented a practical method to interpret the injection-falloff test response for a CBM reservoir in connection with an aquifer to quantify the connection factor between the aquifer and the reservoir using transient-pressure tests, which are usually conducted during the field appraisal/exploration phase. In addition to complementing the conventional straight-line method for determining the leakage factor, Escobar et al. (2015a) also provided a solution using characteristic points found on the pressure, pressure derivative and second pressure derivative log-log plot of a "leaky aquifer" reservoir model (Cox and Onsager, 2002) which allowed relationships to be developed for accurately estimating the leakage factor. An extremely useful application of the second pressure derivative was also included to estimate the unknown reservoir permeability for cases in which the radial flow regime is completely masked by other flow regimes.

Unconventional gas reservoirs have become an integral part of the energy supply basket due to an ever-increasing demand for oil and gas and a decrease in new, conventional discoveries. In the future, these unconventional reservoirs are expected to become more significant, as the era of conventional/easy oil and gas comes to an end. Thus, it is vital to devise new methods for characterizing these complex reservoirs, as accurate knowledge about their production mechanisms and performance over time does not yet exist. Coalbed methane (CBM) has developed into an important part of these unconventional resources. At the time they are discovered, nearly all hydrocarbon reservoirs are surrounded by porous rock containing water, according to Schafer, Hower and Ownes (1993). CBM reservoirs mostly contain natural fractures called cleats, which are in most cases filled with water. As CBM reservoirs operate on the principle of desorption of gas from the coal seam surface due to depressurization, most CBM reservoirs require efficient dewatering before they can produce a commercial volume of gas (Onsager and Cox, 2000). In some instances, water influx from other aquifer units can inhibit the dewatering of the coal and thereby limit CBM recovery (Onsager and Cox, 2000). It becomes extremely important, then, to come up with new methods to characterize the degree of connection between these coalbeds and other units early in field life. The groundwater industry has long recognized that many aquifers have imperfect seals and are in connection with other aquifer units through a low-permeability confining layer (Cox and Onsage, 2002). Shallow reservoirs like small CBM aquifers can be in communication with other aquifer units through low-permeability ("leaky") confining layers, which implies that the aquifer's connection with the reservoir is not perfect ($\Delta P_{res.} \neq \Delta P_{aquifer}$).

Hantush and Jacob (1955) published a methodology known in the groundwater industry as the "leaky aquifer model." The idealized Hantush-Jacob leaky aquifer model assumes a constant pressure boundary at the top or bottom of the confining layer. Such a model more accurately describes bottom/top water-drive CBM reservoirs in which coal is only a producing layer. This case can also be modeled with methods given by Neuman and Witherspoon (1972) and by Guo, Stewart and Toro (2002), which are essentially the same as that of Hantush and Jacob (1955). Hantush and Jacob (1955) developed a solution for the non-steady distribution of drawdown caused by pumping a well at a constant rate from an effectively infinite and perfectly elastic aquifer of uniform thickness, in which leakage takes place in proportion to the drawdown. Hantush and Jacob (1955) came up with the parameter *b*, or "leakage factor," which is a function of hydraulic conductivity and the thickness of the confining bed through which leakage occurs. The leakage coefficient is defined as "the quantity of flow that crosses a unit area of the interface between the main aquifer and its semi- confining bed, if the difference between the head in the main aquifer and in that supplying leakage is unity" (Hantush, 1956).

MATHEMATICAL MODELING

Onsager and Cox (2002) provided the solution for the diffusivity equation with wellbore storage and skin in Laplace space for a reservoir with an underlying leaky aquifer:

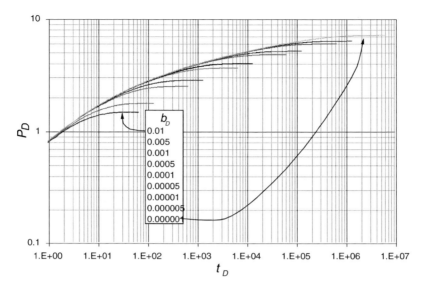

Figure 14.6. Dimensionless pressure versus time log-log behavior for several values of dimensionless leakage factors; adapted from Escobar et al. (2015a).

$$\overline{P_D} = \frac{1}{\ell} \frac{K_o(\sqrt{u}) + s\sqrt{u}K_1(\sqrt{u})}{\sqrt{u}\,K_1(\sqrt{u}) + \ell^2 C_D\left[K_o(\sqrt{u}) + s\sqrt{u}K_1(\sqrt{u})\right]} \quad (14.18)$$

in which:

$$u = \ell + b_D \quad (14.19)$$

$$b = \frac{k_{v,conf}}{h_{conf}} \quad (14.20)$$

$$b_D = \frac{k_{v,conf}\, r_w^2}{kh\, h_{conf}} \quad (14.21)$$

The dimensionless quantities are defined by Equations (1.2), (1.3) and (1.40). The dimensionless second pressure derivative is given by:

$$t_D^2 * P_D{''} = \frac{kh(t^2 * \Delta P'')}{141.2 q\mu B} \quad (14.22)$$

CONVENTIONAL ANALYSIS

Figure 14.6 presents the dimensionless pressure behavior obtained from Equation (1). In this log-log plot, the steady state is reached at a different time depending on the leakage factor value. The pressure at which the steady state takes place is strongly dependent on the leakage factor, as clearly established below:

$$b_D = 1.5338e^{-0.014413\left(\frac{kh\Delta P_{ss}}{q\mu B}\right)} \tag{14.23}$$

Therefore, from a log-log plot of dimensionless pressure drop versus dimensionless time, it is easy to observe that the pressure drop is constant once the steady state is fully developed. However, any conventional plot—for example, semilog or Cartesian—can be used to find the steady-state pressure by drawing a horizontal line on the late steady-state period and finding the intercept on the y-axis. In any case, this is observed by a flat behavior of either pressure or pressure drop. This value is read and placed into Equation (14.23) to easily obtain the leakage factor. It is worth reminding that, unlike the pressure derivative, pressure drop is sensitive to the skin effect; thus, ΔP_{ss} in Equation (14.23) has to be free of skin effects. This means that, for either drawdown (injection) or buildup (falloff):

$$\Delta P_{ss} = P_i - P_{wf} - \Delta P_s \tag{14.24}$$

$$\Delta P_{ss} = P_{ws} - P_{wf} - \Delta P_s \tag{14.25}$$

Additionally, the equivalent time proposed by Agarwal (1980) is recommended for use in buildup tests.

TDS TECHNIQUE

For the case dealt with in the work of Escobar et al. (2015a), refer to Figure 14.7 to observe the dimensionless pressure and pressure derivative behavior. During the radial flow regime, the pressure derivative is governed by a zero-slope straight line with an intercept of 0.5. Tiab (1995) demonstrated that after Equation (1.40) is equalized to 0.5, an expression to find permeability is given by Equation (1.18).

Tiab (1993) also provided an expression to find the skin factor by reading the pressure drop at any arbitrary time during radial flow:

$$s = 0.5\left(\frac{\Delta P_r}{(t*\Delta P')_r} - \ln\left[\frac{k\,t_r}{\phi\mu c_t r_w^2}\right] + 7.43\right) \tag{1.19}$$

However, Escobar et al. (2015a) found that the pressure derivative does not help much, as it has a decaying behavior. In addition to the pressure and pressure derivative, Figure 14.8 also includes the second pressure derivative behavior, in which two characteristic features are clearly seen: (1) a point of intersection between the derivative and second derivative and (2) a maximum point displayed by the pressure derivative, which is given in Figure 14.9 for other cases studies.

Several observations can be drawn from Figure 14.9. The first is that the maximum points are functions of time and that the second derivative is practically constant at a dimensionless value of 0.186:

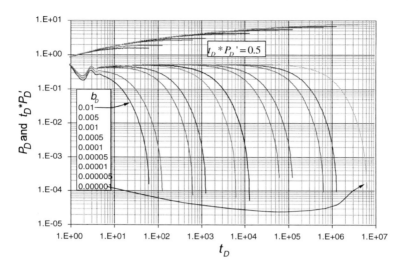

Figure 14.7. Dimensionless pressure and pressure derivative versus time log-log behavior for several values of dimensionless leakage factors; adapted from Escobar et al. (2015a).

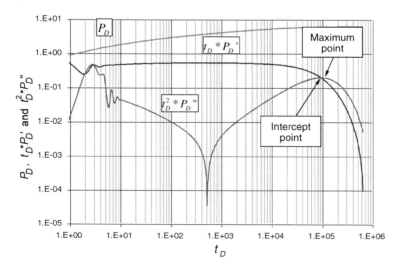

Figure 14.8. Dimensionless pressure, pressure derivative and second pressure derivative versus time log-log plot for $b_D = 0.00001$; adapted from Escobar et al. (2015a).

$$t_D^2 * P_D{}'' = \frac{kh(t^2 * \Delta P'')}{141.2q\mu B} \approx 0.186 \tag{14.26}$$

Solving for permeability:

$$k \approx \frac{26.2632 q\mu B}{h(t^2 * \Delta P'')_{2\max}} \tag{14.26}$$

Equation (14.26) may be practical whenever the radial flow regime is obscured by wellbore storage.

Another observation from Figure 14.9 is that the time of the maximum second pressure derivative, t_{2max}, correlates perfectly with the leakage factor, as given below:

$$b_D = 10^{-1.005\log\left(\frac{0.0002637 k t_{2\max}}{\phi\mu c_t r_w^2}\right) + 0.0185} \tag{14.27}$$

As the third observation on the same plot, the maximum point is difficult to see (probably due to noise), and the intercept between the two derivatives is seen. A perfect correlation is provided below:

$$b_D = 10^{-1.0035\log\left(\frac{0.0002637 k t_{\mathrm{int}2p}}{\phi\mu c_t r_w^2}\right) + 0.0394} \tag{14.28}$$

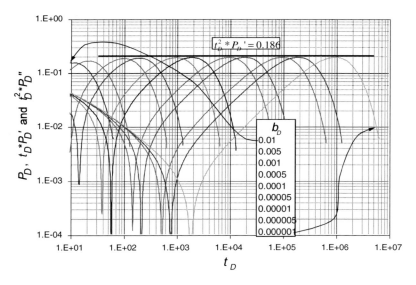

Figure 14.9. Dimensionless pressure, pressure derivative and second pressure derivative versus time log-log plot for several values of dimensionless leakage factors; adapted from Escobar et al. (2015a).

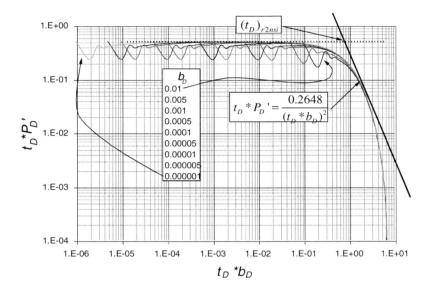

Figure 14.10. Log-log plot of dimensionless pressure derivative versus dimensionless time multiplied by dimensionless leakage factor; adapted from Escobar et al. (2015a).

The fourth observation in Figure 14.9 reveals leakage factors dependence on both time and the second derivative, although the latter is a weak dependence. Equation (14.28) is a correlation of the leakage factor as a function of the pressure derivative and time read at the maximum point. However, for easier manipulation, the second pressure derivative is divided by the pressure derivative during the radial flow regime. If this is unclear, and permeability is known, the pressure derivative can be solved from Equation (1.18).

$$Z = 0.0491070542313 - 2.2954373 F1 - 6.6837964356 \times 10^{-03} F2 \quad (14.29)$$

where:

$$F1 = 1/\log\left(\frac{0.0002637 k t_{2\max}}{\phi \mu c_t r_w^2}\right) \quad (14.30)$$

$$F2 = \left(\frac{(t*\Delta P')_r}{(t^2*\Delta P'')_{2\max}}\right)^2 \quad (14.31)$$

$$b_D = 10^{1/Z} \quad (14.32)$$

A final observation comes from Figure 14.10, which is a log-log plot of dimensionless pressure derivative versus the product of dimensionless time multiplied by the dimensionless leakage factor. As seen in the plot, the late steady-state period unifies into a single line. Thus, drawing a straight line of a slope of -2 will yield the following fitting:

$$t_D * P_D' = \frac{0.2648}{(t_D * b_D)^2} \tag{14.33}$$

The intercept of Equation (14.33) with the radial flow regime pressure derivative $(t_D*P_D'=0.5)$ provides the following expression:

$$b_D = \frac{2759.713\phi\mu c_t r_w^2}{kt_{r2nsi}} \tag{14.34}$$

where t_{r2nsi} the point of intersect between the radial flow and the lines with slope -2.

Example 14.2

A simulated test was performed using the Onsager and Cox (2002) model with the below information:

$B = 1.005$ bbl/STB	$q = 30$ STB/D	$h = 200$ ft
$\mu = 1$ cp	$r_w = 0.4$ ft	$c_t = 1 \times 10^{-4}$ psi^{-1}
$P_i = 3500$ psi	$\phi = 30\%$	$k = 10$ md
$b_D = 0.00035$	$C_D = 10$	$s = 1$

Pressure, pressure derivative and second pressure derivative versus time are provided in Figure 14.11. Characterizing this test with the sole purpose of estimating the leakage factor is required.

Solution by conventional analysis

Actually, this example cannot be solved by conventional analysis, as both wellbore storage and aquifer effects mask the radial flow regime. Therefore, neither skin factor nor permeability can be determined. However, *assuming* both permeability and skin factor are known, based on the log-log plot of ΔP vs. t (Figure 14.11), the value at which the horizontal line is drawn during the late steady-state period intercepts the y-axis at:

$$\Delta P_{ss} + s = 10.853 \text{ psi}$$

There is a need to find the semilog slope from the classic permeability equation in order to find the pressure drop due to skin factor:

$$k = \frac{162.6q\mu B}{hm} \tag{1.10}$$

Solving for m results in:

Bottom Aquifers
369

$$m = \frac{162.6q\mu B}{hk} = \frac{162.6(30)(1)(1.005)}{(200)(10)} = -2.4512 \text{ psi/cycle}$$

This leads to:

$$\Delta P_s = \pm 0.87ms = \pm 2s(t*\Delta P')_r \qquad (14.35)$$

The pressure drop due to a skin factor of 1 is 2.133. Thus, $\Delta P_{ss}= 8.72$ psi, which, when used in Equation (14.23), will provide:

$$b_D = 1.5338e^{-0.014413\left(\frac{(10)(200)(8.72)}{(30)(1)(1.005)}\right)} = 0.000375$$

Solution by TDS Technique
The following information can be read from Figure 14.11:

$t_{r2nsi} = 4$ hr $\qquad t_{int2p} = 4.4$ hr

$t_{2max} = 4.84$ hr $\qquad (t^2*\Delta P'')_{2max} = 0.4342$ psi

This is a very common case in which the radial flow is obscured by wellbore storage. In addition, the steady-state period also contributes to not having a well-defined radial flow regime line. The second derivative at the maximum point can also be used to estimate permeability from Equation (14.26):

$$k \approx \frac{26.2632(30)(1)(1.005)}{(200)(0.4342)} = 9.2 \approx 10 \text{ md}$$

We need the value of the pressure derivative during the radial flow regime—which can be obtained from Equation (14.35)—to estimate from Equation (1.18):

$$(t*\Delta P')_r = \frac{70.6q\mu B}{hk} = \frac{70.6(30)(1)(1.005)}{(200)(10)} = 1.0643 \text{ psi}$$

The leakage factor can be found from Equations (14.27) and (14.28):

$$b_D = 10^{-1.005\log\left(\frac{0.0002637(10)(4.85)}{(0.3)(1)(0.0001)(0.4^2)}\right)+0.0185} = 0.000372$$

$$b_D = 10^{-1.0035\log\left(\frac{0.0002637(10)(4.4)}{(0.3)(1)(0.0001)(0.4^2)}\right)+0.0394} = 0.00044$$

Finally, using the coordinates of the maximum point of the second derivative in Equations (14.29) through (14.32), the leakage factor is also obtained:

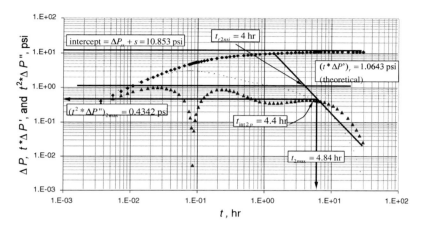

Figure 14.11. Log-log plot of pressure, pressure derivative and second pressure derivative versus time for example 14.2.

Table 14.2. Pressure, pressure derivative and second pressure derivative versus time data for example 14.2

t, hr	ΔP, psi	$t*\Delta P'$, psi	$t^2*\Delta P,''$ psi	t, hr	ΔP, psi	$t*\Delta P'$, psi	$t^2*\Delta P,''$ psi
0.00364	0.40	0.38	0.34	0.39863	8.19	1.52	0.70
0.00546	0.59	0.55	0.48	0.48601	8.48	1.40	0.60
0.00728	0.77	0.70	0.58	0.55882	8.67	1.32	0.53
0.00910	0.94	0.84	0.68	0.63163	8.83	1.26	0.48
0.01092	1.10	0.97	0.75	0.73356	9.01	1.19	0.43
0.01456	1.41	1.20	0.87	0.85006	9.18	1.13	0.40
0.01820	1.70	1.41	0.94	1.025	9.38	1.05	0.37
0.02184	1.98	1.58	0.98	1.200	9.55	1.00	0.36
0.02548	2.23	1.73	0.99	1.374	9.68	0.95	0.35
0.02912	2.47	1.86	0.98	1.578	9.81	0.90	0.36
0.03458	2.81	2.03	0.93	1.811	9.93	0.85	0.36
0.04005	3.11	2.16	0.86	2.161	10.07	0.78	0.38
0.04551	3.40	2.26	0.77	2.510	10.18	0.73	0.39
0.05279	3.74	2.37	0.63	2.860	10.27	0.67	0.41
0.06371	4.20	2.47	0.41	3.267	10.36	0.62	0.42
0.07463	4.59	2.52	0.20	3.675	10.43	0.57	0.43
0.07827	4.71	2.52	0.13	4.258	10.51	0.51	0.43
0.08191	4.83	2.53	0.07	4.840	10.57	0.45	0.43
0.08555	4.94	2.53	0.01	5.423	10.62	0.40	0.43
0.08919	5.04	2.53	0.06	5.889	10.65	0.37	0.42
0.09283	5.14	2.53	0.11	6.354	10.68	0.33	0.42
0.09647	5.24	2.52	0.17	6.937	10.71	0.30	0.40
0.10375	5.42	2.50	0.27	7.869	10.74	0.25	0.38
0.11468	5.67	2.47	0.40	9.034	10.77	0.20	0.34

t, hr	ΔP, psi	$t*\Delta P'$, psi	$t^2*\Delta P,$" psi	t, hr	ΔP, psi	$t*\Delta P'$, psi	$t^2*\Delta P,$" psi
0.12560	5.89	2.43	0.52	10.665	10.80	0.15	0.30
0.14016	6.16	2.37	0.63	12.296	10.82	0.11	0.25
0.16200	6.49	2.26	0.76	14.393	10.83	0.07	0.20
0.19477	6.90	2.11	0.85	17.655	10.84	0.04	0.13
0.23117	7.25	1.97	0.88	20.916	10.85	0.02	0.09
0.26758	7.52	1.84	0.86	23.712	10.85	0.01	0.06
0.30398	7.75	1.73	0.82	26.508	10.85	0.01	0.04
0.34767	7.98	1.62	0.76	29.770	10.85	0.00	0.03

$$F1 = 1/\log\left(\frac{0.0002637(0.1)(1298.48)}{(0.3)(1)(0.0001)(0.4^2)}\right) = 0.127$$

$$F2 = \left(\frac{1.064}{0.432}\right)^2 = 6.25$$

$$Z = 0.0491070542313 - 2.2954373(0.127) - 6.6837964356 \times 10^{-03}(6.25) = -0.284$$

$$b_D = 10^{-1/0.2822} = 0.0003$$

The point of intersection between the radial flow regime and the lines with slope -2 is also used to find the leakage factor from Equation (14.34):

$$b_D = \frac{2759.713(0.3)(1)(0.0001)(0.4^2)}{(10)(4)} = 0.000331$$

Nomenclature

B	Volume factor, rb/STB
b	Partial penetration ratio $= h_p/h$
b	Leakage factor, ft
C	Wellbore storage coefficient, bbl/psi
c_t	Total system compressibility, psi^{-1}
h	Reservoir thickness, ft
h_p	Perforated interval, ft
I_v	Vertical anisotropy (vertical to horizontal permeability ratio)
k	Reservoir horizontal permeability, md
k_z	Reservoir vertical permeability, md
ℓ	Laplace parameter
m_{sps}	Slope of the P vs. $t^{-3/2}$ plot
P	Pressure, psi

P_i	Initial reservoir pressure, psi
P_{wf}	Wellbore flowing pressure, psi
q	Water flow rate, BPD
r_w	Wellbore radius, ft
s	Skin factor
t	Time, days
t_D	Dimensionless time coordinate
$t_D{}^*P_D{}'$	Dimensionless pressure derivative
$t_D{}^2{}^*P_D{}''$	Dimensionless second pressure derivative
$(t{}^*\Delta P')$	Pressure derivative
$(t^2{}^*\Delta P'')$	Second pressure derivative

Greek

ϕ	Porosity, fraction
μ	Viscosity, cp

Suffixes

2max	Maximum of the second pressure derivative
conf	Confining layer
D	Dimensionless
i	Initial
int2p	Intercept of pressure derivative and second pressure derivative
max	Maximum before steady-state regime develops
r	Radial
r2nsi	Intersection of the radial flow line and the line with slope -2
ss	Steady state
sps	Spherical stabilization
spsri	Intersection of the spherical stabilization and radial straight lines
ss	Steady state
v, conf	Vertical in confining layer
wf	Well flowing
ws	Well static

REFERENCES

Cox, D. O. & Onsager, P. R. (2002, January 1). "Application of Leaky Aquifer Type Curves for Coalbed Methane Characterization." *Society of Petroleum Engineers.* doi:10.2118/77333-MS.

Daungkaew, S., Prosser, D. J., Manescu, A. & Morales, M. (2004, January 1). "An Illustration of the Information that can be obtained from Pressure Transient Analysis of

Wireline Formation Test Data." *Society of Petroleum Engineers.* Doi: 10.2118/88560-MS.

Escobar, F. H., Saavedra, N. F., Hernández, C. M., Hernández, Y. A., Pilataxi, J. F. & Pinto, D. A. (2004). *"Pressure and Pressure Derivative Analysis for Linear Homogeneous Reservoirs without Using Type-Curve Matching."* Paper SPE 88874, Proceedings, 28th Annual SPE International Technical Conference and Exhibition to be held in Abuja, Nigeria, Aug. 2-4, 2004.

Escobar, F. H., Muñoz, O. F., Sepulveda, J. A. Montealegre-M, M. & Hernandez, Y. A. (2005a). "Parabolic Flow: A New Flow Regime Found In Long, Narrow Reservoirs." *XI Congreso Colombiano del Petróleo (Colombian Petroleum Symposium).* ISBN, 958-33-8394-5. Oct. 18-21, 2005.

Escobar, F. H., Muñoz, O. F., Sepulveda, J. A. & Montealegre, M. (2005b). "New Finding on Pressure Response In Long, Narrow Reservoirs." *CT&F – Ciencia, Tecnología y Futuro.*, Vol. *2*, No. 6. P. 151-160. ISSN 0122-5383. Dec.

Escobar, F. H. & Montealegre-M., M. (2006). "Effect of Well Stimulation on the Skin Factor in Elongated Reservoirs." *CT&F – Ciencia, Tecnología y Futuro.*, Vol. *3*, No. 2. p. 109-119. ISSN 0122-5383. Dec.

Escobar, F. H., Tiab, D. & Tovar, L.V. (2007a). "Determination of Areal Anisotropy from a single vertical Pressure Test and Geological Data in Elongated Reservoirs." *Journal of Engineering and Applied Sciences.*, *2*(11). ISSN 1816-949X. p. 1627-1639.

Escobar, F. H., Hernández, Y. A. & Hernández, C. M. (2007b). "Pressure Transient Analysis for Long Homogeneous Reservoirs using TDS Technique." *Journal of Petroleum Science and Engineering. ISSN* 0920-4105. Vol. *58*, Issue 1-2, pages 68-82.

Escobar, F. H. (2008). Book "Petroleum Science Research Progress." Nova Publisher. ISBN 978-1-60456-012-1. 2008. Chapter *"Recent Advances in Well Test Analysis for Long and Narrow Reservoirs."*

Escobar, F. H., Hernandez, Y. A. & Tiab, D. (2010a). "Determination of reservoir drainage area for constant-pressure systems using well test data." *CT&F – Ciencia, Tecnología y Futuro.*, Vol. *4*, No. 1. p. 51-72. ISSN 0122-5383. June.

Escobar, F. H., Hernandez, D. P. & Saavedra, J. A. (2010b). "Pressure and Pressure Derivative Analysis For Long Naturally Fractured Reservoirs Using The TDS Technique." *Dyna*, Year 77, Nro. *163*, p. 102-114. ISSN 0012-7353. Sept.

Escobar, F. H., Rojas, M. M. & Bonilla, L. F. (2012a). *"Transient-Rate Analysis for Long Homogeneous and Naturally Fractured Reservoir by The TDS Technique."* Journal of Engineering and Applied Sciences., Vol. *7.* Nro. 3. P. 353-370. March.

Escobar, F. H., Rojas, M. M. & Cantillo, J. H. (2012b). "Straight-Line Conventional Transient Rate Analysis for Long Homogeneous and Heterogeneous Reservoirs." *Dyna.* Year 79, Nro. 172, pp. 153-163. April.

Escobar, F. H., Srivastav, P. & Wu, X. (2015a) "A practical method to determine aquifer leakage factor from well test data in CBM reservoirs." *Journal of Engineering and Applied Sciences.*, Vol. *10.* Nro. 11. p. 4857-4863. June.

Escobar, F. H., Ghisays-Ruiz, A. & Srivastav, P. (2015b). "Characterization of the spherical stabilization flow regime by transient pressure analysis." *Journal of Engineering and Applied Sciences.*, Vol. *10.* Nro. 14. p. 5815-5822. August.

Frimann-Dahl, C., Irvine-Fortescue, J., Rokke, E., Vik, S. & Wahl, O. (1998, January 1.) "Formation Testers vs. DST - The Cost Effective Use of Transient Analysis to Get Reservoir Parameters." *Society of Petroleum Engineers*. Doi:10.2118/48962-MS.

Guo, B., Stewart, G. & Toro, M. (2002). "Linearly Supported Radial Flow-A Flow Regime in Layered Reservoirs." *Society of Petroleum Engineers*. doi:10.2118/77269-PA.

Hantush, M. S. (1956). "Analysis of data from pumping tests in leaky Aquifers." *Transactions of American Geophysics. U.*, *36*, 702.

Neuman, S. P. & Witherspoon, P. A. (1972). "Field Determination of The Hydraulic Properties of Leaky Multiple Aquifer Systems." *Water Resources Research*, vol. *8*, no. 5, pp. 1284-1298.

Stewart, G. & Wittmann, M. (1979, January 1). "Interpretation of The Pressure Response of The Repeat Formation Tester." *Society of Petroleum Engineers*. doi:10.2118/8362-MS.

Tiab, D. (1993, January 1). "Analysis of Pressure and Pressure Derivatives Without Type-Curve Matching: I-Skin and Wellbore Storage." *Society of Petroleum Engineers*. This paper was prepared for presentation at the Production Operations Symposium held in Oklahoma City, OK, U.S.A., March 21-23. doi:10.2118/25426-MS.

Tippie, D. B. & Abbot, W. A. (1978, January 1). "Pressure Transient Analysis In Bottom Water Drive Reservoir With Partial Completion." *Society of Petroleum Engineers*. Doi:10.2118/7487-MS.

Schafer, P. S., Hower, T. & Ownes, R. W. (1993). *"Managing Water-Drive Gas Reservoirs."* *Gas Research Institute*. p. 193

Chapter 15

SHALE RESERVOIRS

BACKGROUND

The actual energetic production is focused on searching for new supply sources that permit the constantly growing need for energy. This necessity and the depletion of conventional resources lead to the goal of finding new oil resources, such as unconventional shale gas reservoirs. Then, an appropriate, accurate and practical way of characterizing these types of reservoirs is needed for better exploitation and managing of these fields. We are currently experiencing a boom in unconventional reservoirs, such as those possessing micro-permeability (deposits of shale gas). These reservoirs are supplying considerable amounts of oil and gas, especially in the United States of America. The key business in the oil and gas industry is "unconventional resources" or "unconventional reservoirs." However, shales are not unique or unconventional reservoirs. Among systems falling into this classification we have: tight, heavy oil; organic-rich shale; oil shale; gas hydrates; gas storage; coalbed methane; tar sands and in-situ conversion. That is why this chapter is not called "unconventional reservoirs."

Most shale reservoirs are developed with horizontal wells. For such wells, several well-pressure behavior models have been presented, but the author's preferred model is the one presented by Goode and Thambynayagam (1987), who also presented conventional analysis for the interpretation of horizontal well pressure tests. The *TDS* technique (Tiab, 1993) for horizontal well pressure tests in both homogeneous and heterogeneous formations was presented by Engler and Tiab (1996a, 1996b). An extension for double-porosity and double-permeability systems in horizontal wells was recently introduced by Lu et al. (2015). Escobar, Zhao and Zhang (2014b) presented a technique for interpreting transient-pressure tests in horizontal wells when the oil requires an onset pressure to flow. Brown et al. (2011) also presented an analytical solution for a system with an outer reservoir, inner reservoir, and hydraulic fracture, coupling the solutions with flux- and pressure-continuity conditions on the interfaces between the regions and then inverting them numerically from the Laplace space. Their mathematical model was used by Escobar, Bernal and Olaya-Marin (2014) to generate a practical methodology for interpretation of pressure tests in shale gas systems. El-Banbi and Wattenbarger (1998) presented a model for a hydraulically fractured horizontal shale gas well, which was modeled as a horizontal well draining a rectangular space containing a network of fractures separated by matrix blocks (a dual-porosity system). While none of these materials are presented in this book for space-saving reasons, the reader should be aware of their existence.

Normally, shale reservoirs are tested under constant well-flowing pressure, but they are sometimes tested under constant-rate conditions. Different reservoir models have been developed for describing either pressure or transient-rate behavior of the different shale plays. In this chapter, only two models will be dealt with: Bello (2009) and Cruz-Fuentes, Gildin and Valkó (2014). Escobar, Rojas and Ghisays-Ruiz (2014d) used Bello's (2009) model to extend the *TDS* technique for transient-rate analysis of shale gas reservoirs drained by a hydraulically fractured horizontal well. On the other hand, Escobar, Montealegre and Bernal (2014a) and Bernal, Escobar and Ghisays-Ruiz (2014) used the model proposed by Cruz-Fuentes et al. (2014) to develop an interpretation methodology based upon both the *TDS* technique for transient-rate analysis and the pressure and pressure derivative analysis.

Since shale gas systems are typically fractured naturally, the initial studies must address the proposal by Warren and Root (1963) to describe a symmetric fractured network limited by cube matrix blocks. Before Cruz-Fuentes et al. (2014), there were other valuable contributions. For instance, Mayerhofer et al. (2006) presented a model for hydraulically fractured well in a shale gas reservoir that represented the hydraulic fracture as an interconnected network of fractures. Their work indicated that drainage did not take place far beyond the stimulated region because of the low matrix permeability. This observation was also stated by Carlson and Mercer (1989). Other important contributions on the transient behavior of shale gas formation include a study by Ozkan, Ohaeri and Raghavan (1987), which presented details of five flow regimes taking place in fractured cylindrical reservoirs. Watson et al. (1989) presented an analytical model for a naturally fractured reservoir with historical matching; this model accounted for variable gas properties. Another model for predicting the production of a dewatered, fractured coal shale gas formation was presented by Spivey and Semmelbeck (1995).

HYDRAULICALLY FRACTURED HORIZONTAL WELLS IN NATURALLY FRACTURED SHALE GAS RESERVOIRS (BELLO MODEL)

Escobar et al. (2014d) developed an interpretation methodology based upon the *TDS* technique (Tiab, 1993), using the analytical solutions proposed by Bello (2009), who used the linear dual-porosity model proposed by El-Banbi (1998) to describe the hydraulically fractured shale gas reservoir system. This consisted of a bounded rectangular reservoir with slab matrix blocks draining into adjoining fractures and subsequently into a horizontal well in the center. The horizontal well fully penetrated the rectangular reservoir. The features of the El-Banbi (1998) model are described below:

- A closed rectangular reservoir containing a network of natural and hydraulic fractures. The fractures do not drain beyond the boundaries of this rectangular space.
- The perforated length of the well, x_e, is the same as the width of the reservoir.
- Flow is towards the well in the center of the rectangular space.
- It is a dual-porosity system consisting of matrix blocks and fractures.
- Both porous media are homogeneous and isotropic.

Shale Reservoirs

- The matrix acts as a uniformly distributed source for the fractures.
- Fluid flows through the fractures to the wellbore.

GOVERNING EQUATIONS

The diffusivity equations for the matrix and the initial and boundary conditions are given by:

$$\frac{\partial^2 P_{DLm}}{\partial z_D^2} = \frac{3}{\lambda_{Ac}}(1-\omega)\frac{\partial P_{Dm}}{\partial t_{DAc}} \tag{15.1}$$

where λ_{Ac} (for the 1D slab matrix) and ω are given by Equations (15.3) and (2.1), respectively:

$$\lambda_{Ac} = \frac{12}{L^2}\frac{k_m}{k_f}A_{cw} \tag{15.2}$$

and:

$$A_{cw} = \frac{L}{2y_e}A_{cm} \tag{15.3}$$

For the 2D or 3D slab matrix, change the constant "2" to "4" in Equation (15.3).

$$\omega = \frac{(\phi c_t)_f}{(\phi c_t)_f + (\phi c_t)_m} \tag{2.1}$$

The dimensionless time and pressure variables for the slightly compressible fluid are:

$$t_{DAc} = \frac{0.0002637 k_f t}{(\phi\mu c_t)_{f+m} A_{cw}} \tag{15.4}$$

$$P_D = \frac{k_f \sqrt{A_{cw}}\left(P_i - P_{wf}\right)}{141.2 q B\mu} \tag{15.5}$$

$$t_D * P_D' = \frac{k_f \sqrt{A_{cw}}(t*\Delta P')}{141.2 q B\mu} \tag{15.6}$$

For the gas case, the variables are:

$$t_{DAc} = \frac{0.0002637 k_f t}{(\phi \mu c_t)_{f+m} A_{cw}} \tag{15.7}$$

For t_{DAch}, the bulk fracture permeability, k_f, is changed by the reservoir permeability, k.

$$m(P)_D = \frac{k_f \sqrt{A_{cw}} \left[m(P_i) - m(P_{wf}) \right]}{1422 q_g T} \tag{15.8}$$

$$t_D * m(P_D)' = \frac{k_f \sqrt{A_{cw}} [t * \Delta m(P)']}{1422 \ q_g T} \tag{15.9}$$

The diffusivity equations for the fracture and both the initial and boundary conditions are given by:

$$\frac{\partial^2 P_{DLf}}{\partial y_D^2} = \omega \frac{\partial P_{DLf}}{\partial t_{DAc}} - \frac{\lambda_{Ac}}{3} \left. \frac{\partial P_{DLm}}{\partial z_D} \right|_{z_D=1} \tag{15.10}$$

CONSTANT PRESSURE INNER BOUNDARY SOLUTION

The solution in the Laplace space to the system presented by Bello (2009) in Equations (15.1) and (15.10) is given by:

$$\overline{P_{wDL}} = \frac{2\pi}{s\sqrt{s f(s)}} \left[\frac{1 + e^{-2\sqrt{s f(s)} y_{De}}}{1 - e^{-2\sqrt{s f(s)} y_{De}}} \right] \tag{15.11}$$

where $f(s)$ depends upon the flow regime as defined in the nomenclature, and:

$$y_{De} = \frac{y_e}{\sqrt{A_{cw}}} \tag{15.12}$$

In Laplace space, the constant P_{wf} case at the wellbore can be found from the solution for the constant rate case using the Van Everdingen and Hurst (1949) relation given by Equation (15.11):

$$\overline{q_{DL}} = \frac{1}{s^2 \overline{P_{wDL}}} \tag{15.13}$$

Equation (15.13) thus becomes the constant P_{wf} case:

$$\frac{1}{q_{DL}} = \frac{2\pi s}{\sqrt{sf(s)}} \left[\frac{1+e^{-2\sqrt{sf(s)}y_{De}}}{1-e^{-2\sqrt{sf(s)}y_{De}}} \right] \qquad (15.14)$$

The dimensionless variables are given for gas production under constant pressure conditions:

$$\frac{1}{q_D} = \frac{k_f \sqrt{A_{cw}} [m(P_i) - m(P_{wf})]}{1422 q_g T} \qquad (15.15)$$

$$t_D *(1/q_D)' = \frac{k_f \sqrt{A_{cw}} [\Delta m(P)]}{1422 T} [t *(1/q)'] \qquad (15.16)$$

TDS FORMULATION FOR RTA (BELLO MODEL)

Escobar et al. (2014d) presented an extension of the *TDS* technique in terms of the reciprocal rate derivative based on the equations presented by Bello (2009) for each flow regime in the case of constant wellbore flowing pressure P_{wf}. New expressions taking into account the interception points between the different flow regimes are proposed. Equivalent equations are presented later in this chapter for the case of a constant flow rate.

EARLY LINEAR FLOW (FRACTURES)

The early-linear flow regime has a slope of 0.5 on the reciprocal rate derivative versus time log-log plot, which reflects the single flow of the fracture system; the matrix does not flow at early times. As seen in Figure 15.1, the flow does not occur at small values of ω; this occurs because this parameter indicates the storage capacity of fractures, and if this is small, the fractures will not be able to affect the response of the reciprocal rate due to low fluid content. We can also observe how the parameter ω does not affect the late response of the reciprocal rate for a particular dimensionless λ_{Ac}. According to the equations presented by Bello (2009), in a log-log plot of the dimensionless reciprocal flow versus dimensionless time, the initial flow regime is governed by the following equation:

$$q_{DL} = \left(2\pi \sqrt{\frac{\pi t_{DAc}}{\omega}} \right)^{-1} \qquad (15.17)$$

Inverting the flow rate and taking the derivative with respect to time yields:

$$t_D *(1/q_{DL})' = \pi \sqrt{\frac{\pi t_{DAc}}{\omega}} \qquad (15.18)$$

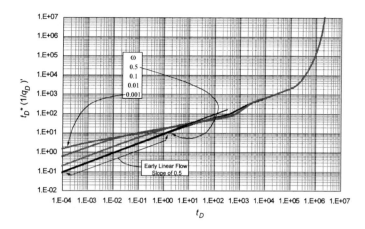

Figure 15.1. Effect storativity, ω, on the early linear flow under the constant interporosity flow parameter, $\lambda_{Ac}=1\times10^{-3}$; $y_{De}=1000$; adapted from Escobar et al. (2014d).

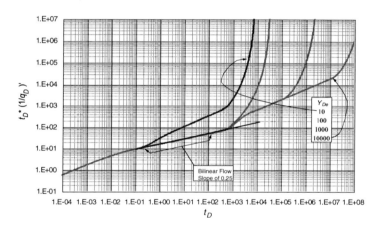

Figure 15.2. Effect of dimensionless reservoir length, y_{De}, on the bilinear flow regime under a constant interporosity flow parameter, $\lambda_{Ac}=1\times10^{-3}$, and a dimensionless storativity ratio, $\omega=1\times10^{-2}$; adapted from Escobar et al. (2014d).

Once the dimensionless quantities given by Equations (15.7) and (15.16) are placed into Equation (15.18), it yields:

$$\frac{k_f\sqrt{A_{CW}}[\Delta m(P)]}{1422T}\left[t*(1/q)'\right]_L = \pi\sqrt{\frac{\pi 0.0002637 k_f t_L}{\omega(\phi\mu c_t)_{f+m} A_{cw}}} \tag{15.19}$$

If the other parameters are known, Equation (15.19) allows one to calculate either the fracture permeability or the dimensionless storativity coefficient:

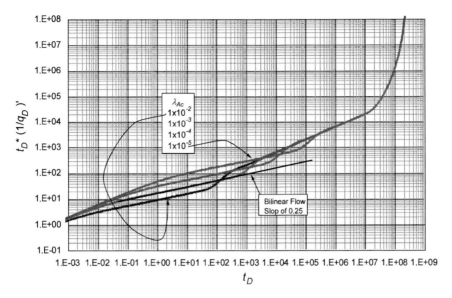

Figure 15.3. Effect of the interporosity flow parameter, λ_{Ac}, on the bilinear flow regime under constant dimensionless reservoir length, y_{De} =1000, and dimensionless storativity ratio, ω =1×10^{-2}; adapted from Escobar et al. (2014d).

$$k_f = \frac{16533.277 t_L}{\omega(\phi\mu c_t)_{f+m}} \left(\frac{T}{A_{cw}[\Delta m(P)][t*(1/q)']_L} \right)^2 \quad (15.20)$$

$$\omega = \frac{16533.277 t_L}{k_f(\phi\mu c_t)_{f+m}} \left(\frac{T}{A_{cw}[\Delta m(P)][t*(1/q)']_L} \right)^2 \quad (15.21)$$

BILINEAR FLOW REGIME

The bilinear flow regime is indicated by a slope of 0.25 for the reciprocal rate derivative vs. time plot (see Figure 15.2). This flow regime is caused by simultaneous transient flow in both fracture and matrix system. In Figure 15.2, this flow regime occurs only in elongated reservoirs—that is, when length y_{De} ($y_e/[A_{cw}]^{0.5} > [3/\lambda_{Ac}]^{0.5}$); otherwise, after the early linear flow regime, the matrix transient linear flow regime will be felt. This is because, if the reservoir is not elongated, the flow in the fracture system is so short that it does not allow the simultaneous flow to occur.

However, from observation of Figures 15.2 and 15.3, this flow regime is only a function of the interporosity flow parameter, λ_{Ac}. Bello (2009), in a log-log plot of the dimensionless rate versus dimensionless time, established that the bilinear flow regime is governed by the following equation:

$$q_{DBL} = \frac{\lambda_{Ac}^{0.25}}{10.133 t_{DAc}^{0.25}} \quad (15.22)$$

Applying the reciprocal rate and taking the derivative with respect to time, the resulting equation yields:

$$t_D * (1/q_{DBL})' = \frac{10.133 t_{DAc}^{0.25}}{4 \lambda_{Ac}^{0.25}} \quad (15.23)$$

Once the dimensionless quantities given by Equations (15.4) and (15.16) are placed into Equation (15.23), it yields:

$$\frac{k_f \sqrt{A_{cw}} [\Delta m(P)]}{1422 T} [t*(1/q)']_{BL} = \frac{10.133}{4} \left(\frac{0.0002637 k_f t_{BL}}{\lambda_{Ac} (\phi \mu c_t)_{f+m} A_{cw}} \right)^{0.25} \quad (15.24)$$

Either fracture permeability or the interporosity flow parameter can be determined from this equation:

$$k_f = \left(\frac{459.045 T}{\sqrt{A_{CW}} [\Delta m(P)][t*(1/q)']_{BL}} \right)^{1/0.75} \left(\frac{t_{BL}}{\lambda_{Ac} (\phi \mu c_t)_{f+m} A_{cw}} \right)^{1/3} \quad (15.25)$$

$$\lambda_{Ac} = \frac{t_{BL}}{k_f^3 (\phi \mu c_t)_{f+m} A_{cw}} \left(\frac{459.045 T}{\sqrt{A_{CW}} [\Delta m(P)][t*(1/q)']_{BL}} \right)^4 \quad (15.26)$$

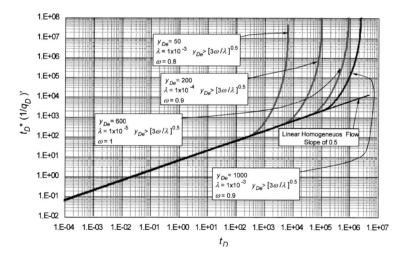

Figure 15.4. Effect of the interporosity flow parameter, λ_{Ac}, storativity, ω, and dimensionless reservoir length, y_{De}, on the linear homogeneous flow regime; adapted from Escobar et al. (2014d).

LINEAR HOMOGENEOUS FLOW

This flow regime occurs with the characteristic slope of 0.5 on the reciprocal rate derivative vs. time plot, similar to the early linear flow regime and the matrix transient linear flow regime. Unlike these, which are accompanied by other flow regimes, the homogeneous linear flow regime is the only one that occurs throughout the test until the pseudosteady state occurs. This behavior reflects the characteristics of a homogeneous reservoir—that is, there is no difference between the permeability of the fracture and the permeability of the matrix; they present a single permeability. In Figure 15.4, this homogeneous flow occurs when $y_e/[A_{cw}]^{0.5} > [3\omega/\lambda_{Ac}]^{0.5}$. For these conditions, the reservoir behaves as a known homogeneous case, in which only one system is unique, and it contains the highest storage capacity for the fluid (in this case, when ω approaches 1). For this linear homogeneous flow regime, the equation representing the behavior of the dimensionless rate versus dimensionless time, as presented by Bello (2009), is:

$$q_{DLh} = \frac{1}{2\pi\sqrt{\pi t_{DAch}}} \tag{15.27}$$

As for the former flow regimes, the derivative is taken after inverting Equation (15.27) to obtain:

$$t_D * (1/q_{Dh})' = \pi\sqrt{\pi t_{DAch}} \tag{15.28}$$

Once the dimensionless quantities given by Equations (15.4) and (15.16) are placed into Equation (15.28), it yields:

$$\frac{k\sqrt{A_{cw}}[\Delta m(P)]}{1422T}[t*(1/q)']_h = \pi\sqrt{\frac{\pi\, 0.0002637 k t_h}{(\phi\mu c_t)_{f+m} A_{cw}}} \tag{15.29}$$

From which the permeability of the homogeneous system is solved:

$$k = \left[\frac{128.5818T}{A_{cw}[\Delta m(P)][t*(1/q)']_h} \sqrt{\frac{t_h}{(\phi\mu c_t)_{f+m}}} \right]^2 \tag{15.30}$$

A MATRIX TRANSIENT LINEAR FLOW REGIME

In this flow, the behavior of dimensionless reciprocal rate is primarily the response caused by matrix drainage from the outer edges towards the matrix block center. A slope of 0.5 on the reciprocal rate vs. time log-log plot is observed. This flow regime is shown in large (y_{De}) elongated fields in Figure 15.5. Additionally, we can see how this flow regime is influenced by the interporosity flow parameter, λ_{Ac}, and dimensionless reservoir length, y_{De}.

Figure 15.5 shows two sets of curves with certain constant parameters. For relatively large values of the interporosity flow parameter ($\lambda_{Ac}=1\times10^{-3}$), the variation of y_{De} affects only the duration of the matrix linear transient flow regime. For elongated reservoirs with large values of y_{De}, this flow regime occurs for a longer test time. The starting time of this flow regime converges at the same point for different y_{De} values, and it is the same in terms of location (i.e., there is no parallel displacement along the time axis).

Conversely, for small values of the interporosity flow parameter ($\lambda_{Ac}=1\times10^{-7}$), the variation of y_{De} and the effect of the duration of this flow regime mean that the starting times for this flow regime does not converge at the same point. These may occur at different times, generating parallel displacement along the time axis. Bello (2009) established the governing equation for the matrix transient linear flow:

$$q_{DLM} = \frac{1}{2\pi\sqrt{\pi t_{DAc}}} \sqrt{\frac{\lambda_{Ac}}{3}} y_{De} \qquad (15.31)$$

Inverting the flow rate and then taking the derivative of the resulting equation leads to:

$$t_D * (1/q_{DLM})' = \frac{\pi\sqrt{3\pi}}{y_{De}} \sqrt{\frac{t_{DAc}}{\lambda_{Ac}}} \qquad (15.32)$$

Once the dimensionless quantities given by Equations (15.4) and (15.16) are placed into Equation (15.32), it yields:

$$\frac{k_f \sqrt{A_{cw}} [\Delta m(P)]}{1422T} [t*(1/q)']_{LM} = \frac{\pi\sqrt{3\pi}}{y_e} \sqrt{\frac{0.0002637 k_f t_{LM}}{\lambda_{Ac}(\phi\mu c_t)_{f+m}}} \qquad (15.33)$$

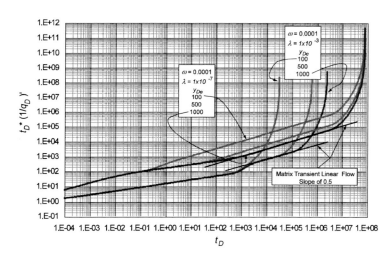

Figure 15.5. Effect of the interporosity flow parameter, λ_{Ac}, and dimensionless reservoir length, y_{De}, on a constant dimensionless storativity ratio, $\omega=1\times10^{-4}$, during a matrix transient linear flow regime; adapted from Escobar et al. (2014d).

Shale Reservoirs 385

Either fracture permeability or the interporosity flow parameter can be solved from Equation (15.33), leading to:

$$k_f = \frac{49599.831 t_{Lm}}{A_{cw} \lambda_{Ac} \left(\phi \mu c_t\right)_{f+m}} \left[\frac{T}{y_e \left[\Delta m(P)\right]\left[t*(1/q)'\right]_{Lm}} \right]^2 \qquad (15.34)$$

$$\lambda_{Ac} = \frac{49599.831 t_{LM}}{k_f A_{cw} \left(\phi \mu c_t\right)_{f+m}} \left[\frac{T}{y_e \left[\Delta m(P)\right]\left[t*(1/q)'\right]_{LM}} \right]^2 \qquad (15.35)$$

For this case (the slab matrix), it is known that:

$$\lambda_{Acw} = \frac{12}{L^2} \frac{k_m}{k_f} A_{cw} \qquad (15.36)$$

$$A_{cw} = \frac{L}{2y_e} A_{cm} \qquad (15.37)$$

Equation (15.31) can be rewritten as:

$$q_{DLM} = \frac{1}{2\pi \sqrt{\pi t_{DAcm}}} \qquad (15.38)$$

Inverting Equation (15.38) and taking the derivative of the resulting equation yields:

$$t_D*(1/q_{DLM})' = \pi \sqrt{\pi t_{DAcm}} \qquad (15.39)$$

As for the former cases, once the dimensionless quantities are placed into the above equation, it yields:

$$\frac{k_m \sqrt{A_{cm}} \left[\Delta m(P)\right]}{1422T} \left[t*(1/q)'\right]_{LM} = \pi \sqrt{\frac{\pi \, 0.0002637 k_m t_{LM}}{\left(\phi \mu c_t\right)_m A_{cm}}} \qquad (15.40)$$

Either the total matrix surface area drained or the matrix permeability can now be solved for:

$$A_{cm} = \frac{128.5818T}{k_m \left[\Delta m(P)\right]\left[t*(1/q)'\right]_{LM}} \sqrt{\frac{k_m t_{LM}}{\left(\phi \mu c_t\right)_m}} \qquad (15.41)$$

$$k_m = \frac{16533.2793 t_{LM}}{(\phi\mu c_t)_m} \left[\frac{T}{A_{cm}[\Delta m(P)][t*(1/q)']_{LM}} \right]^2 \quad (15.42)$$

PSEUDOSTEADY-STATE PERIOD

This flow period represents the transient response of the reciprocal rate when the reservoir boundary begins influencing its behavior. Clearly, this flow period is normally recognized by a unit-slope straight line on a pressure test; however, during the pseudosteady state of a transient reciprocal rate, the flow period does not follow the unit-slope trend. Instead, a constantly increasing curve is displayed; a tangent unit-slope line must then be drawn on the derivative curve during this late time so as to characterize such a regime (see Figure 15.6). As seen in Figure 15.6, this flow period is not influenced by the interporosity flow parameter, λ_{Ac}, which affects the time at which it occurs but not its characteristic equation. A uniform behavior was found by Escobar et al. (2014d), who divided the dimensionless time by the dimensionless square of the reservoir length and the dimensionless reciprocal rate derivative by the dimensionless reservoir length, y_{De} for each case, respectively.

Once a division is made, the dimensionless time and the dimensionless reciprocal rate derivative are multiplied by y_{De}^2 and y_{De}, respectively. The effect of the dimensionless reservoir length on the pseudosteady-state period can be seen in Figures 15.7 and 15.8. As y_{De} changes the tangent points on the curve, the pseudosteady-state period exhibits a linear behavior with the unit-slope. This linear behavior is independent of both the interporosity flow parameter, λ_{Ac}, and the dimensionless storativity ratio, ω. Figure 15.7 shows how these points that are tangent to the pseudosteady state fall on the same line for different values of λ_{Ac}.

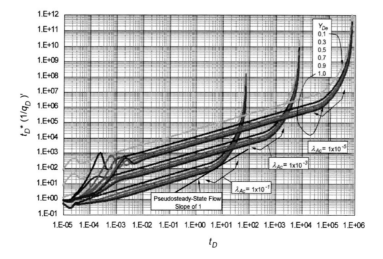

Figure 15.6. Effect of the dimensionless reservoir length, y_{De}, at a constant storativity, $\omega = 0.01$, on pseudosteady-state regime flow; adapted from Escobar et al. (2014d).

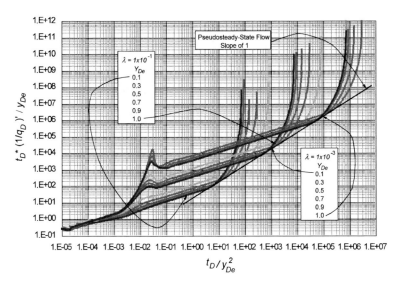

Figure 15.7. Effect of the dimensionless reservoir length, y_{De}, at a constant storativity, $\omega = 0.01$, on pseudosteady-state regime flow; adapted from Escobar et al. (2014d).

In Figure 15.8, the variation of the dimensionless storativity ratio, ω, does not affect the behavior of the pseudosteady-state period. As mention earlier, this parameter affects only the early linear flow regime. Based on these observations, the governing equation during the pseudosteady-state period is given by:

$$\frac{[t*(1/q_D)']_{PSS}}{y_{De}} = 4.76\pi \left(\frac{t_{DAc}}{y_{De}^2}\right)_{PSS} \tag{15.43}$$

Once the dimensionless quantities given by Equations (15.4) and (15.16) are placed into Equation (15.43), it yields:

$$\frac{k_f\sqrt{A_{cw}}[\Delta m(P)]}{1422T}[t*(1/q)']_{PSS} = \frac{4.76\pi\sqrt{A_{cw}}}{y_e}\left(\frac{0.0002637k_f t_{PSS}}{(\phi\mu c_t)_{f+m}A_{cw}}\right) \tag{15.44}$$

Solving for the half-length drainage area from Equation (15.44) results in:

$$y_e = \frac{5.6075T t_{PSS}}{A_{cw}(\phi\mu c_t)_{f+m}[\Delta m(P)][t*(1/q)']_{PSS}} \tag{15.45}$$

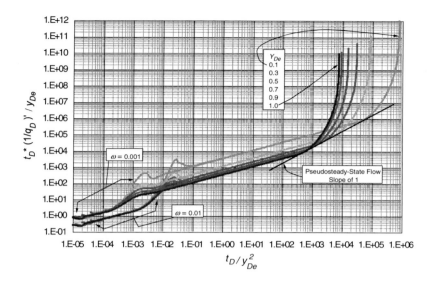

Figure 15.8. Effect of the dimensionless reservoir length, y_{De}, at a constant interporosity flow parameter, $\lambda_{Ac} = 1\times10^{-3}$, on a pseudosteady-state flow regime; adapted from Escobar et al. (2014d).

Once y_e is calculated, then, the total matrix surface area draining in the fracture system can be estimated from:

$$A_{cm} = 2(y_e x_e) \tag{15.46}$$

INTERSECTING POINTS

Based on the above-named flow regimes, the following new expressions are proposed for calculating the interporosity flow parameter, λ_{Ac}, and the dimensionless storativity ratio, ω. Intersections are only meant for reciprocal rate derivatives. The intersection point formed by a line drawn on the early linear flow regime given by Equation (15.18) and the bilinear flow regime line given by Equation (15.23), t_{LBLi}, provides the following equation:

$$t_{DAc_{LBLi}}^{0.25} = \left(\frac{10.133}{4\pi}\right)\left(\frac{\omega}{\pi}\right)^{0.5}\left(\frac{1}{\lambda_{Ac}}\right)^{0.25} \tag{15.47}$$

Once the dimensionless quantities are plugged into Equation (15.47), the storativity ratio, ω, can be solved for:

$$\omega = 0.07846\left(\frac{k_f\, t_{LBLi}\, \lambda_{Ac}}{(\phi\mu c_t)_{f+m}\, A_{cw}}\right)^{0.5} \tag{15.48}$$

The intersection point formed by the line of the bilinear flow regime, Equation (15.23), with the matrix transient linear flow regime line, Equation (15.32), t_{BLLMi}, gives:

$$\sqrt[4]{t_{DAc_{BLLMi}}} = \left(\frac{10.133\, y_{De}}{4\pi} \right) \left(\frac{1}{3\pi} \right)^{0.5} \lambda_{Ac}^{0.25} \tag{15.49}$$

Once the dimensionless quantities given by Equations (15.4) and (15.12) and the definition of λ_{Ac} given by Equation (15.2) are plugged into Equation (15.49), the interporosity flow parameter, λ_{Ac}, can be solved for:

$$\lambda_{Ac} = \frac{0.05541\, A_{cw} k_f\, t_{BLLMi}}{y_e^4 \left(\phi \mu c_t \right)_{f+m}} \tag{15.50}$$

The intersection point formed by line of the early linear flow regime given by Equation (15.18) and the line of the pseudosteady-state period given by Equation (15.43), t_{LPSSi}, is shown in the following equation:

$$\sqrt{t_{DAc_{LPSSi}}} = \frac{y_{De}}{4.76} \sqrt{\frac{\pi}{\omega}} \tag{15.51}$$

Once the dimensionless quantities given by Equations (15.4) and (15.12) are plugged into Equation (15.51), the storativity ratio, ω, can be solved for:

$$\omega = 525.806 \frac{\left(\phi \mu c_t \right)_{f+m} y_e^2}{k_f\, t_{LPSSi}} \tag{15.52}$$

The intersection point formed by the line of the bilinear flow regime from Equation (15.23) and the pseudosteady-state period from Equation (15.43), t_{BLPSSi}, leads to:

$$t_{DAc_{BLPSSi}}^{0.75} = \frac{10.133\, y_{De}}{4 \times 4.76\, \pi \sqrt[4]{\lambda_{Ac}}} \tag{15.53}$$

Once the dimensionless quantities given by Equations (15.4) and (15.14) and the definition of λ_{Ac} given by Equation (15.2) are plugged into Equation (15.53), the storativity ratio, ω, can be solved for:

$$\lambda_{Ac} = 44858022.57\, A_{cw} y_e^4 \left(\frac{\left(\phi \mu c_t \right)_{f+m}}{k_f\, t_{BLPSSi}} \right)^3 \tag{15.54}$$

Finally, the point of intersection resulting from a straight line drawn on the matrix transient linear flow regime given by Equation (15.32) and the pseudosteady-state flow period given by Equation (15.43), t_{LMPSSi}, provides the following expression:

$$\sqrt{t_{DAc_{LMPSSi}}} = \frac{1}{4.76} \sqrt{\frac{3\pi}{\lambda_{Ac}}}$$

(15.55)

As for the former cases, once the dimensionless quantities are plugged into Equation (15.55), solving for the dimensionless storativity ratio ω gives:

$$\lambda_{Ac} = \frac{1577.419 A_{cw} (\phi\mu c_t)_{f+m}}{k_f t_{LMPSSi}}$$

(15.56)

When the bilinear flow regime, the matrix transient linear flow regimes and the pseudosteady-state period are presented, Equations (15.54) and (15.56) can be combined to generate an additional equation that allows for the calculation of fracture permeability regardless of the presence of the interporosity parameter flow. This equation is:

$$k_f = 168.635 \, y_e^2 \, (\phi\mu c_t)_{f+m} \sqrt{\frac{t_{LMPSSi}}{t_{BLPSSi}}}$$

(15.57)

TDS EQUATIONS FOR PRESSURE DERIVATE GAS FLOW (BELLO MODEL)

Escobar et al. (2014d) also presented the constant-rate case. The dimensionless equation representing the first linear flow regime is given by:

$$t_{DL} * m(P_D)'_L = 2\sqrt{\frac{\pi t_{DA}}{\omega}}$$

(15.58)

Once the dimensionless terms given by Equations (15.4) and (15.6) are plugged into Equation (15.58), the fracture permeability can be solved for:

$$k_f = \frac{6700.68 \, t_L}{(\phi\mu c_t)_{f+m} \, \omega} \left(\frac{q_g T}{A_{cw} [t * \Delta m(P')]_L} \right)^2$$

(15.59)

From the above equation, it is possible to know the value of the dimensionless storativity coefficient:

Shale Reservoirs   391

$$\omega = \frac{6700.68\,t_L}{(\phi\mu c_t)_{f+m}\,k_f}\left(\frac{q_g T}{A_{cw}\left[t*\Delta m(P')\right]_L}\right)^2 \tag{15.60}$$

The general equation describing the bilinear flow regime is:

$$t_D * m(P_D)'_{MLL} = \frac{2.28075}{\lambda^{0.25}}(t_{DA})_{BL}^{0.25} \tag{15.61}$$

From Equation (15.61), it is possible to obtain expressions for calculating k_f and λ

$$k_f = \left(\frac{4009.76}{A_{cw}}\right)\left(\frac{q_g T}{\left[t*m(P')\right]}\right)^{\frac{1}{0.75}}\left(\frac{t}{(\phi\mu c_t)_{f+m}\,\lambda_{AC}}\right)^{0.33} \tag{16.62}$$

$$\lambda_{AC} = \left(\frac{6.54 \times 10^{10}\,t}{(\phi\mu c_t)_{f+m}}\right)^{0.33}\left(\frac{q_g T}{\left[t*m(P')\right]k_f^{0.75}A_{cw}^{0.75}}\right)^{\frac{1}{0.25}} \tag{16.63}$$

The dimensionless governing equation representing the second linear flow (matrix transient) is given by:

$$t_{DL} * m(P_D)'_L = 2\left(\frac{3\pi t_{DA}}{\lambda_{AC}}\right)^{\frac{1}{2}}\left(\frac{1}{y_{De}}\right) \tag{16.64}$$

Once the dimensionless terms are plugged into Equation (15.64), the fracture permeability can be solved for:

$$k_f = \frac{60306.44213\,t_{Lt}\,A_{cw}}{(\phi\mu c_t)_{f+m}\,\lambda_{AC}}\left(\frac{q_g T}{y_e\left[t*\Delta m(P')\right]_{Lt}}\right)^2 \tag{15.65}$$

From the above equation, it is possible to know the value of the interporosity flow parameter:

$$\lambda_{AC} = \frac{60306.44213\,A_{cw}\,t_{Lt}}{(\phi\mu c_t)_{f+m}\,k_f}\left(\frac{q_g T}{y_e\left[t*\Delta m(P')\right]}\right)^2 \tag{15.66}$$

The dimensionless governing equation for the second linear flow (homogeneous transient) is:

$$t_{DL} * m(P_D)'_{Lh} = \sqrt{2\pi t_{DH}} \tag{15.66}$$

Once the dimensionless terms are plugged into Equation (15.66), the fracture permeability can be solved for:

$$k_f = \frac{6700.68 \, t_{Lt}}{(\phi \mu c_t)_{f+m}} \left(\frac{q_g T}{A_{cw} [t * \Delta m(P')]} \right)^2$$

(15.67)

The intersection point between early linear and bilinear flow is used to find the interporosity flow parameter:

$$\lambda_{AC} = \frac{k_f t_{LBLi} \, \omega^2}{22130.91 \, (\phi \mu c_t)_{f+m} \, A_{cw}}$$

(15.68)

The intersection point between the second linear and bilinear flow regimes provides the interporosity flow parameter:

$$\lambda_{AC} = \frac{k_f t_{LBLi}}{569632.11 \, y_e^2 (\phi \mu c_t)_{f+m} \, A_{cw}}$$

(15.69)

CONVENTIONAL ANALYSIS FOR RTA (BELLO MODEL)

Along with the governing equations, Bello (2009) also presented conventional analysis for each of the flow regimes. Placing the dimensionless quantities from Equation (15.17) in the early linear flow (fractures) regime yields:

$$\frac{m(P_i) - m(P_{wf})}{q_g} = \frac{1626T\sqrt{t}}{A_{cw}\sqrt{k_f \omega (\phi \mu c_t)_{f+m}}}$$

(15.70)

A plot of $[m(P_i)-m(P_{wf})]/q_g$ vs. $t^{0.5}$ gives a line with slope m_1:

$$A_{cw}\sqrt{k_f} = \frac{1626T}{m_1\sqrt{k_f \omega (\phi \mu c_t)_{f+m}}}$$

(15.71)

Placing also the dimensionless quantities into Equation (15.22) for the bilinear flow regime:

$$\frac{m(P_i) - m(P_{wf})}{q_g} = \frac{4076T\sqrt[4]{t}}{A_{cw}\sqrt{k_f}\sqrt[4]{\sigma k_m (\phi \mu c_t)_{f+m}}}$$

(15.72)

The above expression suggests that a plot of $[m(P_i)-m(P_{wf})]/q_g$ vs. $t^{0.25}$ gives a straight line with slope m_2:

$$A_{cw}\sqrt{k_f} = \frac{4076T}{m_2 \sqrt[4]{\sigma k_m (\phi\mu c_t)_{f+m}}}$$ (15.73)

The linear homogeneous flow is governed by Equation (15.27), which has the dimensional form given below:

$$\frac{m(P_i) - m(P_{wf})}{q_g} = \frac{1262T\sqrt{t}}{A_{cw}\sqrt{k(\phi\mu c_t)_{f+m}}}$$ (15.74)

A plot of $[m(P_i)-m(P_{wf})]/q_g$ vs. $t^{0.5}$ gives a line with slope m_3:

$$A_{cw}\sqrt{k} = \frac{1262T}{m_3\sqrt{(\phi\mu c_t)_{f+m}}}$$ (15.75)

The governing equation for the matrix transient linear flow is given by Equation (15.31), which has a dimensional representation of:

$$\frac{m(P_i) - m(P_{wf})}{q_g} = \frac{1262T\sqrt{t}}{A_{cw}\sqrt{k_m(\phi\mu c_t)_m}}$$ (15.76)

Assuming that $(\phi\mu c_t)_{f+m} \approx (\phi\mu c_t)_m$, a plot of $[m(P_i)-m(P_{wf})]/q_g$ vs. $t^{0.5}$ gives a straight line with slope m_4, which is given in:

$$A_{cw}\sqrt{k_m} = \frac{1262T}{m_4\sqrt{(\phi\mu c_t)_m}}$$ (15.77)

Bello (2009) did not provide a straight-line governing equation for such a period.

Examples 15.1

The reciprocal rate and its derivative for an example reported by Bello (2009) is presented in Figure 15.9 and Table 15.1. Other important data are given below:

$C = 0$ bbl/psi	$s = 0$	$h = 200$ ft
$T = 660$ °R	$\phi = 15\%$	$X_e = 2000$ ft
$c_t = 3.04\times10^{-4}$ psi 1	$L = 50$ ft	$A_{cw}=8\times10^5$ ft^2
$\Delta m(P) = 5.702\times10^8$ psi^2/cp	$\mu = 0.0224$ cp	$k_f = 100$ md
$k_m=1\times10^{-5}$ md		

Solution by TDS Technique

The linear flow regime, linear matrix flow regime and pseudosteady-sate period are presented in this test. The below data are read from Figure 15.9.

$t_L = 9.3 \times 10^{-4}$ hr $[t^*(1/q_g)']_L = 1.7 \times 10^{-8}$ D/Mscf

$t_{LM} = 1$ hr $[t^*(1/q_g)']_{LM} = 2.9 \times 10^{-6}$ D/Mscf

$t_{PSS} = 4.5 \times 10^{4}$ hr $[t^*(1/q_g)']_{PSS} = 7 \times 10^{-4}$ D/Mscf

$t_{LPSSi} = 8 \times 10^{2}$ hr

$t_{MLPSSi} = 3 \times 10^{4}$ hr

Since the matrix permeability is known, the total matrix surface area draining into the fracture system can be calculated by means of Equation (15.41), using the matrix transient linear flow regime, which provides a value of $A_{cm} = 1.61 \times 10^{7}$ ft^2.

$$A_{cm} = \frac{128.5818(660)}{(1 \times 10^{-5})\left[5.702 \times 10^{8}\right](2.9 \times 10^{-6})}\sqrt{\frac{(1 \times 10^{-5})(1)}{(0.15)(0.0224)(3.04 \times 10^{-4})}} = 1.61 \times 10^{7} \text{ ft}^2$$

The reservoir length is found with Equation (15.3):

Table 15.1. Reciprocal rate and its derivative versus time for example 15.1.

t, hr	$1/q$, D/STB	$t^*(1/q)'$, D/STB	t, hr	$1/q$, D/STB	$t^*(1/q)'$, D/STB	t, hr	$1/q$, D/STB	$t^*(1/q)'$, D/STB
0.00031	2.78E-08	6.94E-09	0.427	3.78E-06	1.93E-06	670.21	1.51E-04	7.62E-05
0.00062	3.37E-08	1.08E-08	0.516	4.16E-06	2.13E-06	873.30	1.72E-04	8.66E-05
0.00093	3.86E-08	1.42E-08	0.625	4.58E-06	2.34E-06	1076.40	1.91E-04	9.65E-05
0.00155	4.70E-08	2.16E-08	0.804	5.20E-06	2.64E-06	1340.42	2.13E-04	1.08E-04
0.00217	5.51E-08	3.15E-08	0.982	5.75E-06	2.93E-06	1746.62	2.43E-04	1.22E-04
0.00279	6.35E-08	4.37E-08	1.16	6.26E-06	3.18E-06	2152.81	2.70E-04	1.37E-04
0.00341	7.26E-08	5.88E-08	1.22	6.42E-06	3.27E-06	2680.86	3.02E-04	1.52E-04
0.00403	8.26E-08	7.64E-08	1.62	7.39E-06	3.73E-06	3493.24	3.44E-04	1.73E-04
0.00465	9.34E-08	9.68E-08	2.01	8.25E-06	4.18E-06	4305.62	3.82E-04	1.93E-04
0.00527	1.05E-07	1.19E-07	2.45	9.10E-06	4.63E-06	5361.72	4.26E-04	2.16E-04
0.00589	1.18E-07	1.44E-07	3.24	1.05E-05	5.28E-06	6986.49	4.87E-04	2.46E-04
0.00651	1.32E-07	1.72E-07	4.04	1.17E-05	5.91E-06	8611.26	5.40E-04	2.73E-04
0.00713	1.47E-07	2.01E-07	4.91	1.29E-05	6.54E-06	10723.46	6.03E-04	3.05E-04
0.00775	1.63E-07	2.31E-07	6.50	1.48E-05	7.47E-06	11373.36	6.21E-04	3.13E-04
0.0084	1.80E-07	2.61E-07	8.08	1.66E-05	8.37E-06	14297.94	6.96E-04	3.50E-04
0.0090	1.98E-07	2.91E-07	9.83	1.83E-05	9.25E-06	17222.53	7.64E-04	3.85E-04
0.0099	2.26E-07	3.36E-07	13.00	2.10E-05	1.06E-05	20797.01	8.40E-04	4.25E-04
0.0112	2.67E-07	3.94E-07	13.64	2.15E-05	1.08E-05	26646.18	9.50E-04	4.90E-04
0.0149	4.02E-07	5.37E-07	16.49	2.36E-05	1.19E-05	32495.34	1.05E-03	5.63E-04
0.0192	5.63E-07	6.28E-07	19.67	2.58E-05	1.31E-05	38344.50	1.14E-03	6.49E-04
0.0236	7.08E-07	6.75E-07	25.38	2.93E-05	1.48E-05	49392.92	1.31E-03	8.60E-04
0.0279	8.31E-07	6.94E-07	31.09	3.25E-05	1.64E-05	61091.25	1.50E-03	1.16E-03
0.0366	1.02E-06	6.97E-07	36.80	3.53E-05	1.79E-05	72789.57	1.70E-03	1.54E-03
0.0452	1.16E-06	7.03E-07	46.96	3.99E-05	2.01E-05	90986.97	2.05E-03	2.34E-03
0.0539	1.28E-06	7.33E-07	58.38	4.45E-05	2.24E-05	114383.62	2.61E-03	3.89E-03
0.0626	1.39E-06	7.63E-07	69.80	4.87E-05	2.46E-05	132581.01	3.13E-03	5.69E-03

t, hr	1/q, D/STB	t*(1/q)', D/STB	t, hr	1/q, D/STB	t*(1/q)', D/STB	t, hr	1/q, D/STB	t*(1/q)', D/STB
0.0775	1.55E-06	8.33E-07	86.31	5.41E-05	2.73E-05	153378.04	3.86E-03	8.49E-03
0.0948	1.73E-06	9.17E-07	109.15	6.08E-05	3.06E-05	192372.46	5.71E-03	1.70E-02
0.112	1.89E-06	1.00E-06	134.54	6.75E-05	3.42E-05	233966.50	8.61E-03	3.74E-02
0.115	1.91E-06	1.01E-06	167.54	7.54E-05	3.81E-05	275560.55	1.30E-02	8.29E-02
0.137	2.10E-06	1.11E-06	218.32	8.60E-05	4.33E-05	322353.85	2.05E-02	2.15E+00
0.179	2.41E-06	1.27E-06	269.09	9.55E-05	4.83E-05	405541.94	4.72E-02	8.70E+00
0.224	2.71E-06	1.41E-06	335.10	1.07E-04	5.39E-05	488730.04	1.16E-01	1.53E+01
0.268	2.98E-06	1.55E-06	436.65	1.22E-04	6.12E-05	665504.73	8.56E+00	2.91E+01
0.338	3.35E-06	1.73E-06	538.19	1.35E-04	6.83E-05			

$$y_e = \frac{L}{2A_{cw}} A_{cm} = \frac{50(1.61 \times 10^7)}{2(800000)} = 503.125 \text{ ft}$$

The dimensionless storativity ratio can be calculated using the point of interception between the early linear flow regime and the pseudosteady-state period, with Equation (15.25):

$$\omega = 525.806 \frac{(0.15)(0.0224)(3.04 \times 10^{-4})(503^2)}{(100)(800)} = 0.00168$$

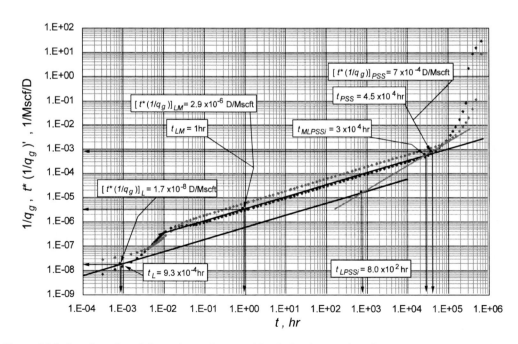

Figure 15.9. Log-log plot of the reciprocal rate and its derivative vs. time for example 3; adapted from Escobar et al. (2014d).

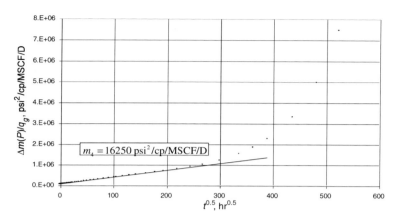

Figure 15.10. Cartesian plot of $\Delta m(P)/q_g$ vs. the square root of time for example 3.

Then, interporosity flow parameter can be estimated with the point of intersection between the matrix transient linear flow regime and the pseudosteady-state period, t_{MLPSSi}, using Equation (15.56):

$$\lambda_{Ac} = \frac{1577.419(800000)(0.15)(0.0224)(3.04\times10^{-4})}{(100)(30000)} = 4.3\times10^{-4}$$

Finally, the half-fracture length is calculated with Equation (15.45), using an arbitrary point from the pseudosteady-state period:

$$y_e = \frac{5.6075(660)(45000)}{(800000)(0.15)(0.0224)(3.04\times10^{-4})(5.702\times10^8)(0.0007)} = 511 \text{ ft}$$

Other available relationships were used to calculate the flow capacity and the dimensionless storativity ratio. The input and calculated parameters are reported in Table 15.2.

Solution by conventional analysis

A slope in the fourth region (the matrix transient linear flow) of 16250 psi^2/cp/MSCF/D was found in Figure 15.11. This slope is then used to find A_{cw} from Equation (15.77):

Table 15.2. Summary of results for example 15.1

Input parameter		Calculated parameter		Equation
y_e (ft)	500	y_e (ft)	511	15.45
A_{cm} (ft^2)	1.6×10^7	A_{cm} (ft^2)	1.61×10^7	15.41
ω	1×10^{-3}	ω	0.00168	15.25
			0.00109	15.21
λ_{Ac}	3.84×10^{-4}	λ_{Ac}	4.3×10^{-4}	15.56
			3.87×10^{-4}	15.35

$$A_{cw}\sqrt{k_m} = \frac{1262T}{m_4\sqrt{(\phi\mu c_t)_m}} = \frac{1262(660)}{16250\sqrt{(0.15)(0.0224)(304.02\times10^{-6})}} = 50700 \text{ md}^{0.5}\text{ft}^2$$

Since A_{cw}=1.6×10^7 ft^2, $k_m = 10^{-5}$ md. The fracture spacing is computed using Equation (15.3):

$$L = \frac{2y_e}{A_{cm}}A_{cw} = \frac{2(500)}{1.6\times10^7}8\times10^{-5} = 50 \text{ ft}$$

HYDRAULICALLY FRACTURED GAS SHALE WELLS USING THE CONCEPT OF INDUCED PERMEABILITY FIELDS (FUENTES-CRUZ ET AL. MODEL)

The permanent search for new hydrocarbon resources is closely related to appropriate reservoir characterization and management, in which well tests have played an important role. Nowadays, gas shale formations are the main target of several oil companies. Since gas shale permeability is ultralow, fracturing the formation is a common strategy for adequate hydrocarbon exploitation. Well test analyses conducted in several gas-producing basins have revealed several flow behaviors as a function of the main fracture plane distance. Transition flow regimes have also been observed, as the propagation radius of the pressure waves does not reach the reservoir boundaries. This implies that evaluation parameters such as reservoir length may be overestimated, depending upon the flow regime used for the calculations.

It is normally expected in ultralow permeability formations that fracturing creates a main fracture plane and a network of microfractures around the well-fracture system. These microfractures may improve the average permeability of the reservoir in zones surrounding the fracture treatment, as stated by Palmer, Moschovidis, and Cameron (2007) and by Ge and Ghassemi (2011). Models such as those presented by Wattenbarger et al. (1998) and by El-Banbi and Wattenbarger (1998) assume uniform permeability in the surroundings of the fracture system, which may not be the case. Recently, Fuentes-Cruz, Gildin and Valko (2014) presented a mathematical model that considered the average effect of the failure of weak planes, leading to a non-uniform permeability distribution, depending on the distance to the hydraulic fracture They performed reservoir characterization by using rate-decline analysis, which was later extended by Escobar, Montealegre and Bernal (2014a) to transient rate analysis by using the reciprocal rate and the reciprocal rate derivative, following the *TDS* philosophy of Tiab (1993).

In their work, Fuentes-Cruz et al. (2014) modeled three cases of permeability variation: uniform (with no variation in permeability), linear and exponential. They used type-curve matching to identify the appropriate permeability model type, which is dealt with in a very different and more practical form in this paper. It was found that, as expected, the uniform model had no permeability variations, and the linear flow was followed by the pseudosteady-state regime. However, for the linear and exponential models, Fuentes-Cruz et al. (2014) reported a transition period between the linear flow regime and the pseudosteady regime, which Escobar et al. (2014a) arbitrarily called the *multilinear flow regime*, which is reflected

398 Freddy Humberto Escobar

as slopes of 0.66 and 0.61 on the reciprocal rate derivative curve for exponential and linear models, respectively.

MATHEMATICAL FORMULATION

The mathematical solution introduced by Cruz-Fuentes et al. (2014) is given below:

$$\bar{P}_D = \frac{\delta\pi}{u\sqrt{u}} \left[\frac{I_1\left(\frac{2y_D^*\sqrt{u}}{\ln(\xi)}\right)K_0\left(\frac{2y_D^*}{\ln(\xi)}\sqrt{\frac{u}{k_D^*}}\right) + I_0\left(\frac{2y_D^*}{\ln(\xi)}\sqrt{\frac{u}{k_D^*}}\right)K_1\left(\frac{2y_D^*\sqrt{u}}{\ln(\xi)}\right)}{I_0\left(\frac{2y_D^*\sqrt{u}}{\ln(\xi)}\right)K_0\left(\frac{2y_D^*}{\ln(\xi)}\sqrt{\frac{u}{k_D^*}}\right) - I_0\left(\frac{2y_D^*}{\ln(\xi)}\sqrt{\frac{u}{k_D^*}}\right)K_0\left(\frac{2y_D^*\sqrt{u}}{\ln(\xi)}\right)} \right] \tag{15.78}$$

(If $k_D^*<1$, $\delta=-1$ and $\xi=1/k_D^*$; if $k_D^*>1$, $\delta=1$ and $\xi=k_D^*$)
The solution for the linear permeability case is:

$$\bar{P}_D = \frac{\delta\pi}{u\sqrt{u}} \left[\frac{I_1\left(\frac{2y_D^*\sqrt{k_D^*u}}{|k_D^*-1|}\right)K_0\left(\frac{2y_D^*\sqrt{u}}{|k_D^*-1|}\right) + I_0\left(\frac{2y_D^*\sqrt{u}}{|k_D^*-1|}\right)K_1\left(\frac{2y_D^*\sqrt{k_D^*u}}{|k_D^*-1|}\right)}{I_1\left(\frac{2y_D^*\sqrt{k_D^*u}}{|k_D^*-1|}\right)K_1\left(\frac{2y_D^*\sqrt{u}}{|k_D^*-1|}\right) - I_1\left(\frac{2y_D^*\sqrt{u}}{|k_D^*-1|}\right)K_1\left(\frac{2y_D^*\sqrt{k_D^*u}}{|k_D^*-1|}\right)} \right] \tag{15.79}$$

(If $k_D^*<1$, $\delta=-1$; $k_D^*>1$, $\delta=1$)
For uniform permeability ($k_D^*=1$), the solution is:

$$\bar{P}_D = \frac{\pi}{u\sqrt{u}}\coth\left(y_D^*\sqrt{u}\right) \tag{15.80}$$

The dimensionless production rate is:

$$\bar{q}_D = \frac{1}{u^2\bar{P}_D} \tag{15.81}$$

The dimensionless time for oil and gas wells in field units is:

$$t_D = \frac{0.0002637k^0 t}{\phi(\mu c_t)_i x_e^2} \tag{15.82}$$

The dimensional length stimulated reservoir volume is:

$$y_D = \frac{y}{x_e} \tag{15.83}$$

The dimensionless permeability quantities for exponential and linear cases, respectively, are:

$$k_D(y_D) = \frac{k(y)}{k^0} = k_D^{*(y_D/y_D^*)} = e^{(\ln k_D^*)^{*(y_D/y_D^*)}}$$

(15.84)

$$k_D(y_D) = \frac{k(y)}{k^0} = 1 + (k_D^* - 1)\left(y_D/y_D^*\right)$$

(15.85)

The dimensionless minimum permeability is given as:

$$k_D^* = \frac{k^*}{k^0}$$

(15.86)

The dimensionless gas flow reciprocal rate and its derivative, respectively, are:

$$\frac{1}{q_D} = \frac{n_f k^0 h\left[m(P_i) - m(P_{wf})\right]}{1424T}\frac{1}{q_g}$$

(15.87)

$$\left[t_D * (1/q_D)'\right] = \frac{n_f kh\left[\Delta m(P)\right]}{1424T}\left[t * (1/q)'\right]$$

(15.88)

The dimensionless oil flow reciprocal rate and its derivative, respectively, are:

$$\frac{1}{q_D} = \frac{n_f k^0 h(P_i - P_{wf})}{141.2B\mu}\frac{1}{q_0}$$

(15.89)

$$t_D * (1/q)'_D = \frac{kh\left(P_i - P_{wf}\right)}{141.2\mu B}(t * (1/q)']$$

(15.90)

Using the concept of stimulated reservoir volume, the length of the hydraulic fracture $(2x_f)$ is equal to the lateral extent of the volume that is stimulated, according to Fuentes-Cruz et al. (2014):

$$2x_f = x_e$$

(15.91)

TDS FORMULATION FOR THE LINEAR FLOW REGIME

This flow regime is presented in the three models: uniform, linear and exponential. It is characterized by the typical half-slope line on the reciprocal rate derivative curve. As expressed by Fuentes-Cruz et al. (2014), at early times, the flow regime is governed by the following equation:

$$\frac{1}{q_D} = \pi^{3/2}\sqrt{t_D} \tag{15.92}$$

This flow regime takes place at about the same period of time for the different y_D, as shown in Figures 15.11 through 15.13. Therefore, its behavior does not depend upon either the variation of the dimensionless reservoir length or the minimum permeability value. Once the dimensionless quantities given by Equations (15.82) and (15.87) are placed into Equation (15.92), taking the derivative of Equation (15.92) and placing it in the dimensionless reciprocal rate derivative, Equation (15.88), yields:

$$(1/q_g)_L = \frac{128.76T\sqrt{t_L}}{n_f h\left[m(P_i) - m(P_{wf})\right]x_e(k^0\phi\mu c_t)^{1/2}} \tag{15.93}$$

$$t*(1/q_g)'_L = \frac{64.38T\sqrt{t_L}}{n_f h\left[m(P_i) - m(P_{wf})\right]x_e(k^0\phi\mu c_t)^{1/2}} \tag{15.94}$$

Since linear flow is independent of the minimum permeability at the end of the main plane of fracture, Equations (15.93) and (15.94) result in the maximum induced permeability, k^0, by reading the values of the reciprocal rate and its derivative at any arbitrary time during the linear flow regime so that:

$$k^0 = \frac{16579.82t_L}{\phi\mu c_t}\left\{\frac{T}{n_f h\left[m(P_i) - m(P_{wf})\right]x_e(1/q_g)_L}\right\}^2 \tag{15.95}$$

$$k^0 = \frac{4144.95t_L}{\phi\mu c_t}\left\{\frac{T}{n_f h\left[m(P_i) - m(P_{wf})\right]x_e[t*(1/q_g)'_L]}\right\}^2 \tag{15.96}$$

Notice that the reservoir length, x_e, can be derived from Equation (15.96):

$$x_e = \frac{64.38T\sqrt{t_L}}{n_f h\left[m(P_i) - m(P_{wf})\right][t*(1/q_g)'_L](k^0\phi\mu c_t)^{1/2}} \tag{15.97}$$

Fuentes-Cruz et al. (2014) introduced an expression to estimate skin factor from any arbitrary point read on the reciprocal rate curve during the linear flow regime. That expression is rewritten here as:

Shale Reservoirs 401

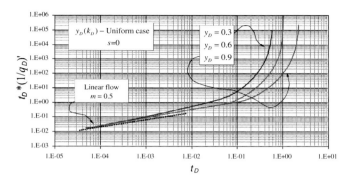

Figure 15.11. Effect of the dimensionless reservoir length (y_D^*) on the flow behavior for the uniform linear case, (k_D^* =0.15); adapted from Escobar et al. (2014a).

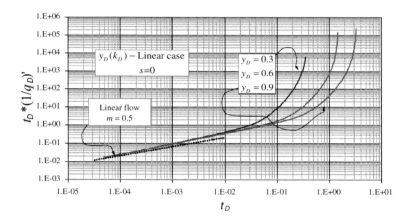

Figure 15.12. Effect of the dimensionless reservoir length (y_D^*) on the flow behavior for the linear flow, (k_D^*=0.1); adapted from Escobar et al. (2014a).

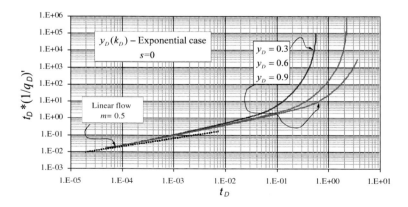

Figure 15.13. Effect of the dimensionless reservoir length (y_D^*) on the flow behavior for the exponential case, (k_D^* =0.1); adapted from Escobar et al. (2014a).

$$S_{initial} = \frac{n_f k^0 h \left[m(P_i) - m(P_{wf}) \right]}{1424T} \left(\frac{1}{q_g} \right)_L$$

(15.98)

TDS Formulation for Multilinear Flow Regime

For both linear and exponential permeability distribution models, Fuentes-Cruz et al. (2014) pointed out the existence of a transition period between the linear flow regime and the boundary-dominated pseudosteady state (BDS). However, Escobar et al. (2014a) found that such transitions may behave as new flow regimes, for the reciprocal rate derivative reflects very characteristic new slopes that are not reported in the literature. They assumed that this flow regime may result from a combination of several flow regimes, which they have arbitrarily called a *multilinear flow regime*. However, they also recommended conducting a simulation study to properly identify the streamlines acting in such case and to identify and name the observed flow regime. For the case of the linear permeability model, the multilinear flow is defined by a slope of 0.61 on a log-log plot of the reciprocal rate derivative. Transient-rate behavior is given in Figures 15.14 through 15.16. As seen in Figure 15.15, the flow behavior is independent of the variation in the dimensionless permeability, k_D, so its representative equation, as developed in this paper, is given below,

$$t_D * (1/q_D)'_{MLL} = \frac{250}{50} (t_D)_{MLL}^{0.6135}$$

(15.99)

The abbreviation *MLL* in Equation (15.99), stands for multilinear flow for the linear permeability model. After placing the dimensionless terms given by Equations (15.82) and (15.88) into Equation (15.99), expressions to estimate either permeability or reservoir length are obtained:

$$k_{MLL} = \left\{ \frac{44.380384T(t)^{0.6135}}{(\phi\mu c_{ti})^{0.6135} x_e^{1.227} n_f h \left[m(P_i) - m(P_{wf}) \right] \left[t * (1/q)'_{MLL} \right]} \right\}^{\frac{1}{0.3865}}$$

(15.100)

$$x_e = \left\{ \frac{44.380384T(t)^{0.6135}}{(\phi\mu c_{ti})^{0.6135} k^{0.3865} n_f h \left[m(P_i) - m(P_{wf}) \right] \left[t * (1/q)'_{MLL} \right]} \right\}^{\frac{1}{1.227}}$$

(15.101)

Notice that the permeability value obtained from Equation (15.100) does not correspond to the initial permeability value, k^0, as the permeability value is undergoing a decay process related to the distance from the fracture system. Once the lateral reservoir length is estimated, the fracture length, x_f, is calculated with Equation (15.91). Fuentes-Cruz et al. (2014) pointed out that the lateral extension of the stimulated reservoir volume is twice the hydraulic fracture length, x_e ($2x_f = x_e$), considering a rectangular reservoir.

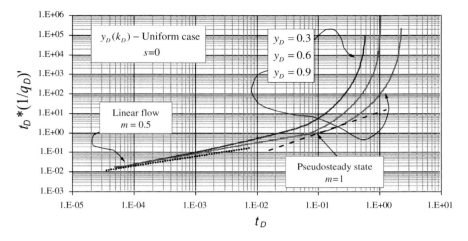

Figure 15.14. Absence of the multilinear case for uniform flow, with variation of permeability in the stimulated reservoir volume (k_D^*) at constant y_D; adapted from Escobar et al. (2014a).

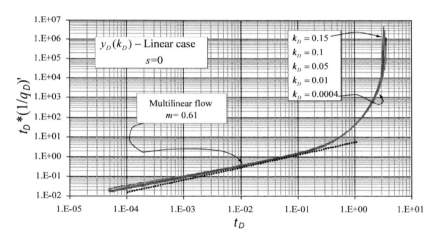

Figure 15.15. Multilinear flow behavior in the linear model with different permeability values (k_D^*) and constant y_D; adapted from Escobar et al. (2014a).

Following the philosophy of the *TDS* technique(Tiab, 1993), the geometric skin factor, s_{MLL}, occurring due to the change from linear to multilinear flow regimes is obtained by taking the ratio between the reciprocal rate—the integral of Equation (15.99)—and the reciprocal rate derivative given by Equation (15.99), and solving for s_{MLL} yields:

$$s_{MLL} = 0.03117 \left(\frac{kt_{MLL}}{\phi \mu c_t x_e^2} \right)^{0.6135} \left[\frac{(1/q)_{MLL}}{t*(1/q)'_{MLL}} - 1.629991 \right] \quad (15.102)$$

The multilinear flow for the case of the exponential permeability model is defined by a slope of 0.66 on a log-log plot of the reciprocal rate derivative. It gives a relationship of three log cycles in the time axis against two log cycles in the reciprocal rate derivative axis. As shown in Figure 15.16, the flow behavior is not uniform relative to the variation of k_D, so it was necessary to determine the most representative mathematical representation of the multilinear flow behavior, which was performed using a probabilistic average:

$$t_D*(1/q_D)'_{MLE} = \frac{405}{50}(t_D)^{0.6612}_{MLE} \tag{15.103}$$

Once again, after plugging the dimensionless terms in Equation (15.103), expressions for permeability and reservoir length can be obtained, such as:

$$k_{MLL} = \left\{ \frac{49.66812320T(t^{0.6612}_{MLL})}{(\phi\mu c_{ti}x)^{0.6612}x_e^{1.3224}n_f h\left[m(P_i)-m(P_{wf})\right]\left[t*(1/q)'_{MLL}\right]} \right\}^{\frac{1}{0.3388}} \tag{15.104}$$

$$x_e = \left\{ \frac{49.66812320T(t^{0.6612}_{MLL})}{(\phi\mu c_{ti}x)^{0.6612}k^{0.3388}n_f h\left[m(P_i)-m(P_{wf})\right]\left[t*(1/q)'_{MLL}\right]} \right\}^{\frac{1}{1.3224}} \tag{15.105}$$

As for the linear permeability model, the geometric skin factor, s_{MLE}, is obtained by dividing the reciprocal rate equation resulting from the integration of Equation (15.103) by the reciprocal rate derivative, Equation (15.103), and solving for the skin factor, like so:

$$s_{MLE} = 0.03488\left(\frac{kt_{MLE}}{\phi\mu c_t x_e^2}\right)^{0.6612}\left[\frac{(1/q)_{MLE}}{t*(1/q)'_{MLE}} - 1.512402\right] \tag{15.106}$$

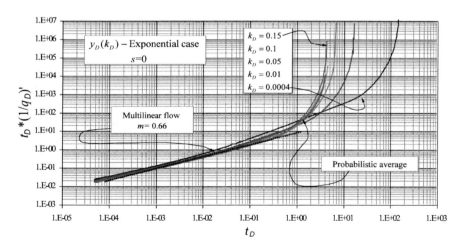

Figure 15.16. Multilinear flow behavior in the exponential model with different permeability values (k_D^*) and constant y_D; adapted from Escobar et al. (2014a).

TDS FORMULATION FOR THE PSEUDOSTEADY-STATE REGIME

The determination of the governing equation for the late pseudosteady period requires a log-log plot of $t_D*(1/q_D)'$ versus t_{DA} using a dimensionless constant permeability (k_D = constant) and varying dimensionless length values (y_D= 0.3, 0.6 and 0.9) for each of the induced permeability models. In each model, a uniform behavior was found by dividing the dimensionless time by the dimensionless length of the stimulated volume reservoir for each case:

$$t_{DA} = \frac{t_D}{y_D^2} \tag{15.107}$$

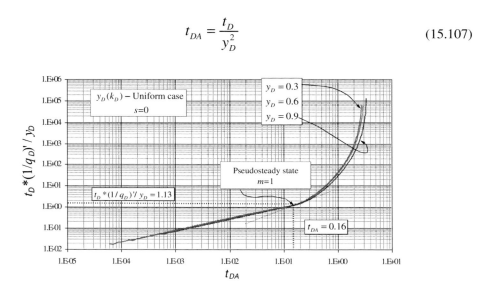

Figure 15.7. Effect of the variation of the dimensionless length on the pseudosteady-state regime for the uniform model, with k_D constant; adapted from Escobar et al. (2014a).

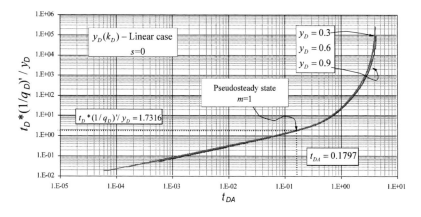

Figure 15.8. Pseudosteady-state behavior on the linear model with constant k_D and varying length of field; adapted from Escobar et al. (2014a).

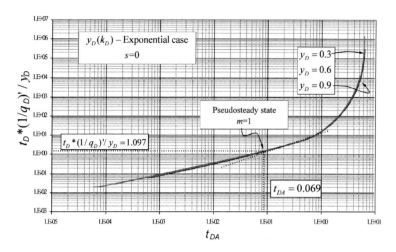

Figure 15.9. Effect of varying the reservoir length in the exponential model with constant k_D; adapted from Escobar et al. (2014a).

As in transient-pressure analysis, the late pseudosteady-state regime is used to calculate the well-drainage area without using the permeability value. Clearly, in transient-rate analysis, the derivative during the pseudosteady state does not follow a unit-slope line but is instead a constantly increasing curve; thus, a tangent unit-slope line must be drawn on the derivative curve to characterize such a regime (see Figures 15.17 through 15.19). For the uniform model in Figure 15.17, the governing equation for the dimensionless reciprocal rate derivative during the pseudosteady-state regime is given below:

$$\frac{[t*(1/q_D)']_{PSSU}}{y_D} = \frac{57}{25}\pi\left(\frac{t_D}{y_D^2}\right)_{PSSU} \tag{15.108}$$

After the dimensionless terms given by Equations (15.82), (15.83) and (15.88) are plugged into Equation (15.108), an expression for the determination of the lateral reservoir length, x_e, is obtained:

$$x_e = \frac{\dfrac{57}{25}\pi(0.0002637)(1424)T}{n_f \phi h (\mu c_t)_i y \left[m(P_i) - m(P_{pwf})\right]} \frac{t_{PSSU}}{[t*(1/q)']_{PSSU}} \tag{15.109}$$

For the linear model, Figure 15.18, the dimensionless reciprocal rate derivative governing equation during the pseudosteady-state regime is given as follows:

$$\frac{[t*(1/q_D)']_{PSSL}}{y_D} = \frac{153}{50}\pi(t_{DA})_{PSSL} \tag{15.110}$$

Shale Reservoirs 407

Table 15.3. Values of alpha for the general dimensionless derivative equation in the pseudosteady-state period

MODEL	α
Exponential	101/20
Linear	153/100
Uniform	57/25

From this, an expression to estimate the lateral reservoir length is developed by replacing the dimensionless parameters in Equation (15.110),

$$x_e = \frac{\frac{153}{100}\pi * 0.0002637 * 1424 * T}{n_f h\phi(\mu c_t)_i y\left[m(P_i) - m(P_{pwf})\right]} \frac{t_{PSSL}}{\left[t*(1/q)'\right]_{PSSL}} \qquad (15.111)$$

For the exponential permeability model in Figure 15.19, the governing equation for the reciprocal rate derivative in dimensionless form takes place during the pseudosteady-state regime and is shown as:

$$\frac{\left[t_D*(1/q_D)'\right]_{PSSE}}{y_D} = \frac{101}{20}\pi\left(\frac{t_D}{y_D^2}\right)_{PSSE} \qquad (15.112)$$

This also leads to the development of an equation to find reservoir length after replacing the dimensionless time, Equation (15.82), and the dimensionless reciprocal rate derivative, Equation (15.87):

$$x_e = \frac{\frac{101}{20}\pi * 0.0002637 * 1424 * T}{n_f h\phi(\mu c_t)_i y\left[m(P_i) - m(P_{pwf})\right]} \frac{t_{PSSE}}{\left[t*(1/q)'\right]_{PSSE}} \qquad (15.113)$$

Finally, it is possible to write a general dimensionless derivative equation for the pseudosteady-state period:

$$\frac{\left[t*(1/q_D)'\right]}{y_D} = \alpha\pi(t_{DA}) \qquad (15.114)$$

The value of α is reported in Table 15.3, depending on the permeability model. The intersection point formed by the line drawn on the linear flow regime, as given by derivative of Equation (15) and the pseudosteady-state period line, t_{LPSSUi}, is provided below:

$$\left(\frac{0.0002637k^{*}t_{LPSSUi}}{\phi(\mu c_{t})_{i}}\right)^{0.5} = 0.388y \qquad (15.115)$$

Equation (15.115) leads to a solution for the maximum induced permeability:

$$k^{\circ} = \left[\frac{23.9225y\left[\phi(\mu c_{t})_{i}\right]^{0.5}}{t_{LPSSU_{i}}^{0.5}}\right]^{2} \qquad (15.116)$$

The point (t_{LPSSLi}) at which the pseudosteady-state period from Equation (15.110) intersects with the linear flow regime from Equation (15.92) is given below:

$$\left(\frac{0.0002637k^{0}t_{LPSSLi}}{\phi(\mu c_{t})_{i}}\right)^{0.5} = 0.28905y \qquad (15.117)$$

The maximum permeability value can be solved for in the above equation to give:

$$k^{0} = \left[\frac{17.7999y\left[\phi(\mu c_{t})_{i}\right]^{0.5}}{t_{LPSSL_{i}}^{0.5}}\right]^{2} \qquad (15.118)$$

The point of intersection, $t_{MLPSSLi}$, between the late pseudosteady-state regime from Equation (15.110) and the multilinear flow regime from Equation (15.99), results in the following equation (in dimensional terms):

$$\left(\frac{0.0002637k^{*}t_{MLPSSL_{i}}}{\phi(\mu c_{t})_{i}x_{e}^{2}}\right)^{0.3865} = 0.5077\frac{y}{x_{e}} \qquad (15.119)$$

This leads to a solution for the low permeability, induced in dimensional terms (k^{*}):

$$k^{*} = \left[12.27\frac{y}{x_{e}^{0.227}}\left(\frac{\phi(\mu c_{t})_{i}}{t_{MLPSSL_{i}}}\right)^{0.3865}\right]^{1/0.3865} \qquad (15.120)$$

The intersection point of the pseudosteady-state regime—Equation (15.112)—and the linear flow regime, t_{LPSSEi}—given by the derivative of Equation (15.92)—provides the following expression:

$$\left(\frac{0.0002637k^0t_{LPSSEi}}{\phi(\mu c_t)_i}\right)^{0.5} = 0.1755y \tag{15.121}$$

Equation (15.122), obtained from the above expression, is useful for recalculating the maximum induced permeability:

$$k^0 = \left[\frac{10.8074y\left[\phi(\mu c_t)_i\right]^{0.5}}{t_{LPSSE_i}^{0.5}}\right]^2 \tag{15.122}$$

The intersection point formed between the pseudosteady-state regime—Equation (15.114)—and the multilinear flow regime, $t_{MLPSSEi}$--Equation (15.103)—provides the following equation:

$$\left(\frac{0.0002637k^*t_{MLPSSL_i}}{\phi(\mu c_t)_i\, x_e^2}\right)^{0.3388} = 0.5111\frac{y}{x_e} \tag{15.123}$$

This is also useful for developing an expression to estimate the minimum induced permeability:

$$k^* = \left[8.3375\frac{y}{x_e^{0.3224}}\left(\frac{\phi(\mu c_t)_i}{t_{MLPSSE_i}}\right)^{0.3388}\right]^{1/0.3388} \tag{15.124}$$

GOVERNING EQUATIONS FOR OIL FLOW (FUENTES-CRUZ ET AL. MODEL)

Linear Flow Regime

The dimensionless equation representing the linear flow, which is independent of the model and the permeability variation, behaves in the manner described by Equation (15.92). Once the dimensionless terms given by Equations (15.82) and (15.89) are plugged into Equation (15.125), the lateral extent of the stimulated reservoir volume can be solved for:

$$x_e = \frac{12.77\mu B}{n_f k^{0.5}h\left[P_i - P_{wf}\right]\left[t*(1/q)'_L\right]}\left[\frac{t_L}{(\phi\mu c_t)}\right]^{0.5} \tag{15.125}$$

From the above equation, it is also possible to know the value of permeability:

$$k = \left[\frac{12.77\mu B}{n_f x_e h \left[P_i - P_{wf} \right]\left[t*(1/q)'_L \right]} \left[\frac{t_L}{(\phi\mu c_t)} \right]^{0.5} \right]^2 \qquad (15.127)$$

MULTILINEAR FLOW REGIME

Linear Model

The general dimensionless equation describing this flow regime for the linear model is given by equation (15.99). Equations for calculating k and x_e can be obtained from Equation (15.125) above:

$$k = \left\{ \frac{4.4(t_{MLL}^{0.6135})}{(\phi\mu c_t)^{0.6135} x_e^{1.227} n_f h \left[P_i - P_{wf} \right]\left[t*(1/q)'_{MLL} \right]} \right\}^{\frac{1}{0.3865}} \qquad (15.128)$$

$$x_e = \left\{ \frac{4.4 t_{MLL}^{0.6135}}{(\phi\mu c_t)^{0.6135} k^{0.3865} n_f h \left[P_i - P_{wf} \right]\left[t*(1/q)'_{MLL} \right]} \right\}^{\frac{1}{1.227}} \qquad (15.129)$$

The geometric skin damage equation obtained for the multilinear flow is:

$$s_{MLL} = 4.8903(t_D)_{MLL}^{0.6135} \left[\frac{(1/q_D)_{MLL}}{t_D*(1/q)'_{MLL}} - 1.629991 \right] \qquad (15.130)$$

Exponential Model

The general dimensionless equation that describes the behavior of linear flow in the exponential model is given by Equation (15.103). After plugging Equations (15.82) and (15.89) into it, solving for both k and x_e gives:

$$k = \left\{ \frac{4.9(t_{MLE})^{0.6612}}{(\phi\mu c_t x)^{0.6612} x_e^{1.3224} n_f h \left[m(P_i) - m(P_{wf}) \right]\left[t*(1/q)'_{MLE} \right]} \right\}^{\frac{1}{0.3388}} \qquad (15.131)$$

$$x_e = \left\{ \frac{4.9(t_{MLE})^{0.6612}}{(\phi\mu c_t x)^{0.6612} k^{0.3388} n_f h \left[m(P_i) - m(P_{wf}) \right]\left[t*(1/q)'_{MLE} \right]} \right\}^{\frac{1}{1.3224}} \qquad (15.132)$$

Shale Reservoirs

The developed equation for calculating the geometric skin factor for the exponential induced permeability model is:

$$s_{MLE} = 0.03488 \left(\frac{kt_{MLE}}{\phi \mu c_t x_e^2} \right)^{0.6612} \left[\frac{(1/q)_{MLE}}{t*(1/q)'_{MLE}} - 1.512402 \right] \qquad (15.132)$$

PSEUDOSTEADY-STATE PERIOD

Uniform Model

After plugging Equations (15.82) and (15.89) into Equation (15.108), solving for the reservoir length gives:

$$x_e = \frac{0.267}{n_f \phi h (\mu c_t)_i \, y \left[P_i - P_{wf} \right]} \frac{t_{PSSU}}{\left[t*(1/q)' \right]_{PSSU}} \qquad (15.133)$$

Linear Model

Similarly, plugging Equations (15.82) and (15.89) into Equation (15.110) yields:

$$x_e = \frac{0.358}{n_f h \phi (\mu c_t)_i \, y \left[P_i - P_{wf} \right]} \frac{t_{PSSL}}{\left[t*(1/q)' \right]_{PSSL}} \qquad (15.134)$$

Exponential Model

Placing Equations (15.82) and (15.89) into Equation (15.112) and solving for the reservoir length leads to:

$$x_e = \frac{0.596}{n_f h \phi (\mu c_t)_i \, y \left[m(P_i) - m(P_{pwf}) \right]} \frac{t_{PSSE}}{\left[t*(1/q)' \right]_{PSSE}} \qquad (15.135)$$

Example 15.2

Reservoir, fluid and well properties are given below for a field case that Fuentes-Cruz et al. (2014) presented and solved using rate-decline analysis.

$h = 306$ ft $\mu_g = 0.018$ cp $\phi = 4.8\%$
$x_e = 800$ ft $T = 633.5$ °R $P_i = 3115$ psi
$B_{gi} = 0.916$ rb/Mscf $P_{wf} = 500$ psi $m(P_i) = 6.83 \times 10^8$ psi^2/cp
$c_t = 2.51 \times 10^{-4}$ psi^{-1} $m(P_{wf}) = 2.08 \times 10^7$ psi^2/cp $y = 552$ ft
$k^0 = 2.8 \times 10^{-3}$ md $n_f = 3$

The rate-time data, which Escobar et al. (2014a) digitized from the work of Fuentes-Cruz et al. (2014), are reported in Figure 15.10. The reciprocal rate derivative is estimated afterwards. This estimate is required to find permeability, damage and reservoir length by transient-rate analysis.

Solution

The following parameters are read from Figure 15.10:

$(t)_{LE} = 852.8151$ hr
$[t *(1/q_g)']_{LE} = 1.06 \times 10^{-4}$ day/Mscf
$(t)_{PSSE} = 13234.6$ hr
$[t *(1/q_g)']_{PSSE} = 9.08 \times 10^{-4}$ day/Mscf
$(t)_{LPSSUi} = 35.6958$ hr

Equation (15.113), applied to the pseudosteady-state regime, is used to calculate the x_e value:

$$x_e = \frac{5.05\pi(0.0002637)(1424)(633.5)}{(3)(306)(0.048)(0.018)(2.51\times10^{-4})(552)\left[6.83\times10^8 - 2.08\times10^7\right]} \frac{(13234.6)}{(9.08\times10^{-4})}$$

$x_e = 755.58$ ft

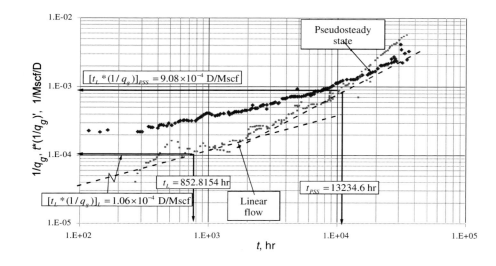

Figure 15.10. Reciprocal rate and its derivative for example 15.2; adapted from Escobar et al. (2014a).

Shale Reservoirs

Table 15.4. Results for example 15.2 against original results

Parameter	Fuentes-Cruz et al. (2014)	Escobar et al. (2014a)
k° (md)	2.8×10^{-3}	2.48×10^{-3}
k^{*} (md)	3.9×10^{-5}	-
x_e (ft)	800	755.58
s	0.31	0.346

Now, applying Equation (15.96) to the linear flow regime yields a maximum induced permeability value of 2.4837×10^{-3} md.

$$k^0 = \frac{4144.95(852.8154)}{(0.048)(0.018)(2.51 \times 10^{-4})} \left\{ \frac{633.5}{(2)(306)\left[6.83 \times 10^8 - 2.08 \times 10^7\right](800)(1.06 \times 10^{-4})} \right\}^2 = 2.4837 \times 10^{-3} \text{ md}$$

Equation (15.98) provides a damage of 0.346.

$$s_{inicial} = \frac{(3)(0.0024837)(306)\left[(6.83 \times 10^8 - 2.08 \times 10^7\right]}{1424(633.5)} 2.07 \times 10^{-4} = 0.346$$

Table 15.4 summarizes the results obtained from example 15.2, compared to the original work of Fuentes-Cruz et al. (2014). Notice that, in spite of the digitization of the data, the results match well.

Nomenclature

A_{cm}	Total matrix surface area draining into fracture system, ft^2
A_{cw}	Well-face cross-sectional area to flow, ft^2
B	Liquid formation volume factor, rB/STB
B_{gi}	Formation volume factor at initial reservoir pressure, rcf/scf
B_g	Volumetric factor, rb/Mscf
c_t	Liquid total compressibility, psi^{-1}
c_{ti}	Total compressibility at initial reservoir pressure, psi^{-1}
$f(s)$	Relation used in Laplace space to distinguish matrix geometry types. In chronological order for each flow it equals ω, $(\lambda A_c/3s)^{0.5}$, 1 and $(\lambda A_c/3s)^{0.5}$.
h	Reservoir thickness, ft
k	Homogeneous reservoir permeability, md
k_f	Bulk fracture permeability of dual-porosity models, md
k_m	Matrix permeability, md
k_V	Vertical permeability, md
k_H	Horizontal permeability, md
k^0	Maximum permeability induced, md
k^*	Minimum permeability induced, md
l	Half of fracture spacing, ft
l^{-1}	Inverse Laplace space operator

L	General fracture spacing, ft
L_w	Horizontal well length, ft
m	Slope; in chronological order, 1 is for early linear flow, 2 for bilinear flow, etc.
$m(P)$	Pseudopressure (gas), psi^2/cp
n_f	Number of main hydraulic-fracture planes
P_i	Initial reservoir pressure, psi
P_{wf}	Wellbore flowing pressure, psi
$\bar{P}$	Laplace space pressure
q_g	Gas rate, Mscf/day
Q	Cumulative production, STB
$\bar{q}$	Laplace space flow rate
$1/q$	Reciprocal flowrate , D/Mscf
r_w	Wellbore radius , ft
s	Laplace space variable
t	Time, days
$t^*(1/q)'$	Reciprocal flow rate derivative, D/Mscf
t_D	Dimensionless time coordinate
t_{DAc}	Dimensionless time based on A_{cw} and k_f (rectangular geometry, dual porosity)
t_{DAch}	Dimensionless time based on A_{cw} and k (rectangular geometry, homogeneous)
t_{DAcm}	Dimensionless time based on matrix A_{cm} and k_m (rectangular geometry)
t_{Drw}	Dimensionless time (radial definition) based on wellbore radius
T	Absolute temperature, °R
u	Laplace space variable
x_e	Drainage area width (rectangular geometry), ft
x_e	Effective reservoir width, ft
x_f	Hydraulic fracture half-length, ft
y_e	Drainage area half-length (rectangular geometry), ft
y_{De}	Dimensionless reservoir length (rectangular geometry)
y^*	Half-length of stimulated reservoirs volume element, ft
z	Coordinate, z-direction (matrix)
z_D	Dimensionless coordinate, z-direction

Greek

γ	Specific gravity
λ	Dimensionless interporosity parameter
ϕ	Porosity, fraction
μ	Viscosity, cp
ω	Dimensionless storativity ratio

Suffixes

BDS	Boundary-dominated state
$BLLMi$	Bilinear flow–matrix transient linear flow intercept

BLPSSi	Bilinear flow–pseudosteady-state flow intercept
D	Dimensionless
DA	Dimensionless based on drainage area
g	Gas
i	Initial
sc	Standard conditions
L	Early linear flow
BL	Bilinear flow
LM	Matrix transient linear flow
h	Linear homogenous flow
LBLi	Early linear flow – bilinear flow intercept
LE	Linear flow, exponential model
LL	Linear flow, linear model
LPSSi	Early linear flow–pseudosteady-state flow intercept
LMPSSi	Matrix transient linear flow–pseudosteady-state flow intercept
LU	Linear flow, uniform model
LPSSUi	Intersection point between linear flow and pseudosteady state, uniform model
LPSSLi	Intersection point between linear flow and pseudosteady state, linear model
LPSSEi	Intersection point between linear flow and pseudosteady state, exponential model
MLL	Multilinear
MLLU	Multilinear flow, uniform model
MLLL	Multilinear flow, linear model
MLLE	Multilinear flow, exponential model
MLLPSSLi	Intersection point between multilinear flow and pseudosteady state, linear model
MLLPSSEi	Intersection point between multilinear flow and pseudosteady state, exponential model
PSS	Pseudosteady-state flow
PSSU	Pseudosteady state, uniform model
PSSL	Pseudosteady state, linear model
PSSE	Pseudosteady state, exponential model
sc	Standard conditions
ST	Short times

REFERENCES

Bello R. O. (2009). *"Rate Transient Analysis in Shale Gas Reservoirs with Transient Linear Behavior."* Texas A&M University. PhD Dissertation.

Bernal, K. M., Escobar, F. H. & Ghisays-Ruiz, A. (2014). "Pressure and Pressure Derivative Analysis for Hydraulically-Fractured Shale Formations Using the Concept of Induced Permeability Field." Journal of Engineering and Applied Sciences., Vol. *9*. No. 10. P. 1952-1958. September.

Brown, M., Ozkan, E., Raghavan, R. & Kazemi, H. (2011, December 1). "Practical Solutions for Pressure-Transient Responses of Fractured Horizontal Wells in Unconventional Shale Reservoirs*." Society of Petroleum Engineers*. doi:10.2118/125043-PA.

Carlson, E. S. & Mercer, J. C. (1991, April 1). "Devonian Shale Gas Production: Mechanisms and Simple Models." *Society of Petroleum Engineers*. doi:10.2118/19311-PA.

El-Banbi, A. H. & Wattenbarger, R. A. (1998, January 1). "Analysis of Linear Flow in Gas Well Production." *Society of Petroleum Engi*neers. doi:10.2118/39972-MS.

Engler, T. W. & Tiab, D. (1996a). "Analysis of Pressure and Pressure Derivatives without Type-Curve Matching. 6- Horizontal Well Tests in Anisotropic Reservoirs." *Journal of Petroleum Science and Engineering.*, *15* (1996), p. 153-168.

Engler, T. W. & Tiab, D. (1996b). "Analysis of Pressure and Pressure Derivatives without Type-Curve Matching. 5- Horizontal Well Tests in Naturally Fractured Reservoirs." *Journal of Petroleum Science and Engineering.*, *15* (1996), p. 139-151.

Escobar, F. H., Montenegro, L. M. & Bernal, K. M. (2014a). "Transient-Rate Analysis For Hydraulically-Fractured Gas Shale Wells Using The Concept Of Induced Permeability Field." *Journal of Engineering and Applied Sciences.*, Vol. *9*. Nro. 8. P. 1244-1254. August.

Escobar, F. H., Zhao, Y. L. & Zhang L. H. (2014b). "Interpretation of Pressure Tests in Horizontal Wells in Homogeneous and Heterogeneous Reservoirs with Threshold Pressure Gradient." *Journal of Engineering and Applied Sciences.*, Vol. *9*. Nro. 11. P. 2220-2228. September.

Escobar, F. H., Bernal, K. M. & Olaya-Marin, G. (2014c) "Pressure and Pressure Derivative Analysis for Fractured Horizontal Wells in Unconventional Shale Reservoirs Using Dual-Porosity Models In The Stimulated Reservoir Volume." *Journal of Engineering and Applied Sciences.*, Vol. *9*. Nro. 12. P. 2650-2669. 2014.

Escobar, F. H., Bernal, K. M. & Olaya-Marin, G. (2014c) "Pressure and Pressure Derivative Analysis for Fractured Horizontal Wells in Unconventional Shale Reservoirs Using Dual-Porosity Models In The Stimulated Reservoir Volume." *Journal of Engineering and Applied Sciences.*, Vol. *9*. Nro. 12. P. 2650-2669. 2014.

Escobar, F. H., Rojas, J. D. & Ghisays-Ruiz, A. (2014d). "Transient-Rate Analysis Hydraulically-Fractured Horizontal Wells in Naturally-Fractured Shale Gas Reservoirs." *Journal of Engineering and Applied Sciences.*, Vol. *10*. Nro. 1. P. 102-114. October.

Fuentes-Cruz, G., Gildin, E. & Valkó, P. P. (2014, May 1). "Analyzing Production Data From Hydraulically Fractured Wells: The Concept of Induced Permeability Field." *Society of Petroleum Engineers*. doi:10.2118/163843-PA.

Ge, J. & Ghassemi, A. (2011, January 1). "Permeability Enhancement In Shale Gas Reservoirs After Stimulation By Hydraulic Fracturing." *American Rock Mechanics Association.*

Goode, P. A. & Thambynayagam, R. K. M. (1987, December 1). "Pressure Drawdown and Buildup Analysis of Horizontal Wells in Anisotropic Media." *Society of Petroleum Engineers*. doi:10.2118/14250-PA.

Lu, J., Zhu, T., Tiab, D. & Escobar, F.H. (2015). "Pressure behavior of horizontal wells in dual-porosity, Dual-permeability naturally-fractured reservoirs." *Journal of Engineering and Applied Sciences.*, Vol. *10*. Nro. 8. p. 3405-3417. May.

Mayerhofer, M. J., Lolon, E. P., Youngblood, J. E. & Heinze, J. R. (2006, January 1). "Integration of Microseismic-Fracture-Mapping Results With Numerical Fracture Network Production Modeling in the Barnett Shale." *Society of Petroleum Engineers*. doi:10.2118/102103-MS.

Ozkan, E., Ohaeri, U. & Raghavan, R. (1987, June 1). "Unsteady Flow to a Well Produced at a Constant Pressure in a Fractured Reservoir." *Society of Petroleum Engineers.* doi:10.2118/9902-PA.

Palmer, I. D., Moschovidis, Z. A. & Cameron, J. R. (2007, January 1). "Modeling Shear Failure and Stimulation of the Barnett Shale After Hydraulic Fracturing." *Society of Petroleum Engineers.* doi:10.2118/106113-MS

Spivey, J. P. & Semmelbeck, M. E. (1995, January 1). "Forecasting Long-Term Gas Production of Dewatered Coal Seams and Fractured Gas Shales." *Society of Petroleum Engineers.* doi:10.2118/29580-MS.

Tiab, D. (1993, January 1). "Analysis of Pressure and Pressure Derivatives Without Type-Curve Matching: I-Skin and Wellbore Storage." *Society of Petroleum Engineers.* This paper was prepared for presentation at the Production Operations Symposium held in Oklahoma City, OK, U.S.A., March 21-23. doi:10.2118/25426-MS.

Van Everdingen, A. F. & Hurst, W. (1949, December 1). "The Application of the Laplace Transformation to Flow Problems in Reservoirs." *Society of Petroleum Engineers.* doi:10.2118/949305-G.

Warren, J. E. & Root. P. J. (1963). "The Behavior of Naturally Fractured Reservoirs." *Society of Petroleum Engineering Journal.* p. 245-255. September.

Watson, A. T., Gatens, J. M., Lee, W. J. & Rahim, Z. (1990, August 1). "An Analytical Model for History Matching Naturally Fractured Reservoir Production Data." *Society of Petroleum Engineers.* doi:10.2118/18856-PA.

Wattenbarger, R. A., El-Banbi, A. H., Villegas, M. E. & Maggard, J. B. (1998, January 1). "Production Analysis of Linear Flow Into Fractured Tight Gas Wells." *Society of Petroleum Engineers.* doi:10.2118/39931-MS.

INDEX

A

adsorption, 213
Alaska, 28
algorithm, 144
alters, 125
amplitude, 71
anisotropy, 13, 352, 353, 355, 356, 371
aquifers, 351, 361, 362
arithmetic, 6
atmosphere, 249

B

behaviors, 153, 164, 168, 204, 298, 308, 397
benefits, xi
Bingham fluids, xiv, 149, 213, 215, 234, 235
bounds, xiv
brothers, v
BTU, 349

C

capillary, 337
chemical, 204
children, v, ix
classification, 149, 203, 375
CO_2, 326, 336
coal, 150, 361, 362, 376
coding, 125
coke, 327
Colombia, 12, 52, 93, 294
combined effect, 215
combustion, xiv, 326, 327, 334, 335, 336, 337
commercial, xiv, 9, 125, 362
communication, 295, 362
compatibility, 188

complement, 225
complexity, 98
compressibility, 25, 37, 39, 42, 49, 86, 87, 88, 89, 102, 122, 144, 183, 200, 211, 244, 276, 291, 316, 337, 348, 371, 413
computer, xiii, xiv, 125
computer software, xiii, xiv
conceptual model, 106
conduction, 319
conductivity, xiv, 55, 56, 57, 58, 59, 60, 61, 62, 65, 67, 69, 70, 89, 92, 179, 180, 254, 255, 265, 266, 270, 271, 272, 291, 295, 296, 297, 298, 299, 301, 302, 304, 306, 307, 308, 311, 312, 313, 315, 316, 317, 362
connectivity, 33
constant rate, 32, 320, 326, 362, 378
convergence, 13, 17, 277
cooperation, ix
correction factors, 180
correlation(s), 7, 35, 38, 56, 59, 74, 77, 79, 80, 88, 98, 99, 109, 118, 139, 142, 167, 191, 197, 206, 207, 211, 220, 221, 222, 229, 235, 242, 258, 263, 269, 283, 284, 286, 288, 311, 366, 367
correlation coefficient, 56, 59, 79, 167
creativity, xi
crude oil, 149, 166, 187, 203, 326
cycles, 226, 231, 233, 330, 403

D

data set, xi
decay, 402
deconvolution, xiv, 204
deposition, 12, 164, 355
deposits, 326, 336, 375
depression, 105, 112, 115, 118, 125, 136, 225
derivatives, 16, 99, 181, 255, 298, 323, 366, 388
desorption, 362

420 Index

detection, 295
deviation, 150, 164, 203, 216, 296, 337
diffusivity(s), 34, 45, 165, 270, 308, 327, 328, 337, 361, 362, 377, 378
discontinuity, 329, 331, 333, 335, 348
displacement, 112, 118, 136, 139, 140, 327, 384
distribution, 30, 215, 320, 327, 362, 397, 402
DOI, 185, 201, 212, 245
double-permeability reservoirs, xiv
double-porosity reservoirs, xiv, 33
drainage, xiv, 5, 8, 21, 27, 43, 56, 80, 84, 181, 204, 205, 207, 211, 212, 234, 238, 241, 275, 281, 290, 296, 333, 373, 376, 383, 387, 406, 415
drawing, 7, 206, 364, 367

E

economic evaluation, 337
economics, 361
energy, 150, 362, 375
energy supply, 362
engineering, xi, 204
Enhanced Oil Recovery, 186
exploitation, 336, 375, 397

F

facies, 22, 326
field tests, 215
fingerprints, 107, 108, 110, 125, 128, 300
flooding, 326, 336, 349
floods, 326
foams, xiv, 149, 164, 203
formation, xiii, 6, 39, 54, 55, 143, 158, 180, 187, 190, 234, 235, 240, 243, 276, 277, 304, 305, 313, 316, 319, 328, 337, 351, 353, 376, 397, 413
formula, 8
fractal dimension, 215
fractures, 29, 30, 31, 33, 39, 53, 54, 60, 65, 88, 95, 105, 106, 107, 113, 115, 123, 125, 126, 145, 215, 225, 286, 362, 375, 376, 377, 379, 392

G

gasification, 150
geometry, 12, 14, 174, 217, 352, 413, 414
God, v
graph, 192, 331
gravity, 54, 235, 328, 337, 414
Greeks, 145, 184, 200, 212, 244
groundwater, 362
grouping, 237

H

heat capacity, 349
heat transfer, 3, 319, 327
heavy oil, xi, xiv, 150, 215, 234, 235, 326, 336, 375
heterogeneity, 287
heterogeneous systems, 32, 249, 282, 286
human, v
hydrocarbon field management, xiii
hydrocarbons, 295, 327, 355
hydrogen, xiii

I

ideal, 30, 218
idealization, 327
identification, 15, 204
income, 44
indirect measure, xiii
industry, xiii, xv, 105, 150, 164, 179, 185, 187, 201, 203, 212, 245, 362, 375
initiation, 71
insertion, 170
institutions, ix
integration, 235, 404
interface, 188, 189, 196, 204, 212, 362
interference, 39
Islam, 327

L

Latin America, xiv
laws, 225
lead, 37, 135, 136, 187, 312, 361, 375
leakage, xv, 351, 361, 362, 363, 364, 365, 366, 367, 368, 369, 371, 373
light, 326, 327
linear model, 398, 403, 405, 406, 410, 415
liquid phase, 328
liquids, 150, 234

M

magnitude, 30, 132, 144, 327, 330
majority, 295
management, xiii, 397
manipulation, 367
masking, 351
mass, 30, 54, 98, 286
materials, 376
mathematics, xiii

Index

matrix, 29, 30, 31, 37, 39, 50, 51, 54, 67, 79, 88, 90, 95, 96, 98, 101, 102, 103, 105, 106, 108, 110, 112, 125, 131, 134, 225, 286, 319, 375, 376, 377, 379, 381, 383, 384, 385, 388, 389, 390, 391, 393, 394, 396, 413, 414
matter, 190, 220
measurement(s), xiii, 229, 352, 361
media, 29, 108, 127, 187, 235
mentor, v
methodology, xiv, 12, 44, 46, 95, 106, 125, 204, 215, 226, 296, 323, 337, 362, 375, 376
Mexico, 105, 143
migration, 295
misuse, xiii
models, xi, xiii, xv, 30, 95, 98, 105, 125, 150, 179, 187, 196, 203, 204, 235, 327, 337, 361, 375, 376, 397, 399, 402, 405, 413
Montenegro, 416

N

Nigeria, 27, 373
nitrogen, 327
non-Newtonian fluids, xiv, 149, 150, 155, 179, 181, 187, 203, 204, 206, 207, 218, 234, 246

O

Oklahoma, ix, 6, 27, 28, 51, 91, 103, 145, 186, 201, 213, 246, 293, 294, 317, 350, 374, 417
operations, 337

P

parallel, xi, 12, 40, 60, 71, 105, 106, 112, 118, 143, 384
parallelism, 226
parents, v
partial completion, 164, 352
PDL, 21, 76
perforation, 353
permit, 375
Petroleum, i, ii, iii, ix, xiii, 27, 28, 51, 52, 91, 92, 93, 103, 123, 124, 145, 149, 179, 185, 186, 201, 203, 204, 213, 245, 246, 293, 294, 317, 349, 350, 372, 373, 374, 415, 416, 417
petroleum engineers, xiii, 149, 179, 203
physical characteristics, 295
plastics, 149, 234
polar, 3
polymer, 149, 159, 187, 203, 213, 326
polymer solutions, 149, 187, 203, 213

porosity, xiv, 29, 30, 31, 32, 33, 34, 36, 37, 38, 39, 42, 44, 53, 54, 62, 76, 88, 95, 98, 99, 100, 105, 106, 107, 108, 125, 126, 128, 150, 216, 217, 218, 221, 224, 226, 229, 375, 376, 413, 414, 416
porous media, xi, 110, 132, 149, 150, 151, 187, 188, 197, 204, 213, 215, 234, 235, 376
pressure gradient, 183, 190, 200, 211, 215, 234, 235, 236, 239, 240, 241, 244, 337
pressure wave traveling, xiii
principles, 105, 249, 352
probe, 352
project, 159, 361
propagation, 397

Q

quartz, 352

R

reading, 12, 43, 61, 63, 144, 169, 191, 197, 228, 229, 275, 364, 400
real time, 276
reality, 105
recall, 87
recognition, 352
recovery, xiv, 159, 164, 187, 326, 327, 336, 337, 362
recovery process(s), 187, 326, 327, 336
regression, xiii, 98, 125, 228, 235, 355, 361
regression analysis, xiii, 98, 228, 235, 355, 361
relatives, v
researchers, xiii, 3, 326
reserves, 215, 326, 336
reservoir information, xiii
reservoir pressure, xiv, 25, 33, 38, 39, 42, 43, 44, 45, 46, 49, 89, 102, 372, 413, 414
resistance, 296, 298, 308, 326, 336
resources, 362, 375, 397
response, xi, 206, 207, 218, 225, 226, 227, 229, 235, 296, 297, 319, 337, 361, 379, 383, 386
rheology, 204
root, xiii, 18, 19, 59, 72, 165, 166, 227, 228, 271, 272, 283, 284, 290, 306, 307, 396

S

saturation, 320, 349
science, xiii, 124
segregation, 9
shape, 5, 12, 25, 30, 43, 44, 45, 46, 48, 49, 108, 127, 188, 197, 226, 241, 251, 291, 352
shear, 149, 150, 190, 203, 216, 234

shear rates, 149, 234
showing, 280
simulation(s), 70, 98, 164, 175, 224, 402
software, xiii, 125
solution, xiii, 3, 5, 29, 32, 34, 45, 106, 152, 153, 155,
165, 166, 179, 204, 206, 216, 235, 250, 251, 253,
256, 270, 295, 308, 319, 320, 321, 326, 329, 337,
353, 361, 362, 375, 378, 398, 408
South America, 81
stabilization, xv, 296, 351, 352, 353, 354, 355, 356,
357, 358, 360, 372, 373
standard error, 79
stimulation, xiv, 149, 187, 203, 327
stress, 30, 149, 150, 184, 190, 203, 216, 234, 235
structure, 143
substitution, 170, 264
surface area, 385, 388, 394, 413
suspensions, 150, 234
symmetry, 169, 170, 171, 174
synthesis, xi, 6, 213

T

tar, 375
target, 397
teachers, xi
techniques, xi, xv, 19, 40, 53, 70, 106, 160, 216, 249,
276, 326, 336, 353, 358
technologies, 185, 201, 212, 245, 351
temperature, 26, 89, 216, 319, 320, 321, 323, 324,
326, 327, 336, 349, 414
test data, xiii, 27, 71, 125, 212, 216, 304, 327, 337,
358, 373
testing, xi, xiii, 3, 30, 53, 149, 150, 179, 187, 249,
259, 327, 328, 351, 361
time pressure, 181, 204
TRA, 270, 278
training, xiv
transference, 286
transformation, 329
transport, 124

transport processes, 124
treatment, 149, 187, 203, 397
trial, xiii, 361

U

uniform, 43, 320, 328, 337, 362, 386, 397, 398, 399,
401, 403, 404, 405, 406, 415
unique features, 319
United States, 375
universe, v

V

variables, 55, 169, 174, 295, 320, 344, 377, 379
variations, 323, 397
Venezuela, 12
viscosity, 25, 54, 86, 87, 88, 150, 151, 161, 162, 176,
184, 188, 193, 196, 197, 199, 200, 203, 208, 210,
212, 216, 244, 276, 319, 326, 336

W

water, xiv, 12, 26, 51, 91, 319, 320, 321, 322, 326,
327, 348, 349, 352, 361, 362
water permeability, 348
wealth, xi
well test interpretation, xi, xiii, xiv, 326
Well testing, xiii, 3
wells, xiv, 9, 12, 39, 43, 45, 53, 60, 88, 89, 92, 102,
150, 164, 179, 180, 187, 203, 215, 234, 246, 249,
270, 274, 276, 291, 319, 327, 352, 361, 375, 398,
416

Y

yield, 19, 40, 55, 59, 74, 75, 149, 150, 203, 234, 235,
252, 254, 280, 367